To Dan

With regards + admiration

¡Keep fighting the good fight!

Chuck

31 October 2021

The Survival Nexus

The Survival Nexus

Science, Technology, and World Affairs

CHARLES WEISS

OXFORD
UNIVERSITY PRESS

OXFORD
UNIVERSITY PRESS

Oxford University Press is a department of the University of Oxford. It furthers the University's objective of excellence in research, scholarship, and education by publishing worldwide. Oxford is a registered trade mark of Oxford University Press in the UK and certain other countries.

Published in the United States of America by Oxford University Press
198 Madison Avenue, New York, NY 10016, United States of America.

Library of Congress Cataloging-in-Publication Data
Names: Weiss, Charles, author.
Title: The Survival Nexus : Science, Technology, and World Affairs/Charles Weiss.
Description: New York, NY : Oxford University Press, 2022 |
Includes bibliographical references and index.
Identifiers: LCCN 2021016619 (print) | LCCN 2021016620 (ebook) |
ISBN 9780190946265 (hardback) | ISBN 9780190946289 (epub)
Subjects: LCSH: Technological innovations—Social aspects. |
Science—Social aspects. | Globalization.
Classification: LCC HM846 .W45 2021 (print) | LCC HM846 (ebook) |
DDC 303.48/3—dc23
LC record available at https://lccn.loc.gov/2021016619
LC ebook record available at https://lccn.loc.gov/2021016620

DOI: 10.1093/oso/9780190946265.001.0001

1 3 5 7 9 8 6 4 2

Printed by Sheridan Books, Inc., United States of America

To Edie

Contents

Author's Preface xi

1. Introduction: Science, Technology, and Survival 1
    - Existential Dangers: Pandemics, Climate Disruption, and Nuclear War 1
    - New Technologies Raise Issues of Ethics and Values 2
    - A Pervasive but Neglected Dimension 4
    - Integrating Science and Technology with World Affairs 6
    - The COVID-19 Pandemic: A Classic Example of the Survival Nexus 7
    - Science and Technology are Not the Same Thing 8
    - Scientific Models and Scientific Uncertainty 9
    - Technological Innovations don't Always Happen by Themselves 10
    - Needed: Strengthened Regimes and New Codes of Conduct 12
    - Global Responses to Global Problems 13
2. Stratospheric Ozone: A Success with Complications 15
    - The Science Behind the Stratospheric Ozone Issue 17
    - The Ozone Hole and the Montreal Protocol 18
    - Clear Scientific Evidence Spurs International Action 20
    - The Scientific Explanation for the Ozone Hole 21
    - Lessons of Success 22
3. Climate Disruption, Not Just Global Warming 24
    - Causes and Effects of Climate Disruption 24
    - The Origins of Climate Change 27
    - The Damage from Global Warming, Degree by Degree 30
    - Greenhouse Gas Budgets and Scenarios 32
    - Climate Negotiations Veer from the Model of Stratospheric Ozone 34
    - The Framework Convention and the Kyoto Protocol 35
    - Market-Based Mechanisms to Reduce Emissions More Cheaply 36
    - Disinformation and Distraction 38
    - Institutionalizing Scientific Advice 40
    - The Paris Agreement: Voluntary, Individual Country Commitments 42
    - The Limitations of Science 45

4. How Do We Address Climate Disruption? 47
Transformation in National Energy Systems 48
Restructuring the Electric Grid 49
Mitigation by Switching Energy Sources 50
Fossil fuels 51
Carbon capture, use, and sequestration 51
Renewable energy 53
Energy storage 53
Nuclear fission 56
Efficient Energy Use 59
Agriculture 59
Urbanization 60
Buildings 61
Transportation 61
Industry 62
Adapting to Climate Disruption: Intervene as Well? 62
So Much to Learn, So Little Time 63

5. Nuclear Issues: Civilization in the Balance 67
Instant, Catastrophic Destruction 67
Mutually Assured Destruction and a Close Brush with Nuclear War 69
Arms Control: Avoiding Mutual Suicide 71
Nonproliferation: Fewer Nuclear Countries Make a Safer World 74
The 1990s: Ups and Downs for Arms Control 77
Iran and North Korea: The Nonproliferation Regime Takes on More than It Can Handle 80
The Second Nuclear Age: Civilization is Still in the Balance 82
Lessons from Nuclear Experience 85
Hypersonic Missiles: Even Shorter Times to React to a Possible Nuclear Attack 86
Fully Autonomous Weapons: Killing Without Human Involvement 88
Why Autonomous Weapons? Why Not? 89
Arms Control for Autonomous Weapons 91
Can We Handle the Nuclear Future? 92

Annex A. A Primer on Nuclear Science and Technology 95

6. Global Health: Security and Inequality 103
Influenza: The Annual Pandemic Zoonosis 105
We're All in this Together 106
Vertical and Horizontal Health Services 108
The World Health Organization and the Politics of International Cooperation 109
"Tropical Health": The Colonial and Postcolonial Background 111
The 1970s: Is Health a Human Right or an Economic Investment? 112

The 1980s: Tough Times for Global Health 115
The 1990s: HIV/AIDS and the Resurgence of Global Health 116
The 2000s: Improved Response to Pandemics 118
The Human Right to Health as a Workable Principle 119
COVID-19 Brings the Survival Nexus into Stark Relief 121
Early Chinese and U.S. Responses to the COVID-19 Pandemic 123
Newly Developed Vaccines Raise Issues of Equity 126

Annex B. Science and Global Health 128

7. Globalization and the Burden of Disease 143
Demographic Transitions and the Quadruple Burden of Disease 144
Health, Globalization, and Trade 145
Bioprospecting and Clinical Testing in Low-income Countries 148
Health and Military Security 149
Who Makes the Decisions Regarding Global Health? 151
Constructive Tension: Health Security and "Health for All" 153

8. The Internet and Social Media: Euphoria and Repression 155
Convenience and Repression 156
The 1970s and 1980s: Early Euphoria Plants the Seeds of Later Issues 157
The 1990s: "The Babe That Roared" 160
2000 and Beyond: The Free Internet Faces Competition 162
The Chinese Government Masters the Internet 163
Democracies have Internet Issues, Too 166
Governance: Who, If Anyone, Will Control the Internet? 168
A Less Open Internet in the Future? 170

9. Cyberwarfare and Cybersecurity 173
Stuxnet and its Aftermath 173
The Snowden Revelations: The Crown Jewels of U.S. Intelligence Spill onto the Table 176
Challenges to Democratic Values and U.S. Dominance 178
Arms Control for Cyberweapons? 183
First Steps Toward Rules of the Road for Cyberconflict 185

10. Frugal Innovations for the "Bottom of the Pyramid" 188
Mobile Finance: The Cell Phone Fills a Banking Vacuum 189
Some Technologies Won't Trickle Down; They Need Special Attention 191
"Appropriate Technology": Philanthropy Plus Engineering 194
Can Frugal Technology be a Business Opportunity? 195
Businesses Without a Profit Motive 197
"Orphan Technology": A Need But No Market 199
The Best You Can Afford: The Ethics of Toilets 200
The CGIAR: Almost $1 Billion a Year for Research for the Small-scale Farmer 202
"Digital Green": Localized Videos on Sustainable Agriculture 210
Technology for the Poor: From Invention to Innovation 210

11. Jobs, Competitiveness, and Inequality: Up the Down Escalators in Manufacturing 213
Global Value Chains and Networks 214
Some Jobs Will be Lost as Productivity Increases, But Others Will be Created 217
The New Jobs Will Demand Different Skills 219
South Korea: A Classic Model of Industrial Development 220
A Worldwide Scramble to Create Good Jobs 223
Manufacturing Employment at the Cutting Edge 224
Manufacturing as a Source of Innovation 225
A Tougher Path for Low- and Middle-income Countries 228
The Rise of China 230
Political Control Versus Technological Innovation in Chinese Industry 233
The Future of Work 234

12. New Technologies, New Issues: Gene Drivers and Geoengineering 237
Gene Drivers: A Technology that Controls Evolution 238
The Pandora's Box Congress 241
The Ethics of Do-it-yourself Biotechnology 242
Negative Emissions Technologies 245
Research Collaboration and Commercial Competition 247
The Ethics of Climatic Intervention 248
Responsible Innovation 249

13. Playing with Fire 251
Stimulating and Guiding Technological Change 252
The Rise of Russia and China 256
Science and Technology as Pervasive Influences on World Affairs 257
Crossing and Uniting Disciplines 259
The Questions We Have Been Asking 260
Norms for Applying Science and Technology to Global Problems 261
Respect for facts and evolving knowledge 262
Cooperation and collaboration 263
Avoiding harm and minimizing risk 263
Equity, sustainability, and inclusivity 263
Adding to knowledge and developing technology 264
Accountability 264
Ethics are Fundamental 264

Acknowledgments 267
Notes 269
Bibliography 317
Index 347

Author's Preface

My introduction to the nexus of science, technology, and world affairs came in 1971, when I had to bone up on the topic of nutrition during my first few weeks as the World Bank's first science and technology adviser. I had come to this job directly from a basic scientific research laboratory on the first light-driven steps of photosynthesis. Each group of experts that I consulted viewed the subject of nutrition through a different lens. To the doctors, malnutrition meant starvation—little children with swollen bellies and matchstick arms and legs, or with vitamin deficiencies that cause rickets or beriberi. To the agriculturists, malnutrition came from insufficient food production due to poor farming methods. The food technologists pointed to crop losses from spoilage, insects, and rodents. The food distribution experts blamed poor roads and poorly run government grain storage monopolies. Political economists told me that farmers in many low-income countries were forced to sell their crops to these government monopolies cheaply at harvest time and to buy them back at high prices during the "hunger time" at off-season, as a way of taxing agriculture to provide money for investments in industry.

This last insight was my introduction to the social science of malnutrition. There was more to come. To economists, malnutrition comes not from any overall shortage of food but, rather, from poverty. Low-income people may not have enough money to buy food. After all, you can starve right outside a grocery store if you haven't any money. To city planners, the problem is urban food deserts where there's no place to buy or grow fruits and vegetables for a balanced diet. To educators, malnutrition is due to lack of information or understanding about what foods to eat. To sociologists, it can be the result of family structure. In some cultures, the father gets to eat first because he is the male breadwinner; the mother and children get whatever he leaves over.

The nexus between science, technology, and world affairs is critical to the survival of humanity. It is fundamental to a broad range of issues: war and peace, environmental sustainability, epidemic disease, the future of the human genome, personal freedom, access to information, economic development, and world poverty and hunger. It's a subject of importance to everyone. That's why I wrote this as a book for everyone, rather than as an academic treatise.

It is an issue for international affairs professionals, too. Despite their importance, the scientific and technological dimensions of these issues tend to be neglected. There is an acute need for people who can bridge the gap between science, technology, and international affairs. As far as I know, there is still no question on science in the examination required for all candidates for the U.S. Foreign Service! And for their part, natural scientists tend to be immersed in their own disciplines, which are demanding enough without the extra work of learning economics and politics.

This gap has implications for university curricula. The links that connect science and technology with world affairs are not generally recognized as having a coherent intellectual framework or as constituting an academic discipline. True, there are many academic programs that deal with these connections as they apply to specific areas: energy, environment, security, health, and many others. But as far as I know, the program in Science, Technology, and International Affairs (STIA), which is now the second most popular major at the Georgetown University School of Foreign Service, where I taught for seventeen years, is still unique in its broad scope.

I am now emeritus from Georgetown and would like to see many more programs like STIA in universities in the United States and abroad. I hope that my book will be useful to this end. This will happen only if institutions of higher learning, which are structured around disciplines, feel pressure from faculty, students, and employers to make room for a new, multidisciplinary endeavor.

I began writing this book with the limited objective of establishing science, technology, and world affairs in the mind of the general reader as an important subject with a clear intellectual framework that cuts across these specific topics. Since the most recent book of comparable scope, Eugene Skolnikoff's excellent work, *The Elusive Transformation: Science, Technology and the Transformation of International Politics*, is now more than twenty-five years old, I adopted a long-term perspective, emphasizing background history and policy issues that are likely still to be of interest many years hence.

As I came to know more about developments in global health, in climate change, in cybersecurity, and especially in nuclear security, I found that the dangers we face are much more urgent and serious than I had realized when I began writing. The outbreak of the COVID-19 pandemic brought at least one of these dangers home to everyone. The tone of the book became much less dispassionate, as I felt the need not only to establish the subject as an important and neglected field of policy, but also, at least as importantly, to sound the alarm that our failure to control technology is taking us to the brink of several worldwide disasters. Even so, we still have time to ensure that science and technology help shape the world in the way we want it to look.

The book guides the reader through the nexus of science, technology, and world affairs as it affects the major global issues of climate change, renewable energy, global health, nuclear weapons, the Internet, cybersecurity and cyberwarfare, competitiveness in global markets, poverty and hunger, gene drivers, geoengineering, autonomous weapons, artificial intelligence, and other new technologies. I chose these issues to illustrate different patterns by which politics, economics, business, law, ethics, psychology, and culture mix with science and technology, and different patterns by which science and technology exert their impact on international affairs. This book is far from the final word on any of the topics I have discussed, and there are many more subjects equally deserving of attention: oceans, water, polar regions, agriculture, space, terrorism, science diplomacy, and technology transfer, to name just a few. I leave these to younger authors than myself, in hopes that the framework I have developed will be useful to them in relating their work to the broader picture of science, technology, and world affairs.

Each chapter provides basic information about the science or technology that has given rise to a problem of global policy. I wove some of this background into the main text and put some into endnotes at the end of the book for each chapter in order not to interrupt the flow of the story. I have introduced and explained technical terms that readers are likely to encounter if they pursue particular subjects in more depth. The scientific questions on global health and on nuclear issues are more numerous than those on other topics, so I have grouped them into annexes at the end of the two relevant chapters.

Each chapter laments the decline of global efforts to deal with an important issue that knows no borders and demands international cooperation, a decline that has typically taken place over several decades. I have done my best to hew closely to the facts as I understand them and to avoid discussion of personalities, so that my words will be accessible to readers of all political persuasions. Understanding of the critical role of science and technology in world affairs, and that of world affairs in the advancement of science and technology, is particularly important at a time when the legitimacy of scientific fact—and, indeed, of facts of any kind—is under systematic attack in many parts of the world.

Charles Weiss

1
Introduction
Science, Technology, and Survival

Science and technology and have given us longer and healthier lives, opened new intellectual horizons, put information and entertainment at our fingertips, lifted hundreds of millions of people out of poverty, enabled us to buy products from all over the world, and made it easy to spread ideas, to launch new businesses, to travel all over the world, and to keep in touch with almost anyone anywhere.

At the same time, science and technology lie at the heart of some of the greatest dangers to the survival of humanity and the sustainability of the Earth: climate change, pandemics, and nuclear war. These dangers are real, affect everyone, know no borders, and demand our attention. Science and technology are central to issues of competitiveness, jobs, inequality, cyberconflict, and the freedom of the Internet. All these issues involve a tangled mix of science and technology with politics, economics, business, law, psychology, ethics, security, and culture—the Survival Nexus that is the subject of this book.

Existential Dangers: Pandemics, Climate Disruption, and Nuclear War

The COVID-19 pandemic that began in December 2019 gave the world a taste of the costs of neglect of the Survival Nexus. One government after another ignored the warnings of scientific experts, downplayed the seriousness of the virus, and squandered critical weeks that could have been used to limit its spread and to save many of the millions of lives that were ultimately lost, in addition to the trillions of dollars in economic losses. The exponential spread of COVID-19 forced political leaders around the world to order belated lockdowns of businesses and people, causing a global economic recession that hit low-income people the hardest. Political leaders, some of whom had taken firm measures of control early in the pandemic, succumbed to political and social pressures to reopen too early, causing still further disease and disruption.

Climate change will cause much more significant long-term damage if it is not stopped, although the damage will take place at a slower pace. It will cause a rise in sea levels that will flood coastal cities, beaches, and low-lying islands around

The Survival Nexus. Charles Weiss, Oxford University Press. © Charles Weiss 2022.
DOI: 10.1093/oso/9780190946265.003.0001

the world. It will intensify droughts, storms, and hurricanes. It will spread mosquito-borne diseases to areas from which they have long been eliminated. It will dry out the Amazon Valley and reduce food production in many parts of the world.

Climate change is better described as climate *disruption,* since its effects are ubiquitous, vary from place to place, and will soon affect almost all aspects of daily life. The offenders here are carbon dioxide and other greenhouse gases that are emitted by all forms of human activity, but especially by industries, vehicles, buildings, and electric utilities that still burn fossil fuels. To limit climate disruption will require a major reorganization of the world economy that will depend, in turn, on a change in attitude, especially in the United States, where many people still do not believe that human-caused climate change is real and dangerous. This is an ethical as well as an environmental issue. Citizens of low-income countries and future generations will suffer the most from climate change even though they have contributed the least to it.

The risks of pandemics and climate disruption pale before yet a third danger, that of nuclear war, which can wipe out millions of lives in just a few minutes and possibly bring an end to human civilization. With no nuclear war in seventy-five years, the prospect has receded from public consciousness. Yet hundreds of nuclear missiles, each capable of wiping out a major city, are on hair-trigger alert in both the United States and Russia and could be fired, intentionally or by mistake, in a moment of crisis. Nuclear war between regional powers like India and Pakistan could also have devastating consequences for everyone.

The world is playing fast and loose with technologies that are taking us into uncharted and dangerous territory. We have come within a hair's breadth of accidental nuclear war several times in recent decades. Nevertheless, political leaders have been increasingly willing to risk nuclear war in order to gain short-term advantage. Countries are competing to be the first to develop and deploy hypersonic missiles and autonomous weapons combined with artificial intelligence. These weapons may be impossible to control and make accidental escalation or a catastrophic accident even more likely. Meanwhile, the safeguards that political leaders, diplomats, and scientists have erected over the decades to address these and other global threats have gradually been undermined and eroded, and technologies like renewable energy that are badly needed face unnecessary obstacles to their development and deployment.

New Technologies Raise Issues of Ethics and Values

The revolution in information and communication technologies, while not an immediate a risk to human life like nuclear war, has given rise to a global conflict over whether these technologies are to continue to embody values of access to information and freedom of thought, expression, and innovation. That such a conflict over values might develop was not at all apparent when the Internet was originally designed by scientific researchers, for whom these freedoms were a given and the pursuit of truth was a way of life. Internet enthusiasts hoped that it would be impossible to control the free transmission of information across borders, and that the Internet by its very nature would undermine undemocratic governments, increase civil liberties, and lead to desirable political change.

A seemingly unstoppable juggernaut when it was first launched, the Internet facilitated globalization, disrupted industries, created new forms of warfare, and at first eluded efforts at government control. The technologies underlying the information revolution benefited from U.S. government support to basic science, and from the much larger research budget of the military, for whom even a small technological advantage—the speed and maneuverability of a fighter airplane in a dogfight, for example—can be the difference between victory and defeat, life and death.[1] Spin-offs from military research were critical to the development of the technologies underlying the computer, the Internet, the smartphone, and the global positioning satellite, to name just a few examples.

The values of an open society that were embedded in the infrastructure of the Internet provoked a counter-reaction from authoritarian governments from North Africa to China that did not share these values and soon became aware of the threat that they posed to their power. After a stumbling start, these governments figured out how to use the new technology to control the flow of information and to keep track of what their people are saying and doing—the exact opposite of the intent of the original innovators.

These cyber-authoritarian regimes then used the Internet and social media as weapons for a worldwide counterattack on liberal and democratic values. China has developed a comprehensive, technology-based system for repression and is exporting it to other authoritarian countries. Russia and its allies use the Internet and social media, along with television and other traditional media, for systematic campaigns of propaganda and disinformation intended to influence democratic elections, undermine Western alliances, erode trust in science, and whittle away at the idea that there is such a thing as objective truth. Terrorist organizations use the new technology for recruiting and propaganda.

The ongoing revolution in manufacturing technology both creates and destroys jobs and in this way gives rise to ethical and social issues of a different kind. The new jobs it creates are likely to be in different places and to require a higher level of education and familiarity with computers than those being lost. This "skill-based technological change" will continue long-standing trends of reduced opportunities for less educated workers in high-income countries, worsening inequities and fueling popular resentment, unless educational systems can be upgraded and adapted to the new requirements.

The consequences of the manufacturing revolution will be even more serious in low-income countries, where it means fewer job opportunities in labor-intensive industries like textiles and clothing. In countries like South Korea, these industries were the first rungs on the ladder of economic development. At the same time, agriculture, another major employer, is being hit by climate change. We may expect the resulting unemployed to try to emigrate in large numbers to higher-income countries, potentially provoking an increased populist and xenophobic backlash.

Strong arguments can be made for both exploiting and banning the application of gene drivers and geoengineering. Applied to human germ cells, gene driver technology offers the promise of reducing the frequency of inherited diseases, but at the risk of permanent and unpredictable changes in the genetic heritage of future generations. Geoengineering technology, if it can be developed and applied at scale, could reduce global warming, but at the risk of tempting the world to undertake a permanent commitment to try to control climate change without reducing emissions of greenhouse gases. These are not simple issues. They require a careful balancing of benefits and risks that can best be addressed with a clear understanding of the scientific and technological concepts at play, and how these relate to the broader context of world affairs.

A Pervasive but Neglected Dimension

Science and technology are ubiquitous dimensions of world affairs. Nuclear weapons have upended geopolitics, redefined great power status, and made obsolete the idea of winning an all-out war. Information, communication, and transportation technology have made possible the globalization of trade and investment and the global circulation of people and ideas. Information technology has also given rise to a new form of warfare and has become the centerpiece of a new kind of technological geo-economic and geopolitical conflict. Scientific researchers bring us a constant flow of information demanding urgent international action on environment, climate change, and health. Many more international issues have science and technology at their core: investment, finance, narcotics, terrorism, pollution, oceans, water, fisheries, and the polar regions, to name just a few.

Developments in science and technology raise a host of global policy issues. Some of these are new, like the international governance of the Internet, gene drivers, and geoengineering. Others are familiar from other areas of international policy: How do political leaders get expert advice and what do they do with it? How are decisions made when the facts are still unclear? How are new ideas and new disciplines introduced and integrated into long-established "legacy" institutions? Who gains and who loses from any given action? Who bears its costs and risks? Who is to make the decisions? How and to whom are they to be accountable? How are the public and civil society to be informed and involved?

International policymakers, business leaders, diplomats, military personnel, and experts need to understand the scientific and technological dimensions of the problems they are addressing. This requires disciplinary knowledge that is not usually familiar to international affairs professionals and that often suffers from neglect on their part. Those who do have this understanding have often been hived off from the professional mainstream and relegated to the role of specialists who devote themselves to subjects they find interesting, challenging, and important, even at the risk of losing opportunities for professional advancement. This disciplinary divide between the two cultures of science and technology, on the one hand, and world affairs, on the other, is anomalous and even dangerous at a time when science and technology are a critical dimension in foreign policy and international business.

The public, too, needs a basic understanding of science and technology and how they relate to the broader context of world affairs. There is, therefore, every reason for people interested or expert in what is happening internationally, and for people interested or expert in science and technology, to become familiar with each other's concerns and with the basic concepts of each other's fields. This has important implications for our high schools and colleges.

To be sure, there is no way for any one person, whatever their professional qualifications, to become an expert in all of the many fields of science and engineering that have important ramifications for international affairs: molecular biology, epidemiology, nuclear physics, geophysics, seismology, computer science, climatology, stratospheric photochemistry, solid-state physics, aerodynamics, and manufacturing engineering, to name only some of the fields touched on in this book. But neither can they be expert in all the fields of social science and the humanities that influence world affairs, a list that includes economics, politics, law, business, history, sociology, psychology, anthropology, linguistics, ethics, and religion—and that's just to start. In natural as in social sciences, we can learn enough to understand basic facts, to relate these facts to the general context of world affairs, and to detect obvious falsehoods.

The meshing of science with world affairs involves not only collaboration among people trained in different disciplines, but also a mutual understanding on the part of people accustomed to very different roles, disciplines, and professional cultures. Science is a search for truth about the world around us, sometimes motivated by sheer curiosity and sometimes by the hope of practical application (applied science).[2] In science, until an assertion is confirmed by empirical evidence, it is a mere hypothesis or conjecture. Professionals in the field authenticate research results as being backed by convincing evidence, a process known as peer review. In the world of science, controversy is normal and knowledge is always developing. A new theory is accepted when there emerges a consensus of the scientific community that it explains the available evidence better than the theory that it claims to replace. (In science, a theory is not a speculation, as it is in informal conversation but, rather, is a comprehensive explanation of apparently unrelated phenomena, like the theories of gravitation, evolution, or relativity).

Once they have earned the consensus of the scientific community, scientific facts are facts—or at least as close to facts as humans can make them—and will outlast any efforts to ignore them. Up to that point, to most scientists at least, inadequate understanding means that there is a need for more research and for more funding with which to do it. They are by and large instinctively reluctant to speculate in public on the answer to an unsettled question. They expect themselves and their colleagues to stick to the science and can look down on those who venture into popular education or advocacy. In contrast, policymakers need to make a decision here and now, based on the information available, however incomplete it may be. They are typically trained in the social sciences, find science and technology to be alien and forbidding, are unfamiliar with the sociology and cultures of the scientific and engineering community, and do not instinctively turn to scientific research and technological innovation as a way to help to achieve their objectives.

Scientific advisory bodies play a key role in explaining and interpreting scientific findings, including existing uncertainties, in a form that policymakers can understand. These bodies can come under fire, especially when the science is still uncertain, either because of genuine disagreement among researchers or (as in the case of climate and vaccine deniers) because of deliberate efforts to obscure an existing consensus.

Integrating Science and Technology with World Affairs

The world's experience with the integration of scientific and technological issues into broad international strategy has been mixed. A particularly important example on the positive side is in the area of nuclear arms control, in which U.S. nuclear scientists, political scientists, and diplomats worked closely with their counterparts in the Soviet Union and in many other countries in the 1950s and 1960s to think through the technical and political implications of the idea—radical at the time—that the world would be safer with fewer nuclear weapons and fewer missiles with which to deliver them. The resulting regimes for arms limitation and nonproliferation have been successful for many decades, although they are now under stress.

The story of the international effort to protect the stratospheric ozone layer, which shields ocean ecosystems from disruption and protects us from skin cancer, which we will discuss in chapter 2, is another successful example. After research scientists discovered the ozone hole and brought it to the public's attention, governments across the world worked with scientists, nongovernmental organizations (NGOs), and business to agree on a solution. The international regime that emerged was grounded in facts, was equitable to lower-income countries, and protected the rights of future generations. Chapter 3 recounts how decision makers tried to apply this model to climate change, with much less success.

The COVID-19 Pandemic: A Classic Example of the Survival Nexus

The early stages of the COVID-19 pandemic provide a more complicated example of the interaction of science with politics and economics. As we shall see in chapter 7, critical decisions needed to be made and communicated to the public in real time even as scientific understanding of the virus and the ways in which it spread was still unfolding, so that the recommendations of public health experts for action were constantly changing. In response, scientists from all over the world collaborated to unravel the secrets of the virus and to develop more than a hundred candidate vaccines in record time. Governments and private foundations responded quickly with financial support, both to basic research and to vaccine development. In order to encourage innovation and save time, they made unprecedented advance financial commitments to purchase hundreds of millions of doses of promising vaccines, recognizing that the resulting production facilities would be useless if the vaccines did not pass tests of efficacy and safety.

At the same time, the classic problems inherent to the links between science, technology, and politics emerged. Political and religious leaders in several countries made public statements minimizing the importance of the pandemic, suppressing scientific evidence and epidemiological data or cherry-picking it to fit their preferences, hyping or hawking unproved "cures," suppressing or censoring the remarks of professional experts and attacking them ad hominem, and pressing for reopening the economy long before it was safe to do so. There were political pressures to award contracts to politically favored companies or to speed vaccine development by cutting corners, as well as ignorant or politically inspired campaigns against wearing masks, social distancing, and vaccination.

Scientists, too, made important mistakes. Epidemiological models paid insufficient attention to the economic and political costs of the shutdowns they were recommending. There were also mistakes in the United States in managing the development of technologies for testing people who may or may not have been infected with the virus—a key measure for controlling its spread.

Science and Technology are Not the Same Thing

If we are to understand how science and technology relate to world affairs, we need to be clear about how science and technology differ from each other and how they relate to one another, how scientists express their understanding of the world through theories and models, and how economics and politics influence the direction of research and the dynamics of technological innovation.

We begin with some definitions. When I speak of technology, I mean any technical knowledge that is organized for a practical purpose—not just computers, not just hardware, and not just high-tech. By this definition, technology ranges in sophistication from a pit latrine for basic sanitation to space technology for planetary exploration.[3] Both are of value to humanity. When I speak of science, I mean knowledge of the human and natural world, or the way in which scientists derive this knowledge from a combination of theory and experiment, or the way in which scientists communicate with each other and arrive at a consensus as to what is a fact. Science and technology are not the same, and "scienceandtechnology," as it is often pronounced, is a plural, not a singular noun.

Science, technology, and world affairs influence one another. Science leads to technology. The laser, for example, emerged from basic scientific studies of how ammonia absorbs and emits microwaves. It made possible the technology of the CD and the scanners in supermarket checkout stands, both of which were far from the minds of its inventors. Conversely, technological advance leads to advances in basic science. As an example, the Hubble Telescope, a triumph of space technology, transformed cosmology, the study of the beginning of the universe, from a subject of philosophical and religious speculation to a field of empirical scientific research.

Completing the triangular relationship, the speed and direction of advances in both science and technology are influenced by world affairs. As examples on the technology side, international treaties on arms control, environment, and human rights gave rise to numerous innovations, like using remote sensing to identify nuclear test sites, to measure forest loss and pollutant emissions, and to detect the locations of concentration camps and mass burial sites. In science, seismology grew in importance and sophistication when seismologists were challenged to distinguish earthquakes from atomic explosions. The sciences of meteorology, climatology, and human geography likewise flourished when they were asked to support the understanding of climate disruption. World affairs influence even the financing and conduct of basic scientific research. During the Cold War, for example, U.S. research support tilted strongly to physics and engineering. When the Soviet Union fell apart, public support shifted to the health sciences.

Scientific Models and Scientific Uncertainty

In both the natural and social sciences, knowledge typically takes the form of a model, an intellectual construct that helps us to understand some portion of the world, usually, but not always, involving mathematical tools. Models help bridge the gap between scientific experts and government policymakers and negotiators. In the natural sciences, for example, models help predict the course of epidemics and the consequences of climate change. In the social sciences, they help to predict whether a new weapon will make it easier or harder to prevent a geopolitical crisis from escalating into accidental war.

In a scientific model, data and scientific principles go in, and predictions come out. No model is a perfect representation of reality; a lot of things can go wrong. If the inputs are faulty, the outputs will be, too. Model builders may leave out things that they assume to be unimportant for their particular purpose, or because they lack understanding of matters outside their own discipline. Critical data may not be available when it is needed most—or may depend on future human actions about which there can only be speculation.

Still, a model can be useful even if it does not give accurate predictions, depending on how it is used. As an example from the COVID-19 pandemic, the key determinants of the rate of the spread of virus infection depended not only on biological and physical factors that were not clear in early stages of the pandemic, but also on people's willingness to abide by mask wearing, social distancing, and other measures to protect public health. This, too, was also impossible to predict at these early stages. The result was that even the best initial predictions of when the pandemic would peak in different places were essentially educated guesses, at a time when important strategic decisions had to be made regarding the allocation of life-saving protective equipment and medical devices and the shutting down of important parts of the economy. Even so, early models of the epidemic were sufficient to allow experts to make an important contribution to public policy, namely, to advise that efforts be focused on "flattening the curve"—i.e., on stretching out the duration of the epidemic and reducing the height of the peak rate of expected casualties—to reduce the total number of victims and to prevent healthcare systems from being overwhelmed with critically ill patients.

Part of the job of model builders is to explain the limitations of their models. In principle, politicians and policymakers can then decide whether or not to rely on them. Less happily, they have been known to use this uncertainty as a handy excuse to choose the model that gives the results most to their liking, so as to reach a more politically convenient policy conclusion. This happened in the early stages of the COVID-19 pandemic and of the debates over stratospheric ozone depletion and climate change. In the latter two cases, better models, fed with better data and run on more powerful computers, gradually resolved the uncertainties and confirmed the seriousness of the threats. In the ozone case, the level of international regulation increased as the level of scientific understanding improved. In the climate case, however, increased scientific understanding has not yet led to sufficient international support for effective measures to address the problem.

Technological Innovations don't Always Happen by Themselves

Some of the most revolutionary innovations in history, from nuclear power to lasers to tomographic (CAT) scanners, have resulted from applying the results of basic scientific research. Others have resulted from accidents, as when Alexander Fleming discovered penicillin by noticing that bacteria did not grow near the mold in a neglected petri dish. But most innovations, from incandescent light bulbs to precooked dinners to social media, have been the work of entrepreneurs who sensed that there would be a market for their new and different ideas, sometimes with new technology and sometimes by putting technology that already existed together in a new way.

In other words, most innovations are products of the "magic of the marketplace." This magic works only when there is a commercial market for the innovation. A compelling need is not enough. Potential users must be willing to pay for the innovation and must have the money to buy it. Otherwise, there will be no market to attract private entrepreneurs to invest in the necessary research, new product development, and manufacturing. If there is a need but no market for a particular innovation, only government, private philanthropy, or some other nonprofit entity can make it happen.

This directly affects the prospects for innovations in energy technology. Emitters of carbon dioxide and other greenhouse gases are not forced to cover the costs that they impose on others by way of pollution and climate change (economists call these externalities). As a result, the price users pay for a barrel of oil or a ton of coal does not represent its true cost, and greenhouse gas emitters have no financial incentive to invest in ways to reduce their emissions. There may even be "perverse subsidies" that advantage those who damage the environment. The government subsidies and tax advantages available to coal, oil, and natural gas all over the world, for example, dwarf those for renewable sources, a reflection of the political and economic power of the companies that produce them. This stacks the odds against renewables, despite the fact that solar and wind are not only better for the environment but are now the cheapest and most convenient sources of energy in many situations.

Similar logic applies to the protection of the stratospheric ozone layer, which protects us from skin cancer. As long as refrigerators and aerosol cans dependent on ozone-destroying chlorofluorocarbons (CFCs) could be sold without restriction and as long as their manufacturers and users did not have to pay for the damage that these chemicals were causing to the stratospheric ozone layer, they felt little need to invest in developing substitute technologies. When CFC production and sale was prohibited by international agreement, manufacturers quickly developed and marketed suitable replacements. Technological innovation was critical to solving the stratospheric ozone problem, but it took place only in response to major changes in policy.

Innovations addressing the special needs of low-income people provide more examples of orphan technologies: compelling needs that do not constitute profitable markets. It takes a lot of time and money—and an appetite for risk—to develop a new product or service and launch it on the market, and most innovators want to recoup their investment quickly. This usually means that a high-end product is first developed and marketed to high-income people, who like to have the latest and the best, and is scaled down for lower-income buyers only later. If there is no high-end market, the innovation may never be developed.

This dynamic shows up most dramatically in global health. Diseases like malaria that afflict hundreds of millions of low-income people attract little investment from private business, unlike non-life-threatening illnesses that affect high-income people. Public investments in research, development, and service infrastructure are likewise largely aimed at the problems of high-income people with more political clout. The good news here is that technologies developed for high-income customers sometimes trickle down to low-income people with dramatic speed. Here an example is the cell phone, the fastest spreading technology in human history, which has become so cheap that just about everyone can afford it—and, as a bonus, can be used for mobile finance and for spreading information on nutrition, public health, and improved agricultural practices.

Still, targeted efforts by governments or nonprofit organizations are often needed to solve specific problems unique to low-income people. Happily, malaria and other parasitic diseases that affect hundreds of millions of people have attracted support from governments and private philanthropic foundations, and several cures and vaccines are now under testing for these and other previously neglected diseases. Other technologies to meet needs specific to low-income people are low-cost, "frugal" innovations, which we explore in chapter 10. Examples include an emergency food for the severely malnourished, fly- and odor-resistant latrines, village-level systems for cataract surgery, a simple and cheap artificial foot, and rats trained to detect land mines and diagnose diseases. These worthy innovations do, however, raise the ethical question of why low-income people should have to settle for such "second-best" solutions, when better, although more expensive, technologies are available to people who have more money.

Needed: Strengthened Regimes and New Codes of Conduct

The global issues we are addressing in this book—stratospheric ozone depletion, climate change, nuclear arms control, global health, Internet governance, cybersecurity, information technology, geoengineering, and gene drivers—demand a global response that integrates science and technology with politics, economics, law, business, ethics, and culture, in what we have called the Survival Nexus.

To deal with some of these issues, the world has developed global regimes, usually enshrined in formal treaties among governments. These regimes are made up of established norms of international behavior, intergovernmental institutions, scientific advisory committees, national targets and reporting requirements, inspections to ensure compliance, mechanisms for dispute settlement, hotlines for crisis communication, and much more. Of these regimes, the most important are those for nuclear arms control and nonproliferation, which have been instrumental in preventing nuclear war but which are now eroding and in danger of unraveling. This is in part because of new weapons systems and changed geopolitical alignments, but mostly because of growing distrust of multilateral arrangements and organizations.

The COVID-19 pandemic exposed the consequences of the long-standing neglect of another important, long-standing international regime, namely, the one charged with responsibility for promoting and protecting the world's health. At the apex of this regime is the World Health Organization (WHO), the United Nations technical agency responsible for expert global leadership and coordination of the global response to pandemics. The WHO has long been hampered by inadequate and inflexible funding and a management and staff constrained and undermined by political controls and commercial pressures, again reflecting a more general distrust of multilateral organizations. Bilateral agencies and private foundations have stepped in to complement its work, but the resulting ad hoc system is still struggling with its important global mandate.

Yet another need for norms arises from the advent of gene drivers, cyberwarfare, hypersonic missiles, and autonomous weapons, all of which raise knotty technical and ethical issues that concern not only governments, but also businesses, nongovernmental and religious organizations, foundations, and even individuals all over the world. Academies of science and interdisciplinary international forums of various kinds are beginning the process of drafting proposed norms and codes of conduct for these technologies. Ideally, these will in time become widely accepted among people of different cultures and eventually form the basis for binding agreements among governments of different political orientations.

Global Responses to Global Problems

The stories we have chosen to tell in this book show how technology, properly managed, can be mobilized to address global problems. They also sound a series of alarms. Left to evolve on their own in response to commercial or military demands, technological innovation can lead to accidental nuclear war, rampant climate disruption, or out-of-control pandemics. All countries and all peoples, regardless of their culture, religion, or political philosophy, have an interest in avoiding these shared calamities.

We also need to defend against technology-based attacks on the values of freedom of thought, innovation, and access to information. We need agreed norms to limit the use of cyber technologies that can cripple critical health, water, financial, transportation, and financial infrastructure. We need to develop and put in place controls on technologies like gene drivers and geoengineering that are capable of both wonderful benefits and great environmental or genetic damage. We need to enable everyone to participate in the economic benefits of technological change, not just those well positioned to take advantage of it, and to help low-income countries generate jobs, adapt to climate disruption, provide basic health services, and fight infectious diseases that can spread beyond their borders.

These are global problems that demand global responses. They cannot be solved by the magic of the marketplace or individual countries acting alone or in pairs. They require international cooperation, involving governments, businesses, universities, nongovernmental organizations, religious institutions, researchers, and ordinary citizens, regardless of background or affiliation. They require agreement on international norms of behavior, so that all may benefit from advances in science and technology and so that new technologies will not get out of hand. They require confronting facts and complexities at a time when the very idea of scientific facts is under attack in many parts of the world.

In sum, these problems require attention to the Survival Nexus: the interweaving of science and technology with politics, economics, law, business, psychology, culture, and ethics.

2

Stratospheric Ozone

A Success with Complications

The successful defense of the stratospheric ozone layer is a template from what seems like long ago for how the international community can take timely measures to solve a global environmental problem brought to its attention by research scientists. There is much to learn, both from this success and from the difficulties that were encountered by the efforts to mobilize global action on stratospheric ozone. These difficulties will reappear when we consider climate change and other science-intensive global issues.

Two larger-than-life figures dominate the early phases of the ozone story: one a scientist and the other a politician. The first, a brilliant chemist named Thomas Midgley, had a greater influence on the chemistry of the Earth's atmosphere than anyone else, either before or since. Midgley's first invention was tetraethyl lead, the molecule that made possible leaded gasoline. His second was chlorofluorocarbons (familiarly known as CFCs), which were models of chemical non-reactivity, or inertness, at least when used at ground level.[1]

Hailed as miracle compounds, CFCs found dozens of applications as refrigerants, propellants for aerosol cans, foaming agents for plastics, and cleaning solvents for delicate electronics. A scientific triumph—until a series of scientific discoveries showed that just about every molecule of CFCs ever manufactured that wasn't inside an aerosol can, a refrigerator, or an air conditioner had ended up in the atmosphere. There, they catalyzed chemical reactions that damage the thin layer of ozone in the stratosphere that protects us from skin cancer by absorbing ultraviolet radiation from the sun.

The second man, Ted Hallock, was a powerful and legendarily profane state legislator from Oregon, a state that, then as now, prides itself on its reputation as an environmental pioneer. Hallock rammed through the state legislature a law prohibiting "nonessential" uses of CFCs, arguing that the convenience of CFCs as an aerosol propellant for hair spray and other consumer items did not justify even an uncertain risk of skin cancer. A few states followed suit.

Industry took the opposite view. Dupont, the major U.S. producer of CFCs, promised to stop making and selling CFCs if there was proof—by which it meant rigorous proof, the kind you can publish in a peer-reviewed scientific

The Survival Nexus. Charles Weiss, Oxford University Press. © Charles Weiss 2022.
DOI: 10.1093/oso/9780190946265.003.0002

journal—that they cause damage to public health. But if the science was still uncertain, they argued through a trade association that CFCs should be allowed to be manufactured and used. Dupont had provided the first serious financial support to basic research on the photochemistry of the stratosphere, which at the time was a neglected field of science, but this support faded in the deregulationist atmosphere of the 1980s.

The uncertainty expressed by Dupont was real enough. Clearly the ozone layer was in trouble—but how much trouble? Answering that question required a reliable computer model of the chemistry and the wind patterns of the stratosphere, which the computers of the day were unable to create. Skeptical newspaper editorials mocked the wide range of ever-changing estimates of damage to the ozone layer, as scientists tried to model complicated chemistry using uncertain empirical data on the interactions of upper atmosphere molecules with sunlight.[2]

These skeptics raised an important policy question: should a useful product be banned because of a risk of uncertain magnitude to the environment and to public health? Given the expected benefits of the manufacture and use of CFCs, what level of scientific certainty and what level of risk were required to justify regulatory action? Environmentalists urged a ban on CFCs even though the facts were still uncertain. Industry argued the opposite: act only when the evidence is sufficient to prove damage when judged by rigorous scientific standards, so that more research was needed before action was required. Later, when this argument was used to forestall action on climate change, this philosophical disagreement led to the development of the precautionary principle, which states that the lack of full scientific certainty should not be used as a reason to delay action to deal with threats of serious or irreversible environmental damage.[3]

But let us return to the story. A new, informal "epistemic community" of internationally minded scientists, journalists, researchers, nongovernmental organizations (NGOs), government officials, and international organizations arose to educate governments about a new kind of political problem that required a knowledge of both stratospheric photochemistry and international politics. Under the leadership of the Scandinavian countries and the prodding of the NGOs, a framework convention, called the Vienna Convention on the Protection of the Stratospheric Ozone Layer, was negotiated in 1985, by which countries agreed, in principle, that ozone destruction was a global problem and that the parties to the treaty would agree to a limit on CFC production in a subsequent "protocol," diplomat-speak for a more detailed follow-up. The meeting to negotiate the first such protocol was set for September 1986 in Montreal.

The Science Behind the Stratospheric Ozone Issue

The stratospheric ozone layer is formed
from the action of sunlight on oxygen.

Most of us are familiar with ozone. Those of us who live in cities encounter it on sunny days as a nuisance component of smog. In the stratosphere, however, fifteen miles or more above the Earth's surface, a small concentration of ozone is important to our health because it absorbs dangerous ultraviolet light that would otherwise reach the surface of the Earth and cause skin cancer. Stratospheric ozone also protects plankton—tiny transparent plants and animals that are a major source of food for fish and even whales—from lethal damage to their DNA and, hence, guards marine ecosystems against disruption.

An ozone molecule (chemical formula O_3) consists of three oxygen atoms linked to each other in a broken line. It is formed in the stratosphere when a free oxygen atom (O) combines with a molecule of oxygen—the kind we breathe, chemical formula O_2—which, in turn, consists of two oxygen atoms tightly linked to each other in the shape of a dumbbell, so tightly that it takes a lot of energy to split them apart into single oxygen atoms.

In the stratosphere, this energy can come from light. Light exists at the same time as a collection of waves and a collection of particles called photons (never mind how or why—quantum mechanics says that's the way it is). The shorter the wavelength of the light, the more energy each photon has. A gamma ray photon has more energy than an ultraviolet photon, which, in turn, has more energy than a photon of visible light, and so on down the electromagnetic spectrum to infrared radiation (heat), radio waves, and microwaves. All of these are forms of electromagnetic radiation. This means that a photon can have a wavelength even though it is a particle. Quantum mechanics. Go figure.

The sun puts out electromagnetic radiation across a broad range of wavelengths, from ultraviolet to infrared. This radiation gives rise to chemical reactions that first form and then destroy stratospheric ozone. The most energetic of these solar photons, ultraviolet photons of wavelength less than 230 nanometers, have enough energy to break the bond connecting the two atoms of an O_2 molecule and form two free oxygen atoms. (A nanometer, abbreviated "nm," is a billionth of a meter). This happens in the upper stratosphere. The newly formed oxygen atoms (O) collide with molecules of oxygen (O_2) and attach themselves to them to make molecules of ozone (O_3). These ozone molecules can absorb photons of slightly lower energy and longer wavelength (up to 270 nm), splitting them back into free oxygen atoms and molecular oxygen, rereleasing atomic oxygen, liberating energy, and warming the stratosphere.

The overall result of these reactions is to absorb the sun's ultraviolet radiation. The balance between the formation and the destruction of ozone in the stratosphere results in a stable concentration of ozone. (The analogy is to a bathtub with the faucet on and the drain open, resulting in a layer of water of unchanging depth in the bottom of the tub.) Before the invention of CFCs in 1930, the stratosphere had about three ozone molecules for every eight million molecules of nitrogen and every two million molecules of oxygen. This small but valuable concentration of ozone is known as the stratospheric ozone layer.

A single CFC molecule can catalyze the destruction
of one hundred thousand molecules of ozone.

In the stratosphere, CFCs are exposed to photons of ultraviolet light, each of which has sufficient energy to strip off a chlorine atom from the CFC molecule. The free chlorine (Cl) grabs the third oxygen atom off an O_3 molecule, destroying it, creating a molecule known as chlorine oxide (ClO) and recreating ordinary oxygen, O_2. But alas for the ozone layer, the process is not finished. The ClO collides with a free atom of oxygen (which had been created when an ultraviolet photon split an O_2 molecule into two atoms, as we described earlier). The free oxygen atom grabs the oxygen atom off the ClO molecule to reform O_2, leaving behind a chlorine atom for another go-around. To summarize, the cycle started with an O_3 molecule, an oxygen atom, and a free chlorine atom derived from a CFC molecule but ends with two O_2 molecules and the same chlorine atom, which is freed up to destroy another O_3 molecule, and so on. The cycle repeats itself one hundred thousand times, each time destroying an ozone molecule.

The upshot is that one ultraviolet photon from the sun splits a free atom of chlorine off a CFC molecule in the stratosphere, and the free Cl atom ends up destroying one hundred thousand molecules of ozone. There is enough CFC in the stratosphere for this process to drastically reduce the amount of stratospheric ozone and its ability to absorb harmful ultraviolet radiation. In this way, atomic chlorine facilitates the destruction of ozone without itself being transformed, acting as what chemists call a catalyst. The cycle continues until the atomic chlorine chances upon a molecule of nitrogen dioxide (NO_2), which changes it into a chemical (chlorine nitrate, or ClONO) that can no longer do any damage. The whole series of reactions is known as a catalytic chain reaction, and the nitrogen dioxide is known as a chain terminator. We'll need these terms later when we explain the chemistry of the ozone hole.

The Ozone Hole and the Montreal Protocol

Back to the diplomacy. The Vienna Convention on the Protection of the Stratospheric Ozone Layer set the stage for a second scientific shocker. Scientists discovered an ozone hole—an area of much-reduced stratospheric ozone the size of the United States, which appeared every spring in the southern hemisphere, centered over Antarctica but reaching as far as the population centers of Tasmania and southern Australia. Space-based instruments had measured these reduced ozone concentrations four years earlier, but the measurements had been ignored because they were so far from what had been expected that they were assumed to have been due to instrumental error!

The ozone hole was definitive evidence of a danger to public health, but were CFCs to blame? Two groups convened to deal with the issue. The first were the diplomats convened in Montreal; the second, a team of scientists huddled in an airplane taking air samples of the stratosphere over Antarctica, seeking telltale molecules of ClO that would pin definitive blame for the ozone hole on CFCs.

The diplomats in Montreal decided not to wait for the results from Antarctica. Scared by discovery of the ozone hole and under pressure to act despite the admitted lack of definitive proof that CFCs were to blame, the negotiators split the difference. The resulting agreement, called the Montreal Protocol, called for a 50% reduction in CFC production as compared with 1986 levels by the year 2000, together with a ban on CFC trade with countries that were not parties to the Protocol. This last restriction was to ensure that parties could not defeat the purpose of the treaty by importing CFCs from nonparties or by exporting CFCs to them. The trade ban was also intended to encourage countries to join the treaty and not to be "free riders" that benefited from the defense of the ozone layer but did not contribute to its costs. The 50% figure was a purely political compromise with no scientific basis. Science proposes but politics disposes.

The Montreal Protocol set up a regime of rules and institutions that served as the prototype for numerous future agreements dealing with international environmental problems. First was the format: a framework agreement that set forth the internationally agreed nature of the problem and broad strategy. A protocol and five amendments have followed the framework convention, setting forth specific measures.

Second, the parties agreed to provide annual reports on their efforts to phase out CFC production and use, and to meet at regular intervals in a so-called Conference of the Parties (COP) to review progress and to receive a formal technical review. This was not the first time such a technical review had taken place in support of the implementation of a treaty. Scientific advice for diplomatic negotiations went back at least to the 1946 treaty regulating whaling, but the ozone review was by far the highest-profile example of such an advisory system and became a precedent for other negotiations on global environmental problems. In the process, atmospheric scientists, at first a relatively insular and underfunded group, gained political sophistication as the results of their work became more and more important to international environmental diplomacy and as they themselves became more involved in both domestic and international political processes.

Third, the Montreal Protocol provided for an Implementation Committee charged with reviewing cases in which countries had failed to comply with the requirements of the treaty, and with developing practical measures to help them to do so. This replaced the pattern in some previous agreements that relied solely on judicial procedures to settle disputes and to impose penalties for noncompliance.

Fourth, the parties to the Montreal Protocol recognized that countries that consumed only small amounts of CFCs were not responsible for having put CFCs into the atmosphere in the first place, and that requiring these countries to do without them would be expensive and difficult, especially in tropical countries where CFCs were needed for refrigeration and air conditioning. They therefore allowed low-income, low-consuming countries an extra ten years to comply with the requirement of the Protocol to phase out ozone-destroying chemicals.

Fifth, the Protocol made it possible to speed up the phaseout of particular chemicals in response to new scientific evidence, without the need for the cumbersome process of going back to governments to amend the treaty. (Adding new chemicals to the list still requires an amendment to the Protocol.)

Finally, the Protocol made it possible for countries to graduate from low-consuming to high-consuming status, a seemingly obvious feature that was not made part of the Kyoto Protocol on climate change.

Clear Scientific Evidence Spurs International Action

Nine months after the adoption of the Montreal Protocol, scientists announced that CFCs in the Antarctic stratosphere were, indeed, the culprits that were giving rise to the ozone hole, thus resolving this uncertainty.[4] There was no further need to invoke precautionary action to forestall an uncertain risk. On the contrary, the danger was unquestionably real and demanded action. The first COP, held in London in 1990, therefore called for a complete end to the manufacture and consumption of CFCs.

Two further problems had become apparent by the time of the London COP. First, several other ozone-destroying chemicals needed to be added to the list of those to be phased out, notably a group of firefighting chemicals called halons. Second, high-income, high-consuming countries promised to contribute to a special fund to help low-consuming, low- and medium-income countries adjust to the new situation. Brazil, China, and India, all of which are warm-climate countries with large potential markets for CFCs, would not have joined without this provision. Diplomacy. These measures came to be known as the London Amendment.

Faced with incontrovertible scientific evidence that CFCs constituted a danger to human health and after vigorous internal debate, the Dupont Company stood by its earlier commitment to cease their manufacture, honorably resisting the very real temptation to stonewall and squeeze out a few more years of profit from sales of CFCs. Despite earlier fears that it would be difficult to replace CFCs in refrigeration and their other applications, Dupont's expertise enabled it to develop substitute chemicals—even plain water was found to be an adequate substitute in some applications. Indeed, its confidence that its superior expertise in this branch of chemistry would translate into profitable opportunity was a factor in its decision to discontinue CFC manufacture and to support international regulations that would level the playing field between it and its competitors.

The first substitutes for CFCs, hydrochlorofluorocarbons (HCFCs), can also be ozone destroyers but for chemical reasons were thought to be less likely than CFCs to reach the stratosphere. They, too, began to be phased out in 1992 under the provisions of the Copenhagen Amendment, both for their ozone-destroying capacity and because they are greenhouse gases that contribute to global warming. The Kigali Amendment to the Montreal Protocol, which came into force in 2019, calls for a rapid reduction in the production and consumption of hydrofluorocarbons (HFCs), chemical relatives of HCFCs that are also substitutes for CFCs and potent greenhouse gases.

The Scientific Explanation for the Ozone Hole

The catalytic chain reaction by which CFCs destroy ozone continues for much longer in the stratosphere over the Antarctic than in the rest of the stratosphere.

Antarctica is very cold and very dark in winter. It is even colder than the Arctic because it is a continent-sized island whose atmosphere is surrounded by a vortex of circular winds that prevent warmer air from lower latitudes from penetrating it. The Antarctic stratosphere in winter is just as dark as the land under it and is even colder, going down to −180°F. It is so cold that it gives rise to a special kind of cloud, called (logically) a polar stratospheric cloud (PSC).

PSCs have a pearly opalescent beauty. They consist of tiny ice particles that, like anything small, have lots of surface area compared with their volume. CFCs interact with PSCs to give rise to a bevy of chlorine-containing compounds, especially diatomic chlorine (Cl_2) and hypochlorous acid (HOCl). These compounds do not dissolve in the ice but, rather, accumulate on the cold, dark surfaces of the PSC crystals. At the same time, NO_2, which we have met before as the terminator of the ozone-destroying chain reaction, dissolves throughout the volume of the PSC crystals and stays there all winter.

Come the beginning of spring and the Antarctic dawn, the chlorine compounds are freed from the crystal surface and are released into the stratosphere, while the NO_2 is left behind in the volume of the crystal. Ultraviolet light from the sun splits off atoms of chlorine from their compounds, freeing the atoms to destroy ozone by a variant of the catalytic chain reaction that we described earlier. In contrast to the situation in the rest of the stratosphere, the NO_2 is trapped inside the ice crystals of the PSCs, out of action and unavailable to perform its usual role of terminating the chain. The result is that atomic chlorine is free to run amok, destroying ozone, until the ice crystals melt and free the NO_2. In this way, much more ozone is destroyed in the Antarctic spring than anywhere else in the world, a phenomenon known to the world as the ozone hole. A smaller hole sometimes forms over the Arctic.

Lessons of Success

The Montreal Protocol has achieved real results and is among the most successful of all global environmental initiatives. The ozone hole has stopped growing and, according to NASA, has begun to recede. It is expected to return to pre-CFC levels by about 2060. Outside of the Arctic and Antarctic, stratospheric ozone is expected to recover to 1980 levels by the 2030s in the northern hemisphere, and by "mid-century" in the southern hemisphere.[5] What is more, since CFCs, HCFCs, and HFCs are greenhouse gases, the success of the Montreal Protocol in reducing their production and emission into the atmosphere has done more to mitigate climate change than all the other international efforts in this direction! So far, so good.

To be sure, there are still problems. The illegal manufacture of CFCs and HCFCs is still an issue, especially in China.[6] So is the smuggling of refrigerants, especially of HCFCs, into Europe and the United States. CFCs still leak into the atmosphere from old refrigerators and air conditioners.[7] And the Montreal Protocol regime will have to operate for many decades if the ozone problem is to be fully resolved. But these are the problems of success.

The ozone story illustrates some typical general features of the interaction between science, technology, and international affairs that form the themes of this book. Scientists can put an issue onto the international agenda, but only governments and political processes can manage them. Ideally, the two systems coevolve. Governments gain increasing understanding of the scientific facts and their importance, and scientists gain increasing understanding of what governments need to know in order to find political solutions to the problems that the scientists have uncovered. At the same time, multinational corporations make investments in alternative technologies in anticipation of international regulations and in the process become less resistant to their adoption.

Governments and scientists were not the only actors involved in the ozone story. Epistemic communities of scientists, NGOs, journalists, and sympathetic officials were needed to educate governments and the public about the newly discovered issues. The public must be involved in the choice of whatever measures must be adopted and must support them as they are carried out. Business and government must confront the same global issues, although from different points of view. The ozone story would have been quite different had Dupont not stood by its commitment to end production of CFCs when definitive proof appeared that they were harming human health. The line between public and private issues has in this way become so blurred as to be almost meaningless.

The treaty process itself has come to play an important role in raising awareness. The negotiation of an environmental treaty educates governments, businesses, and the public. It prods the recalcitrant and exposes spurious scientific arguments to peer review. Ratification and implementation of an environmental treaty can build confidence, establish commercial standards, stimulate new technical solutions, and build technical and administrative capacity. Sounds great. But as will become apparent when we discuss climate, this process of confidence building may collapse when the costs of a new regime are much greater than they are with stratospheric ozone depletion, so that those threatened by the new knowledge mount a determined counterattack.

The defense of the stratospheric ozone layer was not immune to these challenges. A well-financed, media-based campaign, backed by right-wing activists opposed to government regulation, attacked both the scientific findings on CFCs and stratospheric ozone depletion and the honesty and integrity of the researchers who produced them. Foreshadowing the later tactics of climate denialists, the campaign publicized the views of authoritative-sounding scientists whose expertise lay in other fields, who spread misinformation and continued to repeat scientific arguments that had already been tested and disproven by the normal processes of science.[8] The campaign was unsuccessful, in part because of the agreement of industry to stop producing CFCs, but also because of the solidity of the science underlying the international agreements, which won a Nobel Prize in Chemistry for Sherwood Rowland, Mario Molina, and Paul Crutzen.

The role of science on the way to the Montreal Protocol has a parallel in virtually all the global environmental issues faced by the international community: forests, deserts, oceans, coral reefs, water, and biodiversity, to name only a few. The defense of the stratospheric ozone layer, however, is a relatively simple story, involving clear-cut science; a relatively benign political climate and small economic stakes; a few chemicals for which substitutes could be assumed to be reasonably available; a handful of producing companies in the United States, Europe, and Japan; and a major manufacturer that was proud of its reputation for technical honesty. The problems raised by the ozone story, however, will appear most acutely in our discussion of climate change, to which we now turn.

3

Climate Disruption, Not Just Global Warming

The effects of climate change are already being felt worldwide. Nine of the ten hottest years on record occurred between 2009 and 2019. Scientists now hold climate change largely responsible for more frequent dangerous heat waves; more intense rainstorms; stronger hurricanes; increased coastal flooding; widespread bleaching of coral reefs; record-breaking wildfire seasons in California, Australia, and Siberia; droughts in the Mediterranean area; thinning polar ice; and retreating glaciers in the Alps, Himalayas, Andes, and Africa.[1] Climate change goes far beyond mere global warming. It disrupts the Earth's climate, geography, geopolitics, ecosystems, economy, energy systems, and demography.[2] Avoiding catastrophic changes to the Earth we bequeath to our descendants will require urgent, far-reaching changes in technology, economics, law, politics, business, and lifestyles—the very archetype of the Survival Nexus that is the theme of this book.

Causes and Effects of Climate Disruption

Just to list the causes and longer-range effects of climate disruption will take us several pages.[3] Oceans are warming and hence expanding. This works along with melting glaciers to raise sea levels, threatening to submerge low-lying islands and coastal cities. Sea levels are expected to rise by up to a meter (three feet) by 2100, flooding areas where over two hundred million people live.[4] Rising sea levels may also press seawater into groundwater aquifers on which over 2.5 billion people depend for drinking water.[5] The extra warmth of the oceans strengthens tornadoes and hurricanes, stresses fish populations, reduces the amount of oxygen and carbon dioxide dissolved in ocean water, and enlarges the habitat of the germs that cause cholera. Higher temperatures on land allow the mosquitos that carry malaria, dengue fever, and other diseases to spread into new areas. Oceans absorb carbon dioxide from the atmosphere, making them more acidic, destroying coral reefs, and weakening the chalky shells of shellfish and the skeletons of the microscopic plankton at the base of oceanic food chains.

The Survival Nexus. Charles Weiss, Oxford University Press. © Charles Weiss 2022.
DOI: 10.1093/oso/9780190946265.003.0003

Glaciers and ice fields are especially sensitive to climate disruption because water from melted ice absorbs heat from sunlight rather than reflect it the way solid ice does, creating a positive feedback loop that doubles the rate of warming in the Arctic and Antarctic. The more the ice melts, the more the newly exposed land or ocean absorbs energy; the more the newly exposed land or ocean absorbs energy, the more ice melts, and so on. The floating ice of the Arctic Ocean is receding, disrupting the traditional lifestyles of indigenous people and threatening local animals, including polar bears, with extinction. Even under an optimistic (2°C) warming scenario, the Arctic Ocean is likely to be free of ice in September up to once every three years by the year 2100, freeing up navigation and natural resource exploitation but exposing this fragile area to pollution.[6] Frozen soils (permafrost) in the Arctic are melting, setting up another feedback loop—and a possible tipping point—by releasing both carbon dioxide and methane, a greenhouse gas 30–70 times as powerful as CO_2, molecule for molecule.[7] (The ratio depends on the time scale, as methane has a shorter residence time in the atmosphere than does CO_2.)

There's more. Meltwater from land-based ice, especially in Greenland and Antarctica, also contributes to sea level rise, even more than does the thermal expansion of the ocean water. Global warming speeds the "glacial" pace of land-based ice creeping toward the ocean: first, by melting the floating ice that blocks its path, and second, by warming ocean water that seeps inland and lubricates the flow of a glacier by melting its underside. Should these processes make the Texas-sized West Antarctic ice sheet disintegrate and fall into the ocean, global sea level would rise by up to an additional 1 meter. To put these numbers into perspective, a complete melting of both polar ice caps, including all the Greenland ice sheet, would raise sea levels by almost 18 meters (60 feet), flooding coastal areas all over the world, not to mention most of Florida. Some models suggest that this could happen in as little as five hundred years (although several thousand years is more likely) if nothing is done to control carbon dioxide emissions.[8]

As the Earth gets warmer, an estimated one million species of plants and animals will go extinct, in part because their habitats will have moved to higher latitudes faster than they could adjust.[9] A typical summer will be hotter than the hottest summers we have experienced in the last several decades. This will increase human deaths from heat stress and other heat-related illnesses, not only in low-income countries but also among low-income people in high-income countries who cannot afford air conditioning. Dry subtropical regions like much of sub-Saharan Africa will experience conditions comparable to those of the Dust Bowl in the western United States during the 1930s. The resulting loss of agricultural productivity is likely to stimulate widespread migration, leading to political and social tensions that dwarf those we are seeing today.

Most of these environmental changes will persist for centuries after all net carbon dioxide emissions have ended. This is because the carbon dioxide that has already been emitted will remain in the atmosphere for about a century before it is absorbed into the land or dissolved in the ocean, and the heat absorbed in the upper layers of the ocean will take a similar time to diffuse to greater depths.

Worse yet, the world may be approaching any of a number of tipping points at which huge, irreversible changes happen very fast once they have started, perhaps in a few decades or even in a few years. For example, the Greenland ice sheet is melting at an accelerating rate, and no one knows how much of its cold fresh water would have to pour into the North Atlantic for it to give Western Europe a climate like that of Newfoundland, which is at a similar latitude on the other side of the Atlantic.[10] Other possible tipping points involve the disruption of the Asian monsoon and the drying of the Amazon basin. These are still only scary possibilities. As a final blow, a recent comprehensive review of the geophysical data pretty much extinguishes the hope that the consequences of climate change could turn out to be minor.[11]

The impacts of climate disruption are especially severe in low-income countries, many of which are in tropical areas that do not have much margin for further warming. For these countries, warmer temperatures mean more disease, more droughts, less food, more floods, more severe storms, lower agricultural productivity, and more and bigger wildfires. Smoky wood and dung fires spread huge quantities of black particulate matter (black carbon) over the glaciers and icefields of the Himalayas and other Asian mountain ranges, turning them gray. This contributes to climate disruption by increasing the amount of solar energy that they absorb, reducing the amount that they reflect back into space, and increasing the rate at which they melt. In the long run, melting glaciers in the Himalayas mean less water for the Yangtze, Mekong, Brahmaputra, and Yellow Rivers, on which some 1.4 billion people depend for irrigation and drinking water. The ice covering the Himalayas and other Asian mountain ranges already melts earlier in the season than it used to, disrupting the schedules of farmers who rely on irrigation from rivers fed by meltwater.

Compared with high-income countries, low-income countries have fewer resources to deal with climate disruption. They have fewer well-trained policymakers and managers and less stable political institutions, making it more difficult for them to determine and put into practice adaptive measures like dikes to hold back rising seas, while smaller government budgets make it hard to implement these expensive measures. Low-income countries can be less equipped to embark on the novel and perhaps risky path of development based on "green" technological and institutional innovations. For example, having fewer agricultural researchers makes it harder to develop new crops or crop varieties adapted to the new climatic conditions. This is deeply unfair, since most low-income countries emit far less carbon dioxide than high-income countries and hence have contributed little to the climate disruption from which they are now suffering. Even China, which now tops the list of yearly CO_2 emitters, is still well behind the United States and Europe in total aggregate emissions from the beginning of the modern era. The same holds true for India, which is now also a major emitter. Finally, future generations of humankind will bear even more of the costs of climate disruption than we do today, to which they have contributed nothing at all!

In sum, the disruption of the Earth's climate wrought by greenhouse gases is real, is caused by humans, demands urgent action, and goes far beyond just a rise in temperature. "Climate disruption" is a much more accurate description than "global warming."[12] There are, to be sure, some benefits to humans from this disruption. An ice-free Arctic Ocean would substantially shorten transport times from the east coast of America to Asia via the long-sought Northwest Passage and from Europe to East Asia via the Northern Sea Route and would open vast new areas to natural resource exploitation. Russia and Canada will be warmer, and their agriculture will be more productive—although their forests and grasslands are already prone to disastrous fires due to drought. But the global costs from climate disruption far exceed these benefits. Indeed, the net costs are projected to reach 7% of global gross domestic product (GDP) per capita by the year 2100![13]

The Origins of Climate Change

What is the greenhouse effect?

Imagine, for a moment, that the Earth was a ball of cold, bare rock that tumbles as it orbits the sun in such a way that every part of its surface absorbs the same amount of energy from the sun's rays and warms uniformly. As this hypothetical rock-Earth warms, it would emit an increasing amount of energy into space in the form of infrared heat rays, until the amount of energy it absorbs from the sun becomes equal to the amount of energy it emits into space. Its average temperature would level out at about −18°C (0°F). Such a cold Earth would not be a congenial place for life.

Happily for us living creatures, the Earth is not a ball of bare rock. It has oceans, rivers, continents, mountains, valleys, and, most importantly, an atmosphere that contains not only nitrogen and oxygen, but also small quantities of water and carbon dioxide. These and other so-called greenhouse gases are transparent to visible and ultraviolet sunlight but absorb some of the heat rays emitted by the Earth and send them back to Earth ("reradiate" them), raising its equilibrium average temperature to a much more comfortable 15°C (55°F). This process is known as the "greenhouse effect" because these gases play a role somewhat like that of the glass surrounding a greenhouse. You can also think of greenhouse gases as a sort of blanket surrounding the Earth, keeping its heat in. Either way, life on Earth depends on the greenhouse effect!

The problem is that humans are increasing the greenhouse effect well beyond the level that is best for us. Since the Industrial Revolution of the eighteenth century, we have been burning coal, oil, and natural gas, in effect returning to the atmosphere carbon dioxide that had been stored underground for millions of years. This has raised the concentration of carbon dioxide in the atmosphere from 275 parts per million around the year 1750 to over 400 parts per million in 2016—and very likely, the way things are going, to around 550 parts per million by 2100, or about double the preindustrial level. This increases the amount of heat reradiated back to Earth and hence warms the climate. (By warmer climate we mean a higher temperature for the Earth, averaged over its entire surface and over a period of many years. The weather in any one place and time will not necessarily be warmer.)

How do we know that atmospheric carbon dioxide has increased and that humans are responsible for this increase?

The miles-thick ice fields of Greenland and Antarctica include bubbles of air that were trapped when the ice was originally formed and are now at different depths beneath the surface. Scientists drill deep cores into the ice and carefully extract and analyze these air bubbles to determine the composition of the atmosphere at the time the bubbles were incorporated into the ice. This enables them to trace variations in the chemical composition of the Earth's atmosphere going back millions of years. The percentage of CO_2 was relatively constant for several millennia since the end of the last Ice Age around ten thousand years ago but began to increase sharply after about 1880. That's how we know that humans are responsible for the increased carbon dioxide in the atmosphere. Since 1976, a sensor atop an extinct volcano in Hawaii has tracked the level of atmospheric carbon dioxide day by day. Its measurements show that the long-term increase in CO_2 is continuing, with a small dip each Northern Hemisphere summer as green plants take up CO_2 for photosynthesis, followed by a small rise each winter as photosynthesis slows but respiration continues. (There's more land area, and hence more summer photosynthesis, in the Northern Hemisphere than in the Southern.)

How do we know that increased levels of greenhouse gases are responsible for climate change?

To answer this question, we would ideally need two planets, each the size, geography, atmosphere, and orbit of the Earth, one with 275 and the other with 400 parts per million of carbon dioxide in the atmosphere. But we don't have that and must make do with mathematical models that represent the many physical processes involved in forming our climate, and using them to simulate (backcast) the history of the Earth's climate with and without increased levels of greenhouse gases. Despite their best efforts, model builders have not been able to account for the recent changes in climate without taking into account the effects of increased greenhouse gases. The effects of sunspots, volcanoes, aerosols, and all the other possible candidates are included in their calculations but are insufficient to account for the observations. The match of the computer calculations to the empirical fingerprint of greenhouse gases is not limited to their effect on global average temperature but extends to many other phenomena, such as the variation of temperature by region and altitude. That's how we know that greenhouse gases are causing climate disruption.

Why is there uncertainty in the predictions of how much global warming there will be?

There are two basic sources of uncertainty in predicting global warming: how much greenhouse gas humans will put into the atmosphere, and how the atmosphere will respond to any given concentration of greenhouse gas, the climate sensitivity. Let's start with climate sensitivity.

In principle, it should be possible to calculate the degree of expected global warming if we know the concentration of atmospheric CO_2. The basic geophysics of the Earth's climate is well understood and can be translated into mathematical equations. The problem is that climate involves feedback loops that are difficult to capture mathematically, as well as phenomena whose dimensions range from billionths of a meter (cloud physics) to tens of thousands of kilometers (monsoons and El Niño). Computers can calculate the small-scale events and the large-scale events separately, but not both together.

To calculate the implications of increased carbon dioxide, scientists therefore run hundreds of calculations, each with a slightly different set of assumptions and approximations, and find that they tend to give nearly the same predictions for the Earth's average temperature and other important variables. From this and other evidence, they have concluded that a doubling of atmospheric carbon dioxide to 550 parts per million will raise the average temperature of the Earth by to 2.3–4.5°C (4.1–8.1°F).[14] To put these numbers into perspective, an increase in 2°C is roughly the difference between the annual average temperatures in New York City and in Washington, D.C., and 4.5°C is roughly the difference between New York City and Atlanta, Georgia.[15]

The amount of carbon dioxide that people will put into the atmosphere in the future depends on the larger political, economic, technological, and social context, which will strongly influence the willingness of the world to design and carry out measures to limit greenhouse gas emissions. To take into account this larger context, scientists have built models to represent the consequences of alternative visions of the future attitude of the world toward sustainability, inequality, and national security.[16] The most favorable of these scenarios from the climate point of view imagines a world that gradually shifts to a more environmentally sustainable and less energy- and resource-intensive trajectory with lower economic and social inequality. The more pessimistic scenarios imagine either all-out market- and fossil fuel–based development, or else a world sharply divided between a politically powerful, economically well-off and well-connected world and a fragmented, poorly educated, low-tech, low-income world. These alternative baseline, "business-as-usual" scenarios have different implications, both for the future of global warming and for the prospects of international cooperation on mitigation and adaptation. The fact that we don't know in advance what trajectory the world will follow adds to the uncertainty in predicting future greenhouse gas emissions and hence the uncertainty in the ability of the world to meet warming targets. These scenarios are to be incorporated into the Intergovernmental Panel on Climate Change (IPCC) report due in 2022.

The Damage from Global Warming, Degree by Degree

It's easiest to think about the consequences of global warming if we organize our thoughts according to different levels of possible increase in the temperature of the Earth's surface, averaged over the year and over the globe, compared with the temperature that prevailed before the Industrial Revolution.[17] As a point of reference, the Earth has already warmed by about 1.2°C (2°F) since then. The Earth was last at this temperature about 130,000 years ago.[18]

The most recent international agreement concerned with climate change, the Paris Agreement, calls for international action to limit global warming to 2°C (3.6°F), and for *trying* to reach the more ambitious and demanding target of 1.5°C (2.7°F). In brief, the increase of 1.2°C over preindustrial conditions that we have already caused is bad enough for the Earth. An increase of 1.5°C—a very optimistic target—will be *very* bad. An increase of 2°C will be worse, 3°C (5.4°F) worse yet, and 4–5°C (7.2–9.0°F)—to which we are now headed sometime in the next couple of centuries if nothing is done to stop it—downright catastrophic.[19] The consequences of these increased temperatures will take time to take effect, but they will endure for hundreds or even thousands of years past the time when no net greenhouse gas is being added to the atmosphere.[20]

Let's start with what is very likely to happen when global warming reaches 1.5°C, as it is now on track to do sometime between 2030 and 2050 unless drastic action is taken immediately to stop it. Sea levels will rise by about 46 centimeters (18 inches), exposing some thirty million people each year to the risk of flooding by 2050. The Arctic Ocean will become ice free in September once a century. The area of wildfires in the western United States will increase by a factor of two to four. Coral reefs will decline by 70–90%. Per acre yields of U.S. corn and of Indian and African wheat will decline by 7–22%. Severe drought will threaten

130 million people, and the area of dry land exposed to malaria will increase by a quarter.[21] World gross national product (GNP) will have been cut by about 8% by 2100. This is the *optimistic* scenario!

It's hard to imagine the likely consequences of even 2°C of global warming, the stated goal of international efforts and hence as close as we can come to an "internationally acceptable" level. At this temperature, U.S. corn yields and Indian and African wheat yields per acre are expected to decline by 15–30%. Wildfire areas in the western United States are projected to quadruple or even to *octuple*. The thermohaline ocean circulation that keeps Western Europe relatively warm (see note 10) could decrease by a third, with unpredictable consequences. As mentioned earlier, the Arctic Ocean will be ice free in September once every three to ten years. Sea level is projected to rise by an additional 10 centimeters (4 inches) by 2100 compared with the 1.5°C scenario. This is why small island states—which are typically atolls whose highest point is not more than 2 meters (6.5 feet) above present-day sea levels and which are already regularly flooded by storm surges—have pressed for the more ambitious target of 1.5°C.[22] These predictions are best estimates; the actual figures could be lower, but they could also be higher. Experience to date has tended to track the more pessimistic predictions.

We now turn to 3°C, the likely result around the year 2100 if governments achieve the voluntary targets for limiting greenhouse gas emissions that they announced in connection with the Paris Agreement. The Earth was last at this level three million years ago, a time when there was almost no ice in the world and forests grew in the Arctic.[23] At this temperature, coral reefs are expected to disappear entirely. Nine out of ten summers are expected to be hotter on average than almost all of the hottest summers of the latter part of the twentieth century. The Arctic Ocean will be free of ice at some point during the summer at least every other year.[24]

Now to 4°C.[25] The Earth was last at this temperature fifty-five million years ago.[26] The world could reach this temperature by 2100 if the more pessimistic assumptions about future emissions come to pass. At this temperature, summers in most parts of the world would be expected to be hotter *on average* than the hottest summer *days* of the last decades of the twentieth century. Global crop yields would be expected to decline significantly in most regions, including a decline of almost 60% in the Middle East and North Africa, making it harder to feed an increased world population. Sea level rise will expose more than 30% of the populations of the Philippines, Myanmar, and Vietnam to increased flooding. Storm surges will affect more than forty-three million people in ten major cities across the developing world, with Karachi, Manila, and Jakarta being the hardest hit.[27] Ice in the Arctic Ocean will be reduced by 60% and perhaps disappear entirely each summer. It could be worse. If the world returns to the fossil fuel–intensive trajectories of the early 2000s, it could reach 5°C by 2100, within the lifetime of many of today's children.[28]

Greenhouse Gas Budgets and Scenarios

How much climate disruption can the world accept? This question depends both on scientific information and on how much damage and risk people are willing to tolerate. Most of the discussion, both by researchers and by international negotiators, has been structured around the target of 2°C. This target does not correspond to any known sharp break in the effects of global warming that clearly announces, "this far and no farther." As we have seen in the previous section, it is clear that the greater the global warming, the worse will be the disruption. Scientists are concerned that there may be tipping points or discontinuities that would produce a sudden or irreversible increase in damage. However, they cannot predict the temperatures at which such tipping points would take place, although they are obviously more likely the greater the warming.

The target of 2°C was originally proposed in 1975 by William Nordhaus, then a future Nobel Laureate *economist*, as a tentative standard that lies within the historic variation of global average temperature and corresponds to a doubling of the premodern concentration of atmospheric CO_2.[29] Diplomats, journalists, and environmental advocates have used 2°C as an aspirational goal, as a framework within which to translate scientific findings into concrete targets, as a touchstone for critiques of plans for addressing climate disruption, and as a goad for action, much as they did with the goal of 50% reduction in chlorofluorocarbon (CFC) production and consumption provided by the Montreal Protocol on stratospheric ozone that we discussed in the previous chapter.

For researchers and policy analysts, targets provide a valuable framework for thinking about alternative futures. Researchers can estimate the total amount of carbon dioxide and other greenhouse gases in the atmosphere that would give rise to a given temperature, according to the best available scientific models.[30] They can then subtract the amount that is already in the atmosphere to arrive at a *carbon budget*—the amount of CO_2 that the world can probably still emit without exceeding the targeted temperature. As of the year 2018, for example, this budget was about 570 gigatons of CO_2 for a target of 1.5°C and 1,320 gigatons for a target of 2°C.[31] For 3°C, the budget was about 2,240 gigatons in 2014. By comparison, the world emitted over 40 gigatons of carbon dioxide in 2019. At this rate, then, the world will use its 1.5°C budget in about fifteen years, its 2°C budget in about twenty-nine years, and its 3°C budget in about fifty years.[32] Not very long.

Another way to start to think about what to do to limit global warming is to ask how much greenhouse gas must be "saved"—i.e., *not* emitted, as compared with current trends—in order to achieve a given target temperature. In other words, what's the difference between the carbon budget for a given target temperature and the amount of CO_2 we will emit in a given time if we continue to do so at current rates? Analysis along these lines leads to the clear conclusion that that no one measure—no silver bullet—can save enough CO_2 by itself to reach any reasonable target. Instead, it will take several different measures, taken together: so many gigatons from adopting wind or solar energy, so many from substituting natural gas or nuclear energy for coal, so many from electrifying cars with power from renewable sources, so many from redesigning the growth of cities, so many from measures to remove CO_2 from the atmosphere, and so on. These quantities are known as "wedges" because of what they look like on a graph of business-as-usual and target emissions levels over time.[33]

We can estimate how much CO_2 we need to save from each of these wedges for the total amount of saved CO_2 to be enough to reach a given target. We can also ask how fast these measures need to be put in place so that we reach the target without "overshooting"—i.e., so that the temperature doesn't exceed the target and then have to be ratcheted back down, a scenario that would take extra time and put extra stress on biodiversity and other environmental systems. It will take at least a dozen wedges, each of which by itself would require enormous political, economic, social, legal, and technological effort. As we shall see in the next chapter, the best national strategies integrate the use of all the wedges, touching on all sectors of the economy.

There are a variety of combinations of "mitigation" measures that could, in principle, limit emissions, but they all lead to the same startling conclusions. To limit global warming to 1.5°C, we have only until about 2050, within your lifetime, to reduce net CO_2 emissions to zero, a goal known as "net zero." This means that by then we must have compensated for any CO_2 that we put *into* the atmosphere in a given year with an equal amount that we remove *from* it. We need to be halfway to these targets by about 2030, i.e., in less than a decade. To limit global warming to 2°C, we have a bit more time to get to net zero, say to 2070—when today's little children will be in their fifties.[34]

Both the carbon budget and the carbon-saving formulations, moreover, raise major ethical issues. Who gets to use up the carbon budget: the high-income countries that have put most of the CO_2 into the atmosphere but whose emissions have plateaued? China and India, who have not produced that much CO_2 in the past but who are now leading emitters? Or other low-income countries that aspire to become high-income countries, but whose economic growth could be stifled by stringent limits on emissions?

The same ethical questions apply to deciding which countries are expected to "save" the most CO_2 emissions in the future. Low-income communities, after all, also need energy to study; store and prepare food; communicate and gather information; heat and cool homes; and earn a living. They, too, want a better diet, convenient transportation, labor-saving appliances, and other energy-intensive products. Who are we to say that they should not have them?

Climate Negotiations Veer from the Model of Stratospheric Ozone

The issue of climate change started out on a path parallel to that of stratospheric ozone. Science put the issue on the international agenda. An epistemic community of scientists, academics, journalists, and NGOs arose to educate governments and the public about a complex problem involving science, economics, ethics, politics, law, and business. Diplomats met to address the new global problem, armed with authoritative, up-to-date scientific advice from an officially sanctioned, well-structured, continuing body of credentialed experts.

The UN Framework Convention on Climate Change (UNFCCC) of 1992 committed its parties to the objective of stabilizing greenhouse gas concentrations in the atmosphere, and to the principle that low-income countries would get help to carry out their obligations under the Convention. The 1997 Kyoto Protocol to the Convention specified mandatory targets and timetables for developed countries to reduce their emissions of greenhouse gases. These diplomatic efforts clarified the issues and provided experience with the policies and procedures needed to address them. In the meantime, scientists resolved nearly all the uncertainties that had bedeviled earlier discussions and discovered that the climate disruption problem was even more serious than they had initially feared. The diplomatic process looked promising.

But the parallel between climate and stratospheric ozone broke down. The stakes involved in climate disruption are vastly greater than those involved in stratospheric ozone depletion: no less—and, in fact, considerably more—than a wholesale restructuring of the world's energy economy, its transportation systems, its patterns of agriculture and land use, the design of its cities, and the prospects of people in low-income countries to live decent lives. Perhaps unsurprisingly, countries did not live up to the agreed mandatory targets and timetables for greenhouse gas reduction in the Protocol.

The Paris Agreement that followed the Kyoto Protocol in 2016 took a more permissive approach, in the hope that this would be more effective. It required each country to make its own *voluntary* commitment to reduce emissions, rather than impose a uniform *mandatory* target specified by the agreement. While the aggregate of these commitments falls well short of what is needed to limit global warming to the targets of 1.5°C or even 2°C, the Agreement has focused national attention on the changes that are needed and has led to several promising beginnings. In the United States, however, efforts to limit the size and scope of climate disruption have been stymied by a systematic, well-financed campaign of obfuscation and disinformation by vested business and political interests.

The Framework Convention and the Kyoto Protocol

The idea that people are warming the Earth's climate by emitting carbon dioxide into the atmosphere goes back to the Swedish scientist Svante Arrhenius in 1896 but remained a topic for research and speculation for many decades. As late as the 1970s, scientists were still debating whether the Earth was warming or cooling! By the 1980s, the evidence for dangerous global warming was sufficiently persuasive that scientists warned that the problem required urgent attention.[35]

The first reaction of the international community followed the pattern established for stratospheric ozone a decade earlier. In 1988, the World Meteorological Organization (WMO) and the United Nations Environmental Program (UNEP), two technical arms of the United Nations, formed the IPCC, consisting of thousands of volunteer scientists, to provide authoritative scientific advice to governments. (We shall explore the work of the IPCC later in this chapter.)

In 1992, the IPCC announced its conclusion that "Emissions resulting from human activities are substantially increasing the atmospheric concentrations of the greenhouse gases . . . enhanc[ing] the greenhouse effect, resulting on average in an additional warming of the Earth's surface."[36] Acting on this science-based consensus statement, 166 governments began negotiations for a climate framework agreement. In only fifteen months, a remarkably quick response by international standards, they concluded the UNFCCC, with the objective of "stabiliz[ing] greenhouse gas concentrations in the atmosphere at a level that would prevent dangerous anthropogenic interference with the climate system . . . within a timeframe sufficient to allow ecosystems to adapt naturally to climate change." (The negotiators did not specify targets or timetables and did not define the exact meaning of "dangerous.") The UNFCCC is a treaty binding on its parties.

The obligation to deal with climate change is the most dramatic and important example of the principle of *sustainable development*, defined as "development that meets the needs of the present without compromising the ability of future generations to meet their own needs."[37] This general principle, which finds philosophical resonance with virtually all cultures worldwide, is in conflict with the practical realities of politics, economics, bureaucracy, and business, which operate on shorter timescales dictated by the election cycle, financial discount rates, career tracks, and quarterly earnings statements, respectively.

The UNFCCC included another important environmental principle, that of precaution: that lack of full scientific certainty need not prevent taking action to protect the environment from damage that would otherwise be serious and irreversible.[38] We have already encountered this principle in connection with the Montreal Protocol that we discussed in the previous chapter. The precaution principle doesn't say what you have to do to protect against the risk of environmental damage, or even that you have to do anything at all, just that the absence of firm evidence doesn't close the issue. In a nod to critics concerned about costs, the treaty's text on precaution provides that "policies and measures to deal with climate change should be cost-effective."[39]

Other provisions incorporated into the Convention reflect the concerns of low-income countries, which felt, with much justification, that they were being asked to pay for a global problem to which they had not contributed. The Convention accordingly made a general commitment to the principles of equity and of "common but differentiated responsibility," i.e., that we are all in this together, but that higher-income countries have contributed more to the problem than lower-income ones and must bear more of the costs. Specifically, the Convention did not require low-income countries to commit to adopting national policies to slow the rate of climate change. It also recognized that they would need financial and technical help if they were meet the treaty requirements to collect and report their national inventories of sources of human-caused (anthropogenic) emissions of greenhouse gases *into* the atmosphere, and to report any sinks (like growing forests) that were absorbing greenhouse gases *from* the atmosphere, since expanding sinks could, in principle, be as effective as reducing sources in mitigating climate disruption.

In 1997, 191 countries agreed to the Kyoto Protocol, a more detailed follow-on agreement to the Framework Convention that set quantitative, *binding* targets and timetables that *required* high-income countries to reduce their overall emissions of greenhouse gases generally by 6–8% below 1990 levels in the period of 2008–2012. No targets were set for low-income countries, on the grounds that they were not major contributors to the climate problem and that they needed to continue to emit greenhouse gases if they were to continue to grow. All parties to the Protocol acknowledged that it was only a first step and that more serious and demanding measures for the mitigation of global warming would be required in the future. In retrospect, the Kyoto Protocol was an important step forward and led to important policy lessons. However, its approach of imposing binding national targets for greenhouse gas reduction did not turn out to be a politically sustainable solution.

Market-Based Mechanisms to Reduce Emissions More Cheaply

Whatever its flaws, the Protocol was a first exposure to the teething problems associated with many innovative, market-based mechanisms that enable a country to satisfy its emission reduction requirements cheaply by paying to reduce emissions in another country or by trading permits to emit with other high-income countries. Understanding these mechanisms requires an excursion into the world of environmental economics.

Climate change is a classic negative externality. An emitter pays nothing to put carbon dioxide into the atmosphere but imposes costs on everyone else when it does so. Economists have therefore developed the idea of carbon charges, which make emitters share these costs. (They're really carbon dioxide charges, but everyone has apparently gotten tired of the word "dioxide.") The idea is that if emitters are made to pay for their emissions, they will first figure out how best to reduce them using existing methods and will then invest in research to develop new ways to reduce them even further.

Carbon charges are essential to induce emitters to undertake the transformations needed to limit global warming to any acceptable level. However, the investments in existing technologies that are induced by carbon charges are not sufficient in themselves and need to be accompanied by both public and private expenditures for research, development, demonstration, and deployment of innovative technology. Also necessary are explicit regulations encouraging or requiring the use of climate-friendly technology, such as quotas for the use of renewable energy in electricity generation, energy efficiency standards as elements of building codes, fuel efficiency standards for motor vehicles, disclosure and other requirements on financial and industrial firms, and systematic efforts to eliminate legal, financial, or policy obstacles to a more sustainable economy and lifestyle.

The most straightforward way to charge for carbon dioxide would be to measure the amount of carbon dioxide emission associated with every activity and to require people to pay in proportion to their emissions. This is known as a *carbon tax*. (This idea can be extended to all greenhouse gases, but we'll refer just to carbon dioxide to keep it simple.) This could be made revenue neutral by returning the money collected by the government back to the taxpayer as a "dividend," more or less the way Alaskan oil royalties are returned to Alaskans. Ideally, however, some proportion of the money would be set aside to support research, innovation, and adaptation.

Aside from the practical difficulty of measuring who is emitting what, this mechanism suffers from the political disadvantage that it is called a tax, a notion that is unpopular everywhere and has been especially poisonous in the United States. A tax on carbon would also unfairly burden low-income communities because it would end up being a larger portion of their income, an inequity that would call for some kind of compensation. In part for these reasons, economists have come up with the idea of cap and trade, which would have the same effect of forcing emitters to pay for the environmental damage they are causing. In this mechanism, some governing body arrives at a total amount of carbon dioxide or other greenhouse gas that it will permit to be emitted (the cap) and distributes permits (known as "allocations") to prospective emitters, who are then free to buy or sell them as they wish (the trade). The idea is that emitters who face big clean-up bills will buy allocations from emitters for whom cleanup is cheaper, so that the same amount of CO_2 is saved but the overall cost is lower. In this way, total emissions are limited to the prescribed amount, and the overall cost of emissions reduction is minimized. There are now some sixty ongoing experiments in different parts of the world with various forms of cap and trade, providing a wealth of experience and policy learning.

Disinformation and Distraction

The United States is one of the great anomalies in the worldwide effort to address climate disruption. The world's largest economy has stood aside from, and even attempted to thwart, the worldwide effort of countries of all political orientations and at all levels of economic development to wrestle with the difficult questions of economics, politics, technology, business, law, and ethics involved in climate change mitigation. (As this book goes to press, newly elected President Joe Biden has dramatically reversed this position and has announced major climate initiatives and the ambitious goal of halving nationwide carbon dioxide emissions by 2030.) While a handful of countries followed the United States in official climate denial, most countries—as well as the European Union (EU), which is technically a region rather than a country—have realized that they must take matters into their own hands, with or without traditional U.S. leadership. Governments and public opinion in most of the world no longer ask whether a problem exists or whether they are willing to pay the costs of climate mitigation and adaptation. Rather, they are concerned with finding the most cost-effective way forward, and some are concerned that they may not have the ability to pay.

Several of the largest and most powerful companies in the U.S. energy industry have mounted a well-organized, well-financed, and so far largely successful campaign of obfuscation and disinformation to discredit the idea that humans are changing the Earth's climate, and to argue that even if we are, it would be too expensive and disruptive to do anything about it.[40] Working through industrial advocacy groups like the American Petroleum Institute, the Edison Electric Institute, and the National Coal Association, these companies finance an interlocking network of advocacy organizations that systematically echo each other.

These business interests are joined by political conservatives opposed to government regulation, who realized early on that any effort to mitigate climate change would greatly expand the role of government. Reflecting the increasing dominance of pro-free-market and antigovernment philosophy and the increasing influence of conservative donors, the Republican Party in the United States reversed decades of support to environmental measures—the Environmental Protection Agency (EPA) was established and many of the first anti-pollution laws were passed under the administration of Republican President Richard Nixon—and made the extraordinary strategic decision to deny the validity of the technical consensus of the world scientific community that climate change is a genuine, serious problem .[41] Voters tend to echo the views of their friends and of trusted political leaders on technical issues that they do not themselves understand, so Republican elected officials now risk their political future if they suggest that anthropogenic global warming is real, a harbinger of their minimization of the gravity of the coronavirus pandemic in 2020.

The campaign of climate denial began as a vigorous and legitimate public dispute over the scientific evidence, which cited expert opinions by distinguished scientists with excellent credentials that observed increases in average temperatures could be explained without reference to human activities.[42] The campaign continued long after these alternative explanations had been tested and discounted, and the scientific controversy had been essentially settled. At this point, the scientific arguments of the climate deniers focused on small gaps in the understanding of climate change, deriding it as a mere theory and ignoring the fact that it is supported by an overwhelming body of evidence and that any remaining inconsistencies were marginal to the overall picture.

Here, a small digression regarding the philosophy of science is in order. As we explained in the introduction to this book, a scientific theory is not a speculation, as it might be in informal conversation, but, rather, a comprehensive explanation of apparently unrelated phenomena, like the theories of gravitation, evolution, continental drift, or relativity. Such theories constitute much of the basis of human understanding of how the natural world works. Scientists accept a new theory if it offers a better overall explanation of the evidence than its predecessor, even though some loose ends may remain to be explained. The assumption is that further research will clear up these relatively minor issues.[43]

Once the scientific issues were settled, climate denial morphed into a full-fledged lobbying campaign, one governed by the rules of politics and public relations rather than by those of scientific discussion. Climate deniers raised arguments that had long since been refuted. They took advantage of the fact that scientists are generally reluctant to enter the public arena and are better equipped to debate the fine points of scientific evidence than to enter the rough and tumble of partisan political debate, in which ad hominem attacks are the norm and evidence may be cherry-picked and distorted to suit a predetermined conclusion. Ironically, in ignoring the norms of scientific debate, climate skeptics have shown a sophisticated understanding of the chinks in the armor of science as a human enterprise and of the scientific community as a human institution. For example, they charged that the idea of climate change was just a device to keep money flowing to researchers and to provide things for bureaucrats to regulate. They even hacked the internal communications of climate scientists and quoted out of context remarks they had made to each other.[44]

In large part as a result of their efforts, the idea that human-caused climate change is at best an exaggeration and at worst a hoax has become deeply embedded in U.S. domestic politics, greatly complicating international efforts to address the Earth's most important environmental crisis. Ironically, their campaign has been unintentionally facilitated by the ethical obligation felt by U.S. journalists to present both sides of a presumed controversy. Even if one side is demonstrably false, they do not want to be suppressing the expression of an opinion, or the news that this opinion is, in fact, widely held. Fact checking is insufficient to resolve this dilemma.

Institutionalizing Scientific Advice

Scientific advisers, like other technical experts, like to think of themselves as "speaking truth to power."[45] This turns out to be easier said than done, especially in a high-stakes, high-profile, value-intensive, postmodern issue like climate change. The history of the IPCC, the official, authoritative source of information on global warming, illustrates some of the difficulties.

The IPCC is an intergovernmental organization with a mandate to "prepare, based on available scientific information, a report on all aspects of climate change and its impacts, with a view to formulating realistic response strategies."[46] The most authoritative and comprehensive scientific advisory body ever assembled, the IPCC consists of hundreds of volunteer scientists organized into three working groups: the physical science underlying climate change; impacts, adaptation, and vulnerability; and mitigation of climate change. Every six or seven years, the IPCC has produced an authoritative, three-volume report totaling some three thousand pages. Member governments approve the overall report, and they review and approve the summary for policymakers line by line, a process that tends to produce a relatively conservative statement.

During its history, the IPCC has had to overcome two institutional weaknesses that caused increasing complications as the huge economic and political import of its recommendations became apparent. Both problems led to vulnerabilities that in another context might have gone unnoticed. In the highly politicized context of climate change, however, they were vigorously exploited by critics who attacked the integrity of the IPCC itself, charging that its members suffered from conflicts of interest and that its judgments were insufficiently vetted against the scientific literature. The problems they revealed were relatively minor, and the defects in its rules and procedures relatively easy to repair, but the revelations damaged the reputation of the IPCC as a disinterested scientific body and serve as an alert to future advisory mechanisms.

The first problem concerns the treatment of scientific uncertainty. The IPCC is composed of scientists, for whom facts are facts and ignorance is ignorance. Except as a justification for further research, scientists are generally reluctant to make a public estimate of just how uncertain a given unproven idea might be, or how an unsettled scientific issue is likely to be resolved. But policymakers and political leaders need to know not only what is definitely known, but also the different possibilities for how uncertainties are likely to be resolved, so that they can decide *now*. (In private, to be sure, scientists do ask themselves what the truth is likely to turn out to be. That, after all, helps them to design their next experiment!)

To meet the need for a more precise and systematic approach to uncertainty, a group of climatologists, known informally as the IPCC's "uncertainty police," imposed a standard format for expressing uncertainty—no small job. Thenceforth, each assertion in an IPCC report has been required to be accompanied by a specification of the level of confidence that it is, in fact, correct and an assessment of the quality of the evidence that supports it. [47] The level of confidence was *subjective*—the odds that the scientist-authors would take on a bet that the assertion would turn out to be true when all the facts were known. It was not an *objective* probability derived from empirical measurements, like your chances of getting asthma from breathing a certain concentration of particles of a certain size for a given length of time. The guidelines imposed by the uncertainty police paid off in 2007, when the fourth IPCC Report asserted that it was 90% certain ("very likely") that climate change was due to human activities—a figure that made headlines in newspapers all around the world. This confidence level was raised to 95% six years later.[48]

The second problem concerns the structure of the scientific advisory process. The IPCC is a huge, unwieldy group of independent-minded volunteers with only a thin management structure. This became a source of criticism as IPCC findings became more and more far-reaching and more and more important to domestic and international politics and economics. Unlike longer established advisory bodies like the U.S. National Academy of Sciences, the IPCC as it was first constituted lacked formal rules governing conflicts of interest on the part of its volunteer scientists and the appointment, tenure, and behavior of its chairman and staff. The review procedures for its reports also lacked rigor. These defects in institutional structure were addressed in the run-up to the preparation of the fifth IPCC report in 2014, but they complicated efforts to rebut the criticisms of climate deniers at a critical time when the IPCC was under attack.

The Paris Agreement: Voluntary, Individual Country Commitments

As had been anticipated, the Kyoto Protocol did not reduce the world's aggregate emissions of carbon dioxide and other greenhouse gases. The hope had been to reduce greenhouse gas emissions from high-income countries by at least 5% as compared with 1990 levels by 2012. But by 2018, six years after the Protocol was extended, global emissions of carbon dioxide, far from being reduced, instead reached almost 70% more than 1990 levels, with the United States and China responsible for the largest increases. China overtook the United States to become the world's largest emitter in 2007 and now emits more CO_2 per year than the United States and the EU combined.

Of the major emitters, only the EU (whose members were allowed to meet their targets as a group rather than individually) was able to report a net decrease to below 1990 levels, in part because of the closing of the coal industry in several countries and of inefficient communist-era industries in Eastern Europe, but also because of serious measures taken to encourage energy conservation by more efficient end use.[49] In a major blow to the international climate regime, the United States withdrew its signature from the Protocol in 2001 on the stated grounds that its requirements were both expensive and ineffective. There was at the time little political support in the United States for active measures to limit carbon dioxide emissions.

By 2015, it had become clear that a different approach was needed if international consensus was to be reached. The Paris Agreement of that year gave up the requirement that targets for emissions reductions be legally binding, as they had been under the Kyoto Protocol, and instead requires parties to file individual *voluntary* action plans, with numerical targets for emissions reductions, called "nationally determined contributions," with the objective of "holding the increase in the global average temperature to well below 2°C above preindustrial levels." This is the first statement in an international agreement of a specific target for a limit to global warming.

According to the Paris Agreement, each party is to specify its own target for reducing CO_2 emissions and is to meet its target by 2030 by measures of its own choosing.[50] In response to the fears of low-lying islands, like the Maldives in the Indian Ocean and Vanuatu in the South Pacific, that the target of 2°C leaves them vulnerable to inundation, the parties to the Paris Agreement further committed themselves to "*pursu[e] efforts* [my italics] to limit the temperature increase to 1.5°C above pre-industrial levels."

Some 182 countries have filed these voluntary plans. These vary in their ambition. The EU, where "decarbonization" is a governing philosophy in more or less the same way that antitrust has historically been in the United States, has set quantitative requirements for CO_2 emissions and renewable energy portfolios for 2020 and 2030, which individual countries can meet however they prefer. India has also set demanding targets and put in place policies by which to meet them. In contrast, some parties to the Paris Agreement have stated their intention to reduce only their carbon intensity (CO_2 emissions per unit of GNP), rather than to limit their total CO_2 emissions, so that they can keep on growing and using increasing amounts of energy. Worse yet, the United States withdrew from the Paris Agreement in November 2020. (It has since rejoined.)

The run-up and the follow-up to the Paris Agreement reinvigorated the efforts of governments, businesses, and policy researchers in many countries, especially in Europe, Africa, and Asia, to come to grips with the far-reaching, country-specific changes in policy and technology needed to meet even their relatively limited voluntary commitments. The result has been an outpouring of declarations, action plans, scenarios, roadmaps, targets, briefing documents, research reports, and policy papers, as well as the formation of international and national coalitions uniting stakeholders with common interests. As an example, during the last ten years, mayors from all over the world have set targets, regulations, and investment goals for emissions reduction in their cities and have organized networks to advocate action on climate change, to share experiences, to support research, and to help mobilize funding for climate-oriented municipal projects.[51] These initiatives now include more than five hundred cities with nearly 10% of the world's population.

The EU was the first to announce targets for greenhouse gas reductions by 2030. This announcement was made before the Paris Agreement and set the tone for other countries. It was supported by sub-targets for energy efficiency improvement, renewable energy goals, and stiffened requirements for emissions trading systems, which now cover both energy and non-energy sectors of the economy. EU member countries are to submit national plans, which are to be monitored regularly to create pressure to keep on track. Companies with more than five hundred employees and EU-based asset managers and investment funds are to report yearly the environmental, social, and governance (ESG) impact and risks of their activities.[52] Some sixty-one local, regional, and national governments in forty-six countries have experimented with different kinds of cap-and-trade programs. These programs raised $44 billion for investments in clean energy in 2018 alone.[53] The countries involved in these experiments are responsible for some 22% of worldwide carbon dioxide emissions, providing a rich vein of policy learning as to what works and what does not.

Most energy companies have also begun to integrate environmental issues into their business decisions by using internal accounting or shadow prices for carbon dioxide emissions when they evaluate new investments. These are artificial prices that represent costs and risks that must be borne, either by the business itself or by others, but that are not captured by current market prices. Such internal carbon prices have been shown to influence long-term capital investments in high-emissions industries like mining, utilities, cement, and airlines.[54] From the environmental point of view, shadow prices represent the cost of climate change to the world and should be set so as to create incentives for the needed changes in patterns of investment, production, and consumption, as well as to induce technological progress. A World Bank report estimates that the shadow price of carbon dioxide needed to meet these objectives would be $40–80 per ton in 2020, depending on the industry and the company, rising to $50–100 per ton by 2030.[55] (The report recommends lower prices in the earlier years to reduce short-term adjustment costs.)

From the purely business perspective, on the other hand, shadow prices capture the *business* risks from possible future policies and regulations related to climate change, from possible disruptions of markets and supply chains, and from damage to corporate reputation.[56] In principle, environmental and business rationales may produce very different values for a ton of emissions. In practice, the great majority of both corporations and governments now set the price so low that the induced reductions in emissions would not be enough to meet even the modest goals set by the Paris Agreement, although they would be sufficient to eliminate the most CO_2-intensive investments.

On the research and development front, the Mission Innovation program, which includes governments of countries responsible for 80% of global CO_2, doubled public investments in research and development in clean energy from 2016 to 2020. These governments partnered with local private sectors and with the Breakthrough Energy Initiative spearheaded by Bill Gates and fellow billionaires, who pledged billions of dollars of long-term, "patient" investment for research, development, and commercialization. (Such long-term investment is important because the pay-back period of most large energy investments is too long for the typical venture capitalist.[57]) In addition, a coalition of large investor groups, called Climate Action 100+, prods the largest CO_2-emitting companies to commit to specific actions to "curb emissions, to improve governance, and to strengthen climate-related disclosures."[58] The implied message is that its members' future investments will depend on corporate action to limit and disclose climate risks.

These initiatives on the part of private firms are desirable and constructive. Still, the history of corporate support for research related to stratospheric ozone depletion, which stopped in the early 1980s when regulation of ozone-destroying chemicals seemed less and less likely, suggests that helpful private sector programs are most likely to be strengthened and expanded if pressure from governments, investors, and public opinion continues.

Are the voluntary national commitments, private investments, reinvigorated discussions, and disclosure pressures working? The savings from all the filed voluntary national targets would still allow an increase in average global temperature to at least 3°C by 2100, unless there are very substantial additional reductions after 2030.[59] The projected rates of growth for renewable energy, nuclear energy, and carbon capture and sequestration fall far short of what will be needed to achieve the target of 2°C.

On the positive side, many major emitting countries have announced ambitious goals to become carbon-neutral (net zero emissions) by 2050. These include Japan, the United Kingdom, the European Union, China (by 2060), and the incoming Biden Administration in the United States.[60] The major tech companies in the United States (Microsoft, Facebook, Google, Amazon, and Apple), a few oil companies, and a number of other major multinational companies have announced similar goals. It is still early to say whether these declarations will be followed by major new policies. The Paris Agreement provides for a progress report every two years and a "global stocktake" every five years to assess progress toward reaching the agreed targets. Still, each of these five-year periods takes the world closer to the time when we are supposed to have reached zero net emissions. The time is getting shorter and shorter.

The Limitations of Science

Science proposes and politics disposes. Science can define and highlight a problem. It can assess progress, discover causes and effects, identify impacts, predict consequences, and estimate uncertainties. It can point out areas of ignorance and needs for more extensive monitoring and further research. But it cannot decide what is too risky, who should gain benefits, or who should bear costs and risks. It cannot trade off benefits, costs, and risks in one area against benefits, costs, and risks in another. It cannot make judgments regarding how much certainty is required to justify action in any particular case.

Above all, science cannot dictate policy or force action. Civil society can help prepare the way by educating the public, mobilizing public opinion, and bringing pressure on reluctant governments and companies. The public, once it is convinced of the importance of an issue, can contribute by "thinking globally and acting locally" and by organizing political support for effective action. But in the end, sovereign governments and private business—especially powerful multinational companies—are essential to any effort to address the problem of climate disruption on a sufficient scale. Indeed, public and private issues are now inextricably intertwined, and both are, or at least should be, addressing the same climate issues from different perspectives.

International and domestic issues regarding climate change are also thoroughly intertwined. Carbon dioxide emitted by American or Chinese power plants melts Nepali glaciers and changes the flow of the rivers that feed Indian irrigation waters. Energy policy is no longer a purely domestic issue but, rather, one of thoroughly international concern. Perhaps we should refer to it as an "intermestic" issue.

Climate disruption is also a security issue. Climate change has a direct effect on the military in the narrow sense, as when rising seas threaten the safety of coastal bases. But its effect in security is much broader than that. To take an extreme case, suppose that the worst happens and that much of the polar ice caps melts, raises sea levels, and submerges most of Florida. A lot of taxpayers' money has gone to the Navy and the Coast Guard so that they can defend our shores—to what end? As we have discussed earlier in this chapter, climate disruption may dramatically reduce the availability of critical resources like food and water, and the resulting scarcity can exacerbate existing societal and economic tensions, flaring up into conflict.

Climate science has placed the issue of climate disruption firmly on the global agenda and has taken it about as far as it can. A far-reaching, collaborative, sustained, global effort by government, business, and the public is urgently needed to deal with the implications of that science for economics, politics, law, business, demography, and, indeed, the environmental future of the planet. These links are the Earth's Survival Nexus. We now turn our attention to discussing in more detail what needs to be done.

4
How Do We Address Climate Disruption?

No single action, no silver bullet, will suffice to *mitigate* climate disruption—i.e., to reduce the rate of greenhouse gas emissions and hence of global warming—and to enable the world to *adapt* to the multifarious effects of climate disruption. Both mitigation and adaptation require huge and interrelated changes in energy, transportation, land use, urban systems, infrastructure, and industry, changes with serious environmental, social, and political implications. These, in turn, require multifaceted national strategies, whose conception and implementation involve government, business, financial institutions, educational and research institutions, civil society, and the general public. These strategies will need to be specific to each country, sustained over time, and adjusted with experience through continued monitoring, research, and assessment, combined with a communication strategy to assure public participation, understanding, and buy-in. These country-specific strategies are essential building blocks of the collaborative global effort we need to limit climate disruption.[1] Not a small order.

These measures are at least as much social and political as they are scientific or technological. They require, first, popular acceptance of the reality and importance of climate change and a willingness to make major changes in order to address it. This is a substantial political problem in the United States and in some other countries as well. Second, and at least as important, they require a willingness to address the institutional, legal, political, economic, financial, and bureaucratic obstacles to the rapid and far-reaching changes to energy and transportation systems, business practices, financial instruments, building design, urban land use, and way of life that will be needed if any reasonable target is to be met.

Third, they will require popular tolerance of and participation in a substantial period of policy learning—trying out methods for encouraging and incentivizing the needed changes in both business and government, monitoring results, learning from success and failure, and putting the lessons of experience into practice. European and Asian governments and researchers in particular have begun the process of thinking through how this might be accomplished and are experimenting with ways to put these ideas into practice.

Finally, and perhaps most important, they will require a willingness on our part to accept costs and inconveniences in order to assure future generations a sustainable planet on which to live.

The Survival Nexus. Charles Weiss, Oxford University Press. © Charles Weiss 2022.
DOI: 10.1093/oso/9780190946265.003.0004

Transformation in National Energy Systems

What overall changes in the world energy economy would be needed to make it possible to limit global warming to a 2°C target by the year 2100? To answer this question, researchers have developed a number of model scenarios, all of which agree on the need for substantial reductions in carbon dioxide (CO_2) emissions from buildings, transport, and electricity generation.[2] Each of the three is responsible for about a third of world emissions, each requires a different approach, and each has implications for the others. Reducing emissions from buildings would require greatly improved efficiency in energy use, especially through better insulation and more efficient appliances, and by a shift to electric space heating. Reducing emissions from transportation would require switching away from gasoline and diesel to vehicles run by electricity, natural gas, or biofuels.

According to the scenarios, the increased demand for electricity resulting from the electrification of space heating and transportation would be compensated in part by a massive improvement in end use efficiency, especially in buildings and other long-lived infrastructure. Even with this improved efficiency, there would still be an increase in the demand for electricity, which would be met by greatly expanding wind and solar and by increasing nuclear energy to a smaller extent. This would require, in turn, substantial changes in the structure and operation of the electric grid, which we discuss in the next section of this chapter. Coal-fired generating plants would be gradually phased out and in the meantime would need to be coupled to carbon capture, use, and sequestration (CCUS), a technology that we shall describe later in this chapter. Natural gas, which is less carbon intensive than coal but more so than wind or solar, would first be increased and then phased out.

Some of the technologies needed to effect these changes are mature commodities that can compete without subsidy. Wind turbines, solar photovoltaics, and lithium-ion batteries, for example, have achieved economies of scale and reduced their unit costs by meeting the needs of a series of profitable niche markets ("steppingstones") of increasing size.[3] Other needed technologies do not yet exist, are still at the laboratory stage, or require huge, lumpy, long-term investments at the pilot and demonstration phase at a time of regulatory uncertainty. The U.S. system of venture capital is unsuited to this kind of long-term investment.[4]

For these technologies to reach their potential, both industry and government will need to make much larger investments in research, development, demonstration, and commercialization—in short, in the entire innovation pipeline.[5] The "Mission Innovation" and "Breakthrough Energy" programs described in the previous chapter are only a beginning in this direction. The global recession resulting from the coronavirus pandemic threatens to slow private investments in important clean energy technologies that are key to the mitigation of climate disruption.

In contrast, the huge injection of public funds needed to stimulate post-pandemic recovery offers a golden opportunity to boost economic growth, create new jobs, and build cleaner energy systems. Many of the technologies needed to address climate change can also provide major economic and employment opportunities worldwide, including in low- and middle-income countries. Here the Chinese government has shown the way by pursuing well-funded, long-term, consistent strategies and policies of clean energy technology development, both to reduce carbon dioxide emissions and to position itself as the leader in clean energy. The unmatched Chinese skill at scaling up production technologies at all levels of sophistication, combined with internal competition among regions and companies, a nascent but increasing venture capital industry, and government support to basic research, make China a leader and an essential partner for others in developing mitigation technology.[6] Such partnerships with Chinese firms and government laboratories have long been underway in the private sector. The issues of protectionism and intellectual property that they raise are explored in greater detail in chapter 11.

Restructuring the Electric Grid

The energy sector is responsible for some two-thirds of annual global CO_2 emissions. Electric power now provides about a quarter of final energy consumption. This proportion needs to increase because many parts of the economy, especially transportation and space heating, need to be electrified if they are to improve their energy efficiency. As mentioned earlier, this will require a substantial expansion of the electric grid, along with an increase in the use of renewable sources of energy to feed this expansion, in order to reach any of the targets for mitigation of climate change.

The electric power industry is a low-margin industry that has traditionally emphasized reliability, safety, and low cost and has a justifiably conservative attitude toward innovation. In what we might call the classic period of the electric power industry, the supply of electricity to a given market was managed centrally by utility engineers. In those days, most "baseload" power came from hydroelectric or nuclear energy, or else from burning coal in large-scale, capital-intensive generating plants that ran at nearly full capacity all the time except during maintenance and repairs. Natural gas was used for "peaking" power from oversized aircraft engines that acted as flexible generators that could be turned on and off quickly to meet fluctuations in demand. The rigidity of this system made it difficult or impossible to incorporate more than about 10–15% of intermittent renewable energy sources like wind and solar.

Generating low-carbon electricity by integrating a high proportion of wind and solar resources into electricity systems changes the classical industry model of centralized generators and grid operators in fundamental ways and imposes many more demands on regulators and operators of facilities for energy transmission, distribution, and storage. European planners and operators of national electric grids are thinking through the necessary changes in some detail. Unlike the generators in the classical utility model, wind and solar energy are seasonal, intermittent, and unpredictable. The sun does not shine at night, and daytime clouds can obscure it from minute to minute; the wind does not blow all the time, and its strength can fluctuate rapidly in any given location when it does blow. Rooftop solar panels add another complication. The sun may be shining on a solar-equipped roof in one place and may disappear behind a cloud over another such roof nearby. A solar-equipped building may use the electricity its panels generate at times and at other times may export it into the grid.

This means that there will be fluctuations both in the *supply* of electricity and in the *demand*, unlike the fluctuations in the classical model, which were only in *demand*. These fluctuations can last for milliseconds, seconds, hours, or (in the case of seasonal variations) even months. To manage these fluctuations, utilities must digitize distribution grids and strengthen their capacity to allow electricity to flow efficiently in any direction. This, in turn, will require a much more flexible, higher-capacity way to store energy when the system is generating more power than is needed, and to release it when demand increases. Like the generation of electricity, this storage capacity may be either utility scale or decentralized.

In the new model of the electric grid, if the fluctuation in supply or demand of electricity in a given place is too big to be handled by locally available storage, the system will need to be able to import electricity quickly from somewhere else. In many countries, this means expanding the network of interconnections to form a larger "super-grid," increasing the capacity of the transmission wires so that they can carry the power that needs to be transferred from one place to another and creating a digitized payment system that can keep track of all these transfers. All this adds up to major changes in energy markets, in the business models of electric energy companies, and in the job descriptions of their engineers and managers, as well as the entry of a whole new set of generators and storers of electric energy into the energy system.

Mitigation by Switching Energy Sources

Let's now look at some of the energy and related technologies that would form the building blocks of an energy system based on renewable resources.[7]

Fossil fuels

Our starting point is fossil fuels, which are the basis of the existing energy system. *Coal* is being phased out as a source of baseload electric power—in Europe on environmental grounds, and in the United States because the fracking revolution has made natural gas there much cheaper than coal. Substituting *natural gas* for coal for power generation is a net plus from the environmental point of view, although it is far from a perfect solution. Coal is almost pure elemental carbon (C, not CO_2) and hence produces almost entirely carbon dioxide when burned. (Coal mining is also an important source of methane.) In contrast, natural gas is a mixture of hydrocarbons, compounds that contain both hydrogen and carbon. Burning natural gas therefore produces both H_2O (water vapor) and CO_2 and emits about half the CO_2 for each unit of energy as compared with coal. That makes natural gas a useful bridge to a hoped-for future more dependent on renewable sources, even though the pipes that transport natural gas to market often leak methane, a powerful greenhouse gas.

More immediately, the slapdash methods used by early practitioners of fracking technology in the United States have left a legacy of exploding tap water and contaminated ground water that has spawned worldwide banning movements. The low cost of natural gas from fracking, coupled with the fact that underground reserves are a source of income to the property owner and not to the government, has enabled gas to overcome this opposition in the United States. The resulting natural gas exports have helped make the United States a major energy exporter and has made fuel switching from coal to natural gas for electricity generation there economically and financially attractive.[8] This, however, is not the case in most of the world. Natural gas prices outside of North America are several times the U.S. prices, and fuel switching is financially justifiable only when substantial carbon charges are enforced. Except for the United States and some other major gas-producing countries like the United Kingdom and the Netherlands, natural gas has been traditionally used mainly for industry and heating and not so much for the generation of electricity.

Carbon capture, use, and sequestration

The scenarios we discussed earlier for the effective mitigation of climate disruption all agree on the need for some level of CCUS technology: *capturing* the CO_2 from exhaust gases, compressing it, transporting it, and finally either *using* it for commercial purposes or storing ("*sequestering*") it in a place from which it will not escape for a long time—for example, underground in disused oil fields, in coal or salt mines, in empty aquifers, in deep ocean waters, or beneath the ocean floor.

Technologies for carbon dioxide *capture* from the exhaust streams of fossil fuel–based electricity-generating plants and energy-intensive industries are well known. So are technologies for the *use* of carbon dioxide. It adds the fizz to sodas and sparkling water and can be used in the manufacture of cement and other building materials. Carbon dioxide has long been *sequestered* by being injected underground to force more tightly held secondary oil or gas out of fields that have already undergone exploitation. These applications are commercially profitable without subsidy in situations where the geology of the sequestration site is well understood, the CO_2 supply is plentiful and reliable, paying users are nearby, and the regulatory framework is favorable and predictable.

The widespread application of CCUS technology is an essential element of any strategy for reaching the targets of the Paris Agreement, probably as a complement to biofuels.[9] Energy-intensive industries need CCUS to reduce their carbon dioxide emissions, and the coal and natural gas industries need CCUS to have any long-term future in a climate-constrained world. These companies have long employed many of the needed technologies, although at a scale much smaller than will be necessary for significant climate mitigation. One might therefore have supposed that the interests of governments and businesses would here coincide: governments to protect the climate, and businesses to protect their own long-term future. Surprisingly, the total capacity of existing and planned CCUS projects lags well behind the scale that will be necessary to meet even Paris Agreement targets.[10] Present-day applications of CCUS use only tens of *millions* of tons of CO_2 per year, whereas the world's coal and natural gas plants for electricity generation and energy-intensive industries generate tens of *billions* of tons of CO_2 per year.[11]

Given that CCUS is not now privately profitable except in special circumstances, all new pilot, demonstration, or market introduction projects depend on grants, subsidized long-term financing, or tax benefits from governments to encourage and complement private investment. This may change as costs come down with experience and as improved technologies and business models are developed for exploiting captured carbon dioxide. At present, both public and private financing are discouraged by the long time horizon and lumpiness (large minimum size) of the necessary investments, by the absence of carbon charges in many parts of the world, by technological problems, by uncertainty regarding the reliability of existing policies, and by a variety of other legal and policy obstacles. Since many European countries are relying on other means of decarbonization, it seems likely that the United States will need to support and finance the market introduction of CCUS technology in energy and industry on a consistent and sustainable basis if this technology is to fulfill its potential.[12]

Renewable energy

Unlike natural gas, renewable sources of electric energy like solar, wind, and hydro produce little or no greenhouse gas once their generating facilities are in place.

Solar photovoltaic (PV) cells are now at about 1/250 the cost that they were when first introduced in the late 1970s. The balance of system costs—structural elements, installation, and "soft" costs like customer acquisition and financing—are also substantially reduced from earlier levels, although they vary greatly by country due to differences in business models and regulatory environments. PV is already the cheapest alternative without subsidy in locations with at least average levels of sunlight and can be installed on grid at utility scale, or off grid on roofs, in mini-grids, and in community grids, for markets at all income levels. While PV modules have become a commodity produced primarily in Asia, installation and maintenance offer local employment nearly everywhere. Solar energy is projected to rise as a total of world generating capacity from 2% in 2018 to 25% in 2050.[13]

Wind energy is a mature technology.[14] It is now competitive with coal and natural gas, not only in windy locations, but also in most developing and emerging economies. In 2018, it accounted for only 6% of global generation worldwide, with one-third of global installed capacity concentrated in China. By 2050, wind could contribute up to 35% of global power generation.[15] While a few companies in China, Denmark, Germany, the United States, and India are market leaders for turbines, wind installation creates employment opportunities during construction and operation everywhere.

Energy storage

The variability of solar and wind energy resources means that there will inevitably be times when they generate more electricity than is needed. A classic technology for allowing this excess energy to be stored is *pumped storage*, in which water is pumped uphill into a reservoir when power is being produced in excess of demand and allowed to run downhill again through turbines to generate electricity when it is needed. Pumped storage has many of the environmental and social issues associated with hydroelectric dams, which we discuss below.

Batteries are expensive as energy storage devices. Making them cheaper and increasing their ability to charge and discharge quickly are essential to the expansion of wind and solar energy. These involve problems of applied science and engineering that are subjects of active research but are likely to take time to solve.[16] Once developed, improved batteries will have many important applications in electric cars and other fields besides electric power, so they will be able to scale up and reduce costs in stages by filling a series of niche markets of increasing size.

Extra energy can also be stored by being fed into the wider energy system by "*power-to-x*" technologies, which make it possible for renewable energy resources to be used in transport and heat applications that might otherwise be difficult to convert to sustainable technology. For example, "excess" electricity can be used to electrolyze water (H_2O) to form *hydrogen* (H_2), which can be stored until it is needed, or can be mixed with natural gas and transported to the end user via pipelines specially designed to prevent leaks. Solar energy can be stored in a *hot liquid* that can be used directly for process heat or that can produce steam to run a turbine to generate electricity at night.[17] Alternatively, surplus energy can be used to synthesize any of a number of *chemicals* that can be used directly or that can be saved in the summer and reconverted to electric energy in winter. *Electric vehicles* add both complications and opportunities for energy storage. Ubiquitous charging stations would make it easy and convenient to keep their batteries fully charged and would also make it possible for the vehicle owner to sell electricity back to the utility, in effect making the batteries of a fleet of electric vehicles into a decentralized extension of the energy storage capacity of the national energy system.

Returning to renewable sources of energy, *biofuels* are plant materials derived from trees, plant oils, grasses, sugar cane, algae, or crop wastes, which recycle carbon dioxide by absorbing it from the atmosphere while they are growing and releasing it again when they are burned. They currently account for 2% of world energy production. Biofuels mitigate climate disruption by displacing fossil fuels. Some are burned directly as solid fuels for process heat or electricity generation. Palm oil and ethanol from sugar cane are direct liquid substitutes for diesel fuel and gasoline, respectively, and generate 8–10 times the energy that is invested in their production (their energy ratio). Biogas from human, animal, or agricultural waste is important in low-income countries as a replacement for the charcoal, wood, and untreated animal dung that low-income women and girls burn inefficiently for cooking and space heating, resulting both in lung diseases like asthma and in deforestation that contributes to climate disruption.

Ideally, centralized biofuel production and burning would be complemented by CCUS, once this technology is perfected, so that the carbon dioxide absorbed (*captured*) by the growing plants would be permanently removed from the atmosphere, making biofuels intermediates on the road to *use* or *sequestration*. Most strategies for achieving 1.5°C or 2°C assume that this combination of bioenergy plus CCUS (abbreviated as BECCS) will soon be developed to the point where it removes enough carbon dioxide from the atmosphere ("negative CO_2 emissions") to make a major contribution to mitigating climate change. This will require careful measures to ensure that the people who are to grow the biomass have the incentive to do so and have access to the necessary technical assistance and supplies.

On the negative side, biofuels use a lot of land and scarce water. The projected increase in biofuel production in some scenarios would require planting an area equal to 10–20 times the area of France![18] They cost energy for planting, harvesting, transporting, and processing. Like any monoculture, they would harm biodiversity and be subject to mass attack by pests and diseases. Many proposed biofuels, like corn, have much lower energy ratios than sugar cane and therefore have much less value as replacements for fossil fuels. Moreover, clearing new land for biofuels releases and emits carbon dioxide, in this way incurring an upfront carbon debt that must be "amortized" before the biofuel contributes to mitigation. In an extreme case, it takes over four hundred years to work off the carbon debt incurred by clearing peat soil for a Southeast Asian palm oil plantation.[19]

Many biofuels, like switchgrass, grow well on poor soils with little fertilizer or other inputs. The problem is that they grow even better and more profitably on good soils with lots of inputs. This means that expanding biofuels could lead both to deforestation and to competition with food crops that would raise the price of farmland and of food and would link the prices of food and energy and make the two fluctuate disconcertingly in parallel. Depending on national policies, biofuels could become crops for small farmers in low-income countries or, alternatively, could encourage land grabs—acquisitions by wealthy investors, often foreigners, of lands that have long been occupied by small-scale farmers whose formal title to the land is hazy, often with the connivance of corrupt government officials. None of these problems apply to biofuels derived from agricultural wastes.

Hydroelectric power is attractive from the climate point of view but suffers from major environmental and social obstacles. Hydro currently accounts for about 16% of world electric power generation, and as much as 65% in Brazil and 96% in Norway. However, this figure is expected to rise only slowly if at all. Most of the suitable sites for hydroelectric development in higher-income countries have long since been developed, although there are still many exploitable sites in low-income countries. Hydroelectric dams require displacement and resettlement of the people who live in the area to be flooded—the most dramatic example being the resettlement of over a million people in connection with the Three Gorges Dam on the Yangtze River in China.[20] Hydroelectric projects can also suffer from siltation and from water shortages due to drought, a problem that is likely to be exacerbated by climate change.

Geothermal energy (heat mining) exploits Earth's internal heat, which is, in turn, derived in part from the decay of radioactive elements inside the Earth, and in part from the friction generated by tides created in the liquid rock (magma) in the Earth's core as it flows in response to the changes in the gravitational pull of the moon as it orbits the Earth. Traditional geothermal technology, well known from the installations in Iceland and in California, accounts for less than 1% of energy production worldwide, and is limited to areas with hot water that lies naturally close to subterranean hot rocks and is accessible to drilling.

"Advanced" or hot dry rock technology, if developed and implemented, could greatly expand the use of geothermal energy. In many places all over the world, including much of the United States, hot rocks lie reasonably near the surface but are not naturally close to underground water, so that water must be supplied artificially. To capture this heat, two wells, one for injection and one for production, are drilled down to the hot rocks. Water (or in newer technology, supercritical carbon dioxide) is pumped at high pressure down the injection well to create a path from one well to the other by fracturing the intervening rock. Fluid is then pumped down the injection hole to be heated by the hot rocks, pumped to the surface via the production well, and run through a heat exchanger to create steam that generates electricity.[21]

Drilling deep holes is expensive, and the economics of this process still needs to be proved via expensive pilots and demonstration projects. The technology resembles that for fracking and raises similar environmental risks, most notably the possibility of groundwater contamination and induced earthquakes. The financial issues that we have already discussed in connection with CCUS also apply. As in the case of CCUS, government, industry, and venture capital in the United States are all unlikely to invest in the development and demonstration of an expensive, lumpy, engineering-intensive technology with a long payback period while there is still little prospect of a carbon charge that would level the playing field by imposing costs on emissions from burning fossil fuels.

Nuclear fission

Nuclear fission technology supplies almost three-quarters of the electricity in France and more than a quarter in a dozen other countries but has plateaued at about 14% of world electricity supply.[22] In North America, Western Europe, and Japan, few new nuclear plants are being planned or built, and some existing plants are being shut down as uneconomic or because of public pressure to wind down nuclear energy. This is partly because nuclear technology is too expensive to compete with cheap natural gas in electricity markets, partly due to technical issues, and partly due to a long list of economic, political, social, environmental, and managerial issues. On the other hand, investing in the renovations needed to extend the useful lifetime of existing nuclear plants is one of the most cost-effective ways to reduce the future contribution of energy generation to carbon dioxide emissions and global warming. On balance, overall energy production from nuclear power is expected to remain stable, or to increase marginally because of new installations, primarily in South Korea, China, Russia, India, and the United Arab Emirates.

Of the obstacles facing nuclear power, the technological are probably the most straightforward, at least in principle. Nuclear energy enthusiasts insist that new generations of technology promise a safe, low-cost, low-carbon, high-density source of electricity.[23] The difficult environmental and safety issues that have bedeviled nuclear energy, they argue, have promising technical solutions. Passive, "inherently safe" designs should make accidents like Three Mile Island and Chernobyl much less likely. Technical solutions exist to the problem of isolating nuclear power plants from earthquakes. Nuclear wastes, some of which have half-lives of tens of thousands of years, can be safely stored, first in temporary dry cask storage facilities and eventually in underground repositories in stable geological formations. Safe decommissioning of nuclear reactors that have reached their useful lifetime is technically feasible, although long-term and expensive.[24] The main obstacles to decommissioning are political rather than technical; there is still no authorized repository in the United States for nuclear wastes, for example.

Cost issues can be overcome, nuclear advocates argue, by using standardized, modular designs that could be built more cheaply and safely than the large, one-off installations customary in the United States and many other countries. New designs also make it possible to use nuclear power for process heat or as a dispatchable source of electricity for peaking power, in addition to its current use for baseload power. Substantial cost savings in construction can also be gained from sound project management, using standard practices that are nevertheless often neglected.

All these ideas need to be tested, proven, and demonstrated, a process that will require a good deal of time and billions of dollars. Even if money and engineering solutions are found and proven, financial and political issues will continue to complicate efforts to locate sites, both for new installations and for storing nuclear wastes and decommissioned nuclear plants. Nuclear technology is susceptible to coziness between the promoters of the industry and the government officials who are supposed to regulate it for safety, environmental sustainability, and nonproliferation, issues underlying the unfavorable public image that afflicts nuclear power in many countries. The United States dealt with this problem to some extent by separating the regulatory and promotional functions in 1974, a step that China—one of the few countries that is still making major investments in nuclear energy—has yet to take.

This regulatory capture was a major factor in the 2011 Fukushima accident in Japan, in which a tsunami flooded cooling and emergency systems, including backup generators that had been installed (with regulatory approval) at an elevation known to have been reached by a tsunami centuries earlier. The end result was a meltdown of the nuclear core of one of the four Fukushima reactors and the evacuation of some eighty-eight thousand people from the area surrounding the reactor, with about sixteen hundred deaths from the stress associated with the triple disasters and evacuation (direct human health effects from the radiation were too small to be measured because most of the radiation was blown out to sea or discharged directly into the ocean).[25]

Astonishingly, in retrospect, the need to evacuate up to thirty-five million people from the Tokyo area was avoided only by pure luck. Long after the Fukushima incident, researchers realized that an unsuspected water leak had been the only thing that had kept stored fuel rods covered with water and prevented them from being heated to such a high temperature that their zirconium coating (cladding) would have reacted with steam, generating large quantities of explosive hydrogen. The water had leaked through a crack in the wall that separated the fuel rod pool from an adjacent temporary pool of water, which had been put there by pure chance to help with unrelated maintenance problems. But for this totally fortuitous leak, the hydrogen would have exploded, blowing off the protective roof of the reactor and releasing huge quantities of radioactive cesium, which could have reached downtown Tokyo driven by normal seasonal winds.[26] Whew!

There are other reasons to be concerned about the safety of any possible major expansion of nuclear energy, especially to countries prone to corruption or to regulatory or management lapses. These include the possibility of accidents in the operation of the reactor itself, as happened at Chernobyl in 1986 and at Three Mile Island in 1979, the possibility of terrorist attack, and the risks that reactors will be constructed on geologically unstable sites, that spent fuel will be stored unsafely, or that reactors will be unsafely decommissioned once their useful lifetimes have been exceeded.

The most difficult political problem surrounding nuclear power stems from its role as a steppingstone to nuclear weapons capability and hence to the proliferation of weapons of mass destruction. Such proliferation risks could, in principle, be controlled by entrusting enrichment of nuclear fuels to international fuel banks. But as we discuss in chapter 5, the controls now in place have not prevented Pakistan and North Korea from developing nuclear weapons, and the effectiveness of these controls is likely to decline still further as nuclear export capability expands to more and more countries, some of which may not be fully committed to the principles of nonproliferation. The combination of these many issues has led to widespread opposition to nuclear power in the United States and Western Europe, which, in turn, has led to stringent safety requirements and long delays in permitting and construction that have helped to make nuclear energy uneconomic in these countries when compared with other energy technologies. Germany has taken steps to end all reliance on nuclear power by the year 2022, and several other countries have committed themselves to similar measures or to extensive (and likely negative) safety reviews.

Judged purely from the point of view of climate mitigation, nuclear technology is an option but not a necessity. As a practical matter, investment in nuclear energy is likely to continue, some of it in countries that view it more as a stepping-stone to nuclear weapons capability than as a cost-effective source of electricity. This raises the real possibility of serious problems of environmental pollution and nuclear proliferation. Any substantial expansion of nuclear power would have to be accompanied by international control of uranium enrichment and plutonium production. But any such expansion of nuclear power is unlikely to take place on a sufficient scale to contribute significantly to the level of mitigation needed for the world to reach either the 2°C or the 1.5°C global warming target.[27]

Efficient Energy Use

So far in this chapter, we have been discussing the "supply-side" possibilities of generating energy from renewable sources that do not emit greenhouse gases. Now we turn to the possibilities for increased efficiency in the *use* of energy. There are enormous possibilities on the demand side for energy conservation in agriculture, urban design, buildings, transportation, and industry, and, indeed, these are at least as important to mitigating climate disruption as are the sources of renewable energy that we have been discussing.

Our discussion of efficiency in energy use comes with three caveats. First, measures to increase efficiency often require large changes in public attitudes and policies that typically take a long time to occur. Second, they require changes in physical structures and infrastructure that can take half a century or more before they are replaced in the normal course of events. Third and most important, these changes will, for the most part, not take place in the absence of carbon charges or their equivalent that are sufficient to translate the externality of climate change into costs that influence the business decisions of companies and financial institutions and the planning and policy decisions of government authorities at national, regional, and municipal levels.

Agriculture

Agriculture is a major worldwide source of greenhouse gases.[28] Industrial agriculture in technologically advanced areas is a prolific user of fossil fuels for machinery for plowing, planting, and harvesting; for the manufacture of fertilizer; for the pumping of irrigation water; and for transportation. Mineral fertilizer degrades chemically into nitrous oxide (N_2O), a greenhouse gas three hundred times more potent than CO_2, molecule for molecule. Livestock converts only a fraction of feed into meat, further increasing energy use and consequent emissions. Cow burps and to a lesser extent, cow farts (excuse me, flatus), are a large source of methane, as are the settling ponds and lagoons used to process livestock manure. Far-flung supply chains add still further to agriculture's ecological footprint.

Agriculture contributes to climate disruption in all countries. Clearing forest lands for cropping decreases reflected sunlight and returns carbon dioxide that is stored in trees in the form of organic carbon to the atmosphere as they decompose. Rice paddies produce large amounts of methane, as do peat soils when they are drained to create farmland or palm oil plantations. Finally, when incomes increase, people eat more meat, improving their nutrition but increasing their ecological footprint.

Agricultural soils store a large quantity of organic material, much of which is oxidized to carbon dioxide when it is exposed to air. It is therefore important to spread and improve no-till agricultural technology, which does not involve plowing. This requires a bit of explanation. The major function of plowing is not to aerate the soil, as is commonly believed but, rather, to plow weeds underground so that the crops have a chance to outgrow them. Instead of plowing, no-till farmers use a chemical herbicide to kill the weeds. A seed drill makes a small hole in the soil and pours in the seeds, often accompanied by an individual dose of fertilizer and pesticide. In a refinement of the technology, crops are bred to be resistant to post-emergence herbicides applied after planting, so that these chemicals can continue to kill weeds until harvest time. The overall effect is to save money, save energy, and reduce soil erosion.[29]

One of the most important contributions that agriculture can make to the mitigation of climate change is counterintuitive: namely, to increase the productivity of existing cropland so as to free up acreage for biofuel crops, especially if BECCS technology is developed to capture and sequester the carbon dioxide that results from burning them. Uneaten food also contributes to climate disruption: in low-income countries as losses in the field, after harvest, and in storage, and in wealthier countries as garbage generated in food preparation and in storage before and after meals.

Urbanization

More than three-quarters of the world's population will be living in cities by the year 2050.[30] (The present figure is about half.) Since more than 70% of world carbon dioxide emissions and 65% of world energy demand originate from cities, this urban explosion has enormous consequences for climate disruption.[31] Cities can best minimize their future contribution to climate disruption by guiding their growth to reduce sprawl, to preserve forests and agricultural land, and to make transportation less dependent on private automobiles. LED street lighting can save energy. A covering of microorganism-rich biosolids can greatly reduce methane leakage from landfills. Older cities and towns can help by revitalizing their downtowns and industrial areas, and by encouraging vertical growth--growing in and up—rather than the horizontal suburbanization that is common in North America. Relatively self-contained neighborhoods, with residential, commercial, and office buildings close together, can encourage walking and cycling and minimize long-distance commuting. When unavoidable, urban spread can maximize the number of people along "ride-dense" mass transportation corridors linked to investments in public transportation.

Buildings

Heating and cooling buildings produced some 30% of CO_2 emissions worldwide in 2012. This number could be substantially reduced if new housing stock is built to passive, zero-carbon standards. A number of innovative technologies related to building have already begun contributing to mitigating climate disruption, including use of renewable energy for heating and cooling, more efficient construction techniques, longer-lived and recyclable materials, salvageable and less carbon-intensive building components, on-site renewable energy and cogeneration, LED and natural day lighting, energy-efficient appliances, improved insulation and overall energy efficiency, smart meters, and automated control systems.[32]

Building design and retrofits in warmer countries can further contribute by avoiding the use of air conditioning, an important source of chlorofluorocarbons (CFCs), hydrochlorofluorocarbons (HCFCs), which not only destroy stratospheric ozone but are also potent greenhouse gases. They can also revive traditional architecture that makes occupants comfortable in hot weather by controlling natural airflow. Appliance standards and building codes can overcome the problems of split incentives that arise when builders and landlords cannot pass on the initial costs of improved energy efficiency to prospective buyers and tenants.

Transportation

Transportation accounts for about a quarter of final energy consumption worldwide. Here, major mitigation objectives would be to convert short-distance trips from airplanes to railroads or other surface vehicles; to expand public transportation; to reduce gas-consuming delays from traffic congestion; to encourage walking, cycling, and ride sharing; and to replace as many gasoline-fueled automobiles as possible with cars powered by electricity. (Electric motors are more energy efficient and less CO_2 emitting than internal combustion engines, especially if they are run on electricity generated from renewable sources of energy.) Vehicle manufacturers are increasing the energy efficiency of engines, adapting gasoline and diesel engines to biofuels, and are developing improved batteries and fuel cells for hybrid and electric cars. Even transportation methods that are not easy to electrify can reduce emissions; for example, ships can save a lot of energy by moving at slightly slower speeds.[33]

Industry

The good news is that the energy intensity of economic growth—the amount of energy it takes to produce a given rise in GNP—in high-income countries has been decreasing over the past two decades, thanks to improved energy efficiency, increased use of waste heat, use of renewable sources of process heat, and more recycling and more efficient use of materials.[34] Redesign of industrial products, like enzyme-reinforced detergents that wash clothes in tepid water, also contributes. China, the world's center of manufacturing, has announced plans for a major increase in the energy efficiency of its manufacturing industry.[35] Here the International Organization for Standardization (ISO), an international organization consisting of standard-setting bodies from some 164 countries with a practice of close consultation with industry, has issued standard criteria for energy management, ISO 50001, which "specifies requirements for establishing, implementing, maintaining and improving an energy management system . . . in order to enable an organization to follow a systematic approach in achieving continual improvement of energy performance."[36]

Adapting to Climate Disruption: Intervene as Well?

Thirty years of international agreements and domestic efforts have resulted in only limited progress in *mitigation*, i.e., in reducing emissions of greenhouse gases. This experience has forced attention to the need for *adaptation*, which is to say accepting and dealing with the inevitable impacts of climate disruption. This means, for example, building protective dikes or strengthening natural protections for coastal cities against sea level rise, developing new crop varieties or breeds of livestock suited to higher temperatures, improving systems for storm warnings and disaster management, strengthening wildlife management and insect control, and increasing efficiency in water use. The impacts of climate disruption are likely to fall most heavily on precisely those low-income countries that are weakest in local professional and policymaking capacity and least able to bear the expense and institutional strains that these measures would require.[37]

Discouragement with mitigation measures has also increased interest in the possibility of *interventions* to reverse the effects of climate change. These are also known as "negative emissions" or "geoengineering"[38] and involve reducing the amount of solar energy that reaches the Earth by increasing the proportion of sunlight reflected back into space ("solar radiation management," or SRM) or, alternatively, by removing carbon dioxide from the air and storing it in trees, soils, oceans, or rocks ("carbon dioxide removal," or CDR). This subject was long avoided as being excessively risky and likely to divert attention from the need to mitigate climate disruption, an example of what economists call moral hazard.

Except for reforestation, all these measures are risky, and most are expensive to boot. Even if a geoengineering measure succeeds in slowing global warming, it may induce unexpected and undesirable changes in regional or local weather patterns. SRM in particular would reduce the amount of solar energy that reaches the Earth without changing the rate of carbon dioxide emission or doing anything to limit ocean acidification. A good deal of basic geophysical research will be necessary before these effects are adequately understood. Once begun, moreover, SRM would have to be continued indefinitely. If it were ever to be discontinued, the Earth's temperature would increase rapidly, since the solar energy reaching the Earth would return to pre-SRM levels and be subject to an intensified greenhouse effect from the accumulated atmospheric carbon dioxide.

To the environmentalist, then, geoengineering is at best a measure of perhaps justified desperation. It also raises important issues of international law and governance, since many of its methods can be carried out in international waters, and some are within the means of wealthy individuals or are conducive to crowdsourcing. Who would have the right to authorize such measures, and by what criteria? Who would be responsible if they go wrong? Even research, development, and pilot projects involve environmental risks and must be designed and carried out with great care. Nevertheless, so much time has been lost that the world may have no choice but to turn to these risky interventions. We shall return to these issues in chapter 12.

So Much to Learn, So Little Time

The world has had to absorb a remarkable catalog of radical new ideas and concepts since the issue of climate disruption emerged on the international agenda: that humans can cause serious and irreversible damage to the Earth's climate and its oceans; that environmental change can be an issue of national security; that national energy policies can be international, not just domestic issues; that action to forestall environmental damage is urgently needed even though this damage will not be fully apparent for decades; that this generation must act to avoid damage to future generations; that far-reaching actions must be undertaken despite lingering scientific uncertainties; that scientists may accept a new overarching explanation—a new theory—even if areas of uncertainty and ignorance remain to be clarified; that low-income countries need to contribute to the solution of a global problem, to which they have contributed little or nothing, and need to be compensated somehow for doing so; that diplomats need a regular flow of scientific advice from both official and unofficial sources; and that a solid scientific consensus on a technical issue can be vulnerable to a well-funded, skillfully managed program of obfuscation and disinformation.

There's more to this list: that it may be desirable to allow those who need to damage the environment by emitting greenhouse gases to pay for the right to do so; that needed technological innovations will not take place in response to market forces unless there are policies in place to incentivize them and that such innovations will take time, may face difficult structural obstacles and opposition, and will often require expensive and politically risky direct government involvement and encouragement that must be managed carefully to avoid corruption; that insofar as these innovations are business opportunities, they may raise complicated questions of intellectual property and of the need for simultaneous international collaboration and competition; and finally, that almost all the measures that can be taken to mitigate and adapt to climate change carry with them costs, risks, and trade-offs in addition to their clear benefits.

Each of these propositions affects and challenges the preconceptions of some important interest group. All of them must be absorbed while basic scientific knowledge is still accumulating about the geophysics of climate and about the multifaceted impact of climate change on human life. Each links climate science or energy technology with geopolitics, economics, politics, business, ethics, law, or culture—the Survival Nexus that is the subject of this book. That's a lot for the world to learn, even in fifty years! Small wonder, then, that the model of global cooperation that marked the international response to stratospheric ozone depletion has been overwhelmed by the scale and scope of climate disruption and its many impacts, and that a backlash has appeared from many of the constituencies that will have to bear the costs of the interventions needed to address it. In ordinary circumstances, the world might be congratulating itself for having come as far as it has—intellectually, politically, and as a matter of international law.

In retrospect, the international community has addressed climate change in a manner appropriate to a contentious, high-stakes issue in which scientific understanding was still unfolding. It undertook a reasonably comprehensive application of the principles of adaptive or evolutionary management, an accepted learning-as-you-go approach to problems marked by technical uncertainty, an issue to which we will return when we discuss gene drivers and geoengineering in chapter 12.

The international community acknowledged the problem in an initial framework agreement, the United Nations Framework Convention on Climate Change (UNFCC). It established an ongoing, comprehensive system of regular updates on the underlying science and the economic and social impact and met regularly to update the agreement as information and experience accumulated. Meanwhile, national and state governments experimented with policies to limit greenhouse gas emissions, including direct regulation, emission charges, and cap-and-trade regimes, learning what worked and what didn't. Researchers in both the natural and the social sciences, based in universities, government laboratories, NGOs, research institutes, and intergovernmental organizations like the World Bank and the United Nations technical agencies, expanded knowledge of climate science, the global changes now underway, and those expected from anticipated climate disruption.

An epistemic community conveyed their findings to the public, to policymakers, and to the diplomats negotiating international agreements. Poets, artists, philosophers, theologians, philosophers, journalists, and politicians all participated in public discussions. All this took time, allowing opponents to capitalize on real or imagined scientific uncertainties and irregularities in the scientific advisory process to throw multiple monkey wrenches into what a scientist might imagine should be a straightforward process.

The problem is that we are running out of time. Nature does not wait for the slower processes of social learning and political change. Measures that could have been phased in relatively smoothly and affordably twenty or thirty years ago are now drastic, painful, expensive, and hampered in many countries by populist uprisings. Can we now handle it? The positive scenario (the "bull case," as financial professionals would say) would be that the very process of meeting the voluntary targets required by the Paris Agreement, reinforced by the startling warnings in recent Intergovernmental Panel on Climate Change (IPCC) reports, will lead people, governments, and businesses to accept the fact that climate disruption constitutes a real, existential crisis, and that the world will have to undertake the full-scale, top-speed, all-hands-on-deck effort that is now necessary.

We spoke earlier of possible geophysical tipping points like the large-scale melting of ice in Greenland or Antarctica that could lead to quick climatic disaster. Could there also be political tipping points, like the one that happened in the United States with smoking, that lead to rapid shifts in public sentiment and a positive groundswell of support for decisive international action? Here Western

Europe is clearly well ahead of the United States. There are signs that the young generation is organizing worldwide to shame its elders into urgent, drastic action. Some businesses have instituted internal carbon pricing or cap-and-trade systems. Financial institutions are coming under pressure to incorporate sustainability criteria into their investments and investment advice. Some multinational oil companies, especially those with limited oil reserves, are initiating or resuming hedging investments in green technology and stating their intention to reposition themselves more broadly as energy companies.[39]

Climate disruption is ultimately an ethical and moral issue. We have an ethical, moral and arguably a legal obligation to leave the Earth to our descendants in as good a condition as we received it.[40] To do so, we will need to mobilize both science and technology: science to understand and track the sources and impacts of climate disruption, and technology management and technological innovation to mitigate the level of disruption and to deal with whatever we cannot stop. This is one of the first times in history that scientific research and technological innovation have become moral and ethical obligations. The day may come when they are considered to be legal obligations as well.

We now turn to a danger even greater than climate change, namely, that of nuclear war.

5

Nuclear Issues

Civilization in the Balance

Just after midnight on September 26, 1983, the sirens of the Soviet early warning system screamed an alert that five nuclear-armed U.S. intercontinental ballistic missiles would land somewhere in the Soviet Union within minutes.[1] Lt. Colonel Stanislav Petrov faced a terrible decision. His duty was to inform his superiors of the imminent threat, so that they could launch an immediate counterstrike against the United States and start a catastrophic nuclear war. But something didn't feel right. Why would the United States launch a strategic first strike with only five missiles? If they were going to strike, wouldn't they launch scores or even hundreds of missiles in order to destroy Soviet ability to respond?

Colonel Petrov's gut told him that there was at least a fifty-fifty chance that the warning was a mistake, a glitch in the system. He checked with Soviet ground radar, which reported no sign of any looming attack. He had been right to have been skeptical. The monitoring satellites had sounded the alert because they had mistaken sunlight reflected off of high clouds for the glint of incoming missiles. Petrov informed his superiors of the system malfunction and we didn't have a nuclear war. Close.

It could have been a lot worse. At least Colonel Petrov had fifteen minutes to decide. Someday soon, mid-level officers in the United States, France, or India may have to decide in a couple of minutes whether a hypersonic, nuclear-armed, intermediate-range missile is on its way from a hostile neighbor or whether the alert they have just received is a false alarm. Hypersonic missiles are now under development in several countries. It's easy to imagine situations in which Earth-in-the-balance decisions would have to be made far faster than any human can possibly react.

Instant, Catastrophic Destruction

If an atomic bomb of the size that leveled Hiroshima at the end of World War II exploded during working hours at ground-level in mid-town Manhattan, it would topple skyscrapers for a distance of 500 meters in every direction and kill an estimated five hundred thousand people.[2] Even an unsuccessful explosion—a

The Survival Nexus. Charles Weiss, Oxford University Press. © Charles Weiss 2022.
DOI: 10.1093/oso/9780190946265.003.0005

fizzle--would do major damage at least on the scale of a city block.[3] A 400-kiloton hydrogen bomb—of which the United States and Russia still have thousands—would be about thirty times more powerful. If exploded in midtown Manhattan, it would vaporize or incinerate everything within half a mile and burn buildings and cause third-degree burns on exposed skin out to five miles from Ground Zero, an area extending from lower Manhattan to about 100th Street and from Jersey City to the western parts of Queens and the northern parts of Brooklyn, causing 1.2 million deaths and 2 million injuries.[4]

Nuclear weapons technology is now more than seventy-five years old, long enough to have spread all over the world. From one point of view, it is remarkable how few countries have nuclear weapons. During the Cold War, the major threat of apocalypse came from the possibility of a strategic nuclear exchange between the two superpowers, the United States and the Soviet Union. This possibility still exists.[5] Two thousand missiles with nuclear warheads are still on hair-trigger, "launch-on-warning" alert in the United States and Russia, ready to be fired on a few minutes' notice.[6]

The world is closer to nuclear war today that it has been in several decades. The regimes for arms control and nuclear nonproliferation that were carefully constructed by the two Cold War superpowers to prevent nuclear war have been undermined and deconstructed. Political leaders are increasingly willing to threaten to be the first to use nuclear weapons in support of immediate political objectives. Increasingly accurate missile systems favor the attacker and create the temptation to carry out a strategic attack without warning so as to destroy the enemy's ability to retaliate. Tactical nuclear weapons are blurring the once-clear distinction between atomic and conventional weapons, eroding the taboo against the use of nuclear weapons.

Surprisingly in retrospect, the danger of nuclear warfare received little public attention during the decade between the fall of the Soviet Union and the rise of extremist terrorism. My students, for whom World War II and the Cold War seemed like ancient history, used to regard these issues as historical relics. For better or for worse, popular interest in nuclear issues has been rekindled by the rise of China, the resurgence of Russia, nuclear and missile threats from Iran and North Korea, questioning of the reliability of the nuclear umbrella over U.S. allies, concerns over the possibility of nuclear terrorism, and the possible revival of nuclear power as a low-carbon energy source. But despite the efforts of senior international leaders and arms control professionals, the seriousness and urgency of the nuclear threat are still not widely recognized.

Arms control and nonproliferation are areas of international affairs in which it is essential to understand the scientific and technological underpinning of international issues in some detail.[7] For this reason, the treatment of nuclear science and technology in the text and footnotes in this chapter is more detailed than it is in other chapters. We have also added an annex giving a brief introduction to nuclear science for those readers to whom it is not familiar.

Nuclear issues have the classic characteristics of the Survival Nexus. They had their origin in the discovery of nuclear fission, a revolutionary technological development based on cutting-edge basic science.[8] Scientists pioneered in exploring the implications for geopolitics and military strategy. They succeeded in convincing the international community that nuclear weapons had given rise to a new paradigm of international security, based on arms limitations, controls on nuclear proliferation, deterrence, and mutually assured destruction. They collaborated with diplomats and international affairs specialists to work out details of inspection, verification, and other technical aspects of arms control and nonproliferation, and they kept them up to date on the capabilities and limitations of new technical developments.

Generations of well-trained physicists now understand geopolitical issues, and many diplomats and international affairs specialists understand nuclear science and technology. There are now NGOs, research institutes, and academic departments whose expertise rivals that of the military and civilian government officials concerned with security. Diplomats and international affairs specialists have long appreciated the need to work with technical experts and to negotiate with their counterparts in potentially hostile countries on ways to control potential conflicts so as to prevent mutual suicide. This collaboration gave rise to regimes that have avoided nuclear war for the seventy-five years since World War II but are now in danger of collapse.

Mutually Assured Destruction and a Close Brush with Nuclear War

We begin the story of this regime with the origins of the Manhattan Project at the start of World War II. In 1939, when Albert Einstein wrote the famous letter to President Franklin D. Roosevelt that led to the Manhattan Project and the development of the atomic bomb, nuclear physics was an arcane branch of fundamental science.[9] The basic discovery of nuclear fission that made the atomic bomb possible was less than a year old.[10] The President, it has been said, had "no more knowledge [of nuclear physics] than the corner cop."[11] Six years later, nuclear weapons had decided the most destructive conflict in the history of humankind. Four years later, both the United States and the Soviet Union had both the atomic and the hydrogen bomb, firmly establishing them as the global superpowers in a new, bipolar world and giving rise to a Cold War that lasted some forty-five years.

To the community of physics researchers, the atomic explosions over Hiroshima and Nagasaki were an end to innocence. In the words of J. Robert Oppenheimer, Scientific Director of the Manhattan Project, the physicists who developed the bomb "had known sin."[12] While the war was still underway, a few scientists, led by Nobel Laureate James Franck, wrote a letter to President Truman urging him to invite United Nations officials to a demonstration explosion in the desert, rather than to authorize the explosion of a bomb over a Japanese city.[13] The *Bulletin of the Atomic Scientists*, established after the war by alumni of the Project, became an important forum for the exploration of the technical and political issues concerning nuclear weapons, as it continues to be today.

Despite early attempts to establish international control over atomic weapons, a decade or so of an uncontrolled arms race ensued, during which time the two hostile powers hardly communicated with each other. Each developed fusion-based hydrogen bombs a thousand times more powerful than the fission bombs that had destroyed Hiroshima and Nagasaki. Each stockpiled tens of thousands of weapons, along with the means to deliver them—enough to destroy the Earth many times over. A strategic triad of long-range bombers, submarine-launched ballistic missiles, and land-based missiles in hardened silos resistant to attack assured that each side could respond with devastating force to any possible strategic nuclear attack from the other.[14]

The feeling of danger from nuclear war was omnipresent in both countries. Fallout shelters were built in major cities in both the United States and the Soviet Union. U.S. schoolchildren underwent regular duck and cover drills to teach them what to do in case of a nuclear attack. Budgets and jobs for defense-related research expanded rapidly, as did money for basic research and education in physics and engineering, both of which were thought critical for winning the Cold War—a clear demonstration of the influence of international affairs on the direction of science and technology.

Gradually, communication between the two countries was restored, at first largely through the efforts of the National Academy of Sciences and the Pugwash Conference, an NGO consisting largely of scientist alumni of the Manhattan Project and their counterparts in the Soviet Union.[15] The realization slowly spread that nuclear arms were not usable in any reasonable military scenario for the defense of Western Europe (a conclusion reinforced by the devastation produced by a simulated nuclear battlefield in Western Europe in 1957 and the quick escalation from the use of tactical nuclear weapons to full-scale nuclear war in war games conducted in the 1960s), that more of them did not produce more security, and that their real function was to deter a potential enemy from attacking by making it clear that a strategic nuclear exchange would be suicidal for both sides.[16]

The new conceptual underpinning of world security became mutually assured destruction—MAD for short, an apt acronym for such a mind-blowing foundation for global survival. A strong taboo grew up that has held back national governments from using nuclear weapons in any of the many wars that have taken place since the end of World War II and that is still in effect as these words are written. A substantial cottage industry arose of physicists, political scientists, and international lawyers that projected the likely consequences of a nuclear exchange between the two superpowers, even including imagining the "broken-back warfare" that might then ensue.[17] On a more positive note, these researchers also explored the technical requirements for inspection and verification in any agreement to reduce the number of weapons of mass destruction that each was aiming at the other.

Arms Control: Avoiding Mutual Suicide

The basic doctrines of arms control developed during the Cold War hold that the world is better off with fewer rather than more weapons, and with no defenses against nuclear attack rather than with effective ones. These counterintuitive ideas require us to explore some disagreeable—even unthinkable—issues regarding nuclear war.

Since no one except a nihilist or someone bent on suicide wants a nuclear war, the rational purpose of having nuclear weapons is to deter anyone else from attacking you with *their* nuclear weapons, by convincing them that you can inflict unacceptable damage in a retaliatory strike if they do attack you. (To be sure, this has not stopped people from dreaming of "victory" in a nuclear war, but more on that later.) In order to have a credible deterrent, a country must have enough nuclear weapons that can survive a strategic attack to be able to deliver a retaliatory second strike that will inflict unacceptable damage. Deterrence does not work if one side has first-strike capability, the ability to wipe out enough of the nuclear capability of a defender country in a single first strike that the defender country cannot inflict unacceptable damage in a retaliatory second strike. The need to assure second-strike capability sets off an uncontrollable arms race even if neither of the two countries actually intends to wipe out the other.

To see how this works, let's make it personal. If your country fears that my country, a potential enemy, is building up its nuclear and missile forces and is at or near first-strike capability, your country's military will be strongly motivated to build up its own missile force to assure its ability to retaliate, i.e., its second-strike capability. This leads to a spiral that can easily get out of control. My country's military will not be able to tell whether your country is building up its nuclear forces for offensive or defensive purposes. It will therefore assume the worst—that's its job, after all—and continue to build up my missile force in order to be sure that it can retaliate against a possible first strike by you. Your military will observe my country's buildup and come to the same conclusion and so on, resulting in a positive feedback loop and a classic arms race. This is what happened during the Cold War. It led to the United States and the Soviet Union each building arsenals of tens of thousands of nuclear-armed missiles, wildly more than they could use for any rational purpose.

Missile defense makes things worse. To continue our hypothetical scenario, suppose your country builds a defense against a possible first strike by my country's missiles. Never mind that with present technology, this anti-missile system would not be effective against such a first strike. My country's military would assume the worst: namely, that you will now be able to launch a devastating first strike and that your anti-missile defense will then limit the damage from my country's retaliatory second strike to acceptable levels, so that my country no longer has effective second-strike capability and can no longer deter you from attacking me.

My country's military would be professionally obliged to do this even if you insisted that you have no intention whatsoever of launching such a first strike. My country will now respond by building additional missiles to ensure that at least some of them will get through your country's defenses and cause unacceptable damage after your supposed first strike. This maddening logic leads inexorably to an arms race spiral like the one we have just described. For this reason, anti-missile defenses are said to destabilize—i.e., to create uncontrollable incentives for an unlimited arms race.

But wait. Suppose your missile defense would, in fact, be effective against a first strike from my present arsenal. Yes, that would protect you, but only for a while. It would be a lot cheaper for me to add enough missiles to my arsenal to overcome your defenses than it would be for you to deploy enough defensive missiles to reliably shoot them down. We are thus forced to the paradoxical conclusion that the stablest and safest situation for the world is for major powers *not* to have a defense against the missiles of a potential enemy, as long as they each have sufficient second-strike capability.

This is the basis of the doctrine of mutually assured destruction, fittingly abbreviated as MAD and enshrined in the 1972 bilateral Anti-Ballistic Missile (ABM) treaty, in which the major powers agreed *not* to build anti-missile defenses. (The United States withdrew from this treaty in 2001 in order to be able to build anti-missile systems to defend itself against smaller-scale attacks from lesser powers like Iran and North Korea.) It sounds crazy, but except for the Cuban Missile Crisis, which we are about to describe, and despite numerous near-accidents, the MAD system deserves credit—along with a lot of luck—for helping to prevent nuclear war for seventy-five years.

What is more, it is, in principle, more stabilizing if we both target each other's cities rather than each other's military targets and are open about doing so. From the purely military point of view, it clearly makes more sense to target an enemy's military capabilities, a strategy known as counter-force. However, in another of the maddening paradoxes of mutually assured destruction, a counter-force strategy requires numerous weapons aimed at destroying a defender's retaliatory capability—i.e., an attempt to achieve first-strike capability. For this reason, it is likely to induce an uncontrollable arms race for the reasons we have already discussed. A counter-city strategy would require fewer weapons and in theory be safer for the world, at least up to the point where it is actually implemented, even though holding civilian population centers essentially as hostages resembles terrorism more than it does warfare. Regardless of this logic, there is clear evidence that U.S. war plans (the Strategic Integrated Operations Plan, or SIOP) call for a counter-force attack in case of all-out nuclear war and have done so since the beginning of the Cold War.[18]

It gets even worse. An essential element of the MAD regime as it has evolved is the doctrine of launch-on-warning—i.e., that a defender country is to launch its retaliatory attack when it has received an alert that a strategic first strike was on the way but before it has landed and exploded.[19] This doctrine gives the defender country about fifteen minutes to decide whether to authorize a full-scale nuclear attack. This hair-trigger posture carries with it a substantial risk of catastrophic mistake, as vividly exemplified by the Russian false alarm with which we began this chapter.

In principle, a defender country with second-strike capability should not need to launch a retaliatory attack on warning, since by definition it could launch a retaliatory second strike that would inflict unacceptable damage on the attacker. However, military planners on both sides of the Cold War were reluctant to ride out such a first strike, lest it wipe out key command-and-control systems or kill off political leaders and top military commanders in a "decapitation" attack. They also saw advantages in leaving a prospective attacker guessing as to what their response to a first strike might be. Whatever the rationale, launch on warning is built into the strategic operational planning of both Russia and the United States, despite the inherent hair-raising danger.

MAD doctrine thus leaves the world in what a scientist would call a metastable situation. Imagine a book poised precariously at the edge of a table, from which a slight bump can make it fall to the floor. It's stable until it isn't. MAD has worked so far; the book is still on the table. But there have been close calls. Is it ethical to purchase stability at the price of such a risk? It's a MAD, MAD world.

The ultimate test of the MAD regime and the nuclear taboo came with the Cuban Missile Crisis, which was precipitated when the Soviet Union shipped nuclear missiles to Cuba in 1962. Then-President John F. Kennedy came under strong pressure to carry out a "surgical" (that's what the generals called it) bomb strike to destroy the missiles but elected instead to prevent the missiles from landing by "quarantining" the island. (He actually blockaded it—but to acknowledge this fact would have been to announce that what he was doing was an act of war, so he called it a quarantine.)

The world learned much later that it had come within a hair's breadth of nuclear war. The Soviet government had delegated permission to fire nuclear-tipped missiles to mid-level field commanders, including a submarine commander who was prevented from firing a nuclear missile onto U.S. territory at a critical moment only by the objection of another naval officer, Vasili Arkhipov, whose consent was required by the rules governing the launch.[20]

This story calls into question the whole theory of deterrence, on which the stability of the world against nuclear holocaust was thought to depend under the MAD doctrine. In 1995, in a retrospective discussion of the crisis in Havana, Cuban dictator Fidel Castro told Robert McNamara, who was U.S. Secretary of Defense at the time of the crisis, that unknown to the U.S., Cuba already had short-range nuclear missiles on the island when the Soviet shipment was discovered. Castro told McNamara that he would have used them against U.S. territory had Cuba been bombed, even though he knew that the island would have been completely destroyed in a retaliatory attack.[21] So much for deterrence theory. Soviet dictator Nikita Khrushchev withdrew the nuclear missiles from Cuba when he learned of Castro's attitude several months after the crisis was over.

Nonproliferation: Fewer Nuclear Countries Make a Safer World

By the mid-1960s, France, the United Kingdom, and China had also acquired nuclear weapons. It was widely feared that the number of nuclear powers would rise to many scores by the turn of the century, and that nuclear war of some kind would therefore be a virtual certainty. Complicated negotiations led to a series of agreements that established an international regime—a binding set of principles, rules, and decision-making procedures[22] —to prevent the further spread of nuclear weapons, a goal that became known as nuclear nonproliferation.

The first of these agreements, the Treaty Banning Nuclear Weapons Tests in the Atmosphere, in Outer Space and Under Water of 1963 (the Limited Test Ban Treaty), responded to the worldwide concern about radioactive fallout, like the kind that happened near Bikini Atoll, the 1954 Castle Bravo thermonuclear test that had destroyed three small islands in the South Pacific and poisoned 23

Japanese fishermen. The Threshold Test Ban Treaty of 1974 extended this ban to underground nuclear tests above 150 kilotons, ten times the size of the bomb that destroyed Hiroshima; the enforcement of this treaty depended on the ability to distinguish between an underground test and a natural earthquake.[23] This presented an important challenge to the then-nascent field of seismology (the study of earthquakes), whose leading researchers became influential figures in the next generation of U.S. science policymakers, again showing the close relationship that sometimes holds between international affairs and basic science.

The 1968 Treaty on the Non-Proliferation of Nuclear Weapons (the NPT), the centerpiece of the nonproliferation regime, divided the countries of the world into two categories: nuclear and nonnuclear. To forestall further proliferation, it forbade nonnuclear powers from acquiring nuclear weapons, even though it would otherwise have been their sovereign right to do so. To compensate them for accepting this fundamentally inequitable arrangement, nonnuclear countries were promised access to nuclear technology for peaceful purposes and were assured that the nuclear countries would make progress toward nuclear disarmament—reduction in the total number of nuclear weapons. The indefinite renewal of the NPT in 1995 implicitly accepted this inequity in the name of nuclear stability.

The NPT was followed by two carefully negotiated treaties that emerged from the Strategic Arms Limitation Talks and became known together as SALT I. Signed in 1972, these treaties put limits on an arms race that had led on both sides to vastly excessive arsenals capable of massive overkill. The first of these treaties was the ABM Treaty, which limited the number of *defensive* missiles on each side and required that they be deployed in a manner that would defend only a relatively small area. The second SALT I treaty, known formally as the Interim Agreement and Protocol on Limitation of Strategic Offensive Weapons, froze the total number of land- and submarine-based ballistic missiles at the levels then existing.

The 1979 SALT II treaty limited the total number of long-range strategic weapons, including long-range bombers and land- and submarine-based missiles, including multiple independently targeted vehicles (MIRVs), which are missiles that carry several warheads, each of which can be programmed to hit a different target. SALT II also set forth detailed limits on the development and deployment of new systems for the delivery of nuclear weapons. The SALT II Treaty never came into force—i.e., it was not ratified by the U.S. Senate and hence was never legally binding—but its provisions were generally observed in practice until it was superseded by the START Treaty of 1991, which we will describe shortly.

The arms control regime was augmented by the Intermediate-Range Nuclear Forces (INF) Treaty of 1988, under which the two superpowers destroyed some 2,700 missiles with ranges between 300 and 3,400 miles. These missiles had increased the danger of accidental war because they gave the defender very little time to decide whether a given warning was a real attack or a false alarm. The United States withdrew from this treaty in 2019.

The many conferences and negotiations that led up to these agreements served to educate governments, international organizations, and the world scientific community regarding the intricate connections between nuclear technology and international affairs. The United States established the Arms Control and Disarmament Agency (since reconstituted as the Undersecretariat for Arms Control and International Security in the Department of State), in which physicists, political scientists, economists, international lawyers, and foreign affairs generalists collaborate to address these complex issues. Other countries established smaller official entities for the same purpose. The International Atomic Energy Agency (IAEA), an agency of the United Nations established in 1957, was given the task of monitoring compliance with the NPT rules, and specifically with detecting any diversion of nuclear material away from civilian use by a system of voluntary inspections.[24] In addition, the countries that export nuclear technology organized themselves into the London Suppliers Group and a number of similar voluntary organizations, which laid down rules limiting the export of nuclear equipment and technology, including dual-use technology suitable for both civilian and military applications, to non-nuclear countries.

The IAEA was also given the task of helping non-nuclear countries use nuclear technology for civilian purposes like electric power generation and agricultural and hydrological research. Here, the IAEA assumed a task begun by the earlier Atoms for Peace initiative, which was established by the Eisenhower Administration in 1953 in an effort both to build goodwill and to create a market for U.S. nuclear exports.[25] At the time, the U.S. federal government enthusiastically supported the expansion of nuclear power, in part because of the promise of electricity that would be "too cheap to meter," and in part because of a felt moral imperative to achieve a constructive result from the technology that had destroyed Hiroshima and Nagasaki.[26]

In retrospect, the ensuing rush of private firms to capture market share in what was then an exciting new technology of nuclear power resembled the rush to capture eyeballs during the dot-com boom four decades later (see chapter 8). Governments and their scientists all over the developing world eagerly embraced nuclear technology. Some did so to take advantage of the disproportionate level of support available for research on its peaceful uses, but others did so as a way of building the capacity they would need to develop nuclear weapons as instruments of national security, regional hegemony, or national prestige. IAEA assistance resulted in the establishment of small civilian research reactors all over the developing world, and the much more rapid spread of nuclear expertise than would have otherwise taken place.

All things considered, the nonproliferation effort has been remarkably successful. Only four new nuclear-weapons countries have been added to the five original ones: India, Pakistan, North Korea, and (although this has never been officially acknowledged) Israel, although the four are not parties to the NPT. Brazil, Argentina, Taiwan, Libya, Iraq, South Korea, and South Africa all abandoned nuclear programs at various stages,[27] and newly independent Ukraine, Belarus, and Kazakhstan transferred their nuclear weapons voluntarily to Russia when the Soviet Union dissolved. Many of their nuclear weapons experts were re-employed in more peaceful pursuits, some in U.S. government laboratories under the Nunn-Lugar Act and the Megatons to Megawatts Program.[28] In a first step toward the disarmament promised as part of the negotiations of the NPT, the 1991 Strategic Arms Reduction Treaty (START) set limits on U.S. and Soviet missiles, bombers, and warheads. A New START Treaty in 2011 reduced the total deployed missile count per country from 1,600 to 700—a substantial decrease but still leaving a large number.[29] The treaty has been extended until February 2026.[30]

In effect, nearly all of the non-nuclear countries have more or less reluctantly accepted their subordinate status in the name of global stability and/or in exchange for the U.S. promise to defend its allies against attack with nuclear weapons if need be.[31] After all, many non-nuclear countries could easily develop nuclear weapons of their own—the technology is within the capacity of any reasonably advanced country—and have refrained from doing so because of this commitment, but they might well be tempted to develop them if they came to doubt the security that a U.S. alliance provides. Saudi Arabia, for example, has reason to fear a possible Iranian bomb; Japan and South Korea, a North Korean one. In a crisis, they might ask themselves, would the United States be willing to risk the destruction of New York or Los Angeles to defend Riyadh, Tokyo, or Seoul? Will they still be protected if the United States' recent turn toward isolationism turns out to have been more than a temporary blip? Could the U.S. commitment be taken away on the whim of a future U.S. administration?

The 1990s: Ups and Downs for Arms Control

In the aftermath of the dissolution of the Soviet Union, the physical protection of former Soviet nuclear materials was greatly strengthened. Behind this lies a story. The fall of the Soviet Union and the deterioration of the Russian armed forces ironically created problems in the physical control of nuclear materials. Under the Soviet government, *physical* controls of nuclear weapons and fissile materials were unnecessary; the secret police controlled the *people* who had access, and this was enough.

Once Russia became independent, it worked with the United States to strengthen the physical controls inherited from the Former Soviet Union. These often amounted to little more than unlocked metal storage cabinets (the kind we would use to store office supplies) and wire fences patrolled by unpaid security guards. As part of this cooperation, video images of the cabinets that stored Russian nuclear materials were transmitted to U.S. weapons laboratories every few minutes in order to assure their continued security.[32] Unfortunately, the souring of relations between the two countries in the 2010s led to the cancellation of this extraordinary cooperation. Russian nuclear security has continued to improve, although at a slower pace than before.[33]

The experience of the first Gulf War in 1991 also led to a strengthening of the powers of the IAEA to detect violations of the NPT. At the end of this war, Iraq was discovered to have conducted a covert program that had come within a few years of developing nuclear weapons, even though it was a party to the NPT and had been subject to IAEA inspections. (After the second Gulf War, Iraq was discovered to have cancelled its nuclear program shortly after its defeat in the first one, but this is another story.) This led to the acceptance of "additional protocols" by most IAEA member governments, which toughened IAEA procedures and gave the organization the right to conduct unannounced inspections and to examine undeclared facilities.

The nonproliferation regime, too, has suffered major setbacks. India and Pakistan, neither one a party to the NPT, have tested nuclear weapons since the treaty came into force. North Korea pulled out of the NPT and has tested both nuclear weapons and long-range missiles with which to deliver them. The U.S. Senate refused to ratify the Comprehensive Nuclear Test Ban Treaty (CTBT) of 1996, which would have banned its parties from testing nuclear weapons—although the United States has observed a voluntary moratorium on such tests. In 2001, the United States withdrew from the 1972 ABM Treaty.

This slow deconstruction of the nonproliferation regime forces us to reexamine the minority view that more weapons would actually make for a safer world—in other words, that nuclear proliferation is actually desirable. This idea of "proliferation optimism" was first proposed in 1946 and then reformulated in the 1980s. It was long considered marginal but was endorsed by President Trump in 2016, who remarked "let it be an arms race . . . We will outmatch them at every pass and outlast them all."[34] Proliferation optimists argue that a world with more nuclear-armed states would have *fewer* wars because the governments of all countries, whether small or large, high- or low-income, will not attack an adversary that has even a minimum nuclear deterrent, lest they themselves be annihilated.[35] In other words, rational deterrence ensures that mutually assured destruction is a scalable concept, since decision makers in all countries place just as high a value on their own lives as those of existing nuclear powers.

It is not clear that history has borne out this optimism. Some non-NPT governments—the example of North Korea springs to mind—do seem to display a keen sense of self-preservation, but the same cannot be said of ISIS or other ideologically minded terrorist groups. The heads of state of even some countries that are parties to the NPT have been known to make bellicose statements that could have led to unwanted escalation. We shall later discuss the tendency of non-nuclear conflicts to press close to the edge of the nuclear taboo in the second nuclear age. Increased proliferation would increase the number of nuclear powers that have more than one nuclear-armed neighbor, complicating any effort to limit arms races and to prevent the escalation of conflicts, since measures to defend against one nuclear neighbor might be construed as offensive by the second nuclear neighbor. The possibility that some of these nuclear powers would field low-yield tactical nuclear weapons exacerbates this danger. No calamity yet, but we are getting closer to the edge.

The polar opposite of the arguments for proliferation optimism are those made by advocates of complete nuclear disarmament, i.e., for banning all nuclear weapons and destroying the ones that exist. This case starts with the argument that mutual assured destruction is no longer a path to stability in a world with multiple nuclear powers who do not have the Cold War experience of numerous near-misses. Complete nuclear disarmament would end the immediate threat of nuclear war, whether by accident or by design. It would also end the risks of nuclear accident and of the theft of nuclear weapon by terrorists, nihilists, or Mafiosi, not to mention the expense and trouble of maintaining nuclear stockpiles.[36] To the counterargument that a nuclear-free world would encourage cheating, since the possessor of even a few such weapons would automatically become a superpower, nuclear disarmament advocates argue that the risks of the current world trajectory are greater than the risk that a possible future violator of a nuclear weapons ban might go undetected or unopposed.

The Treaty on the Prohibition of Nuclear Weapons, which came into force in January 2021, incorporates the idea of complete nuclear disarmament into a formal, legally binding agreement. The treaty commits its fifty-four parties (thirty-three more countries have signed but not yet ratified as of May 2021) "never under any circumstances to . . . develop, test, produce, manufacture, otherwise acquire, possess or stockpile nuclear weapons. . to use or threaten to use nuclear weapons or . . . to allow any stationing, installation or deployment of any nuclear weapons or other nuclear explosive devices in its territory."[37] The force of this treaty is, however, limited by the fact that no nuclear weapons state is party to and hence bound by it, although it does have important political support in some Western European countries in which U.S. nuclear weapons are based.

It is clear that today's stockpiles are much larger than the few dozen needed for the "balance of terror" required by the MAD doctrine. An eloquent pair of op-eds in the *Wall Street Journal* by top security officials and experts argued that a stated goal of "a world without nuclear weapons" is needed to orient and focus plans for more immediate steps toward a safer world: increasing the time needed to respond to a possible strategic attack, reducing nuclear stockpiles, eliminating short-range nuclear missiles intended for forward deployment, permanently ending testing, stopping the production of fissile materials, and safeguarding the fissile materials that already exist.[38] These measures make sense, whatever one's views on total nuclear disarmament.

Iran and North Korea: The Nonproliferation Regime Takes on More than It Can Handle

Two important setbacks to the nonproliferation regime came with the failure of ambitious agreements intended to end the nuclear programs of North Korea and Iran. In the case of North Korea, these ultimately unsuccessful diplomatic efforts spanned twenty-four years, beginning with the 1994 Agreed Framework, which was hoped to induce North Korea to give up its effort to develop nuclear weapons, and extending to the 2018 bilateral negotiations between the United States and North Korea for a denuclearized Korean Peninsula.

Under the Agreed Framework, North Korea was to shut down its existing graphite-moderated nuclear reactor, and to stop construction on two more reactors of this kind. (Graphite reactors are well suited to producing weapons grade plutonium, for reasons we explain in Annex A). North Korea was also to agree to a treaty denuclearizing the Korean Peninsula, to submit its nuclear facilities to IAEA inspection, and to remain a member of the NPT, in effect agreeing to stop manufacturing fissile material and to cancel its program to develop nuclear weapons. In return, the United States promised to provide free fuel oil to make up for the energy lost from shutting down the North Korean nuclear reactor, and to finance and commission the construction of light-water nuclear power plants, which are less useful than graphite reactors for the manufacture of weapons-grade material. Most important for the North Koreans, the United States provided formal assurances that it would not mount a nuclear attack on North Korea and promised to move toward ending the economic sanctions that were hurting its economy and toward normalizing relations between the two countries.[39]

The agreement began to break down almost as soon as it was signed, a victim of the long-standing mutual hostility and suspicion between the two signatories and their inattention to the need to devote financial and political resources to make good on its provisions. Each side accused the other, probably with some justification, of deliberately slowing the implementation of their respective promises. In the end, North Korea began to manufacture enriched uranium, which was forbidden under the treaty, withdrew from the NPT in 2003, and conducted nuclear tests of increasing sophistication and yield beginning in 2006. It is now thought to have an arsenal of at least two dozen atomic weapons, perhaps including a hydrogen bomb, and missiles with a range that could reach as far as Hawaii. The Framework Agreement had succeeded in delaying the North Korean nuclear program by about eight years before it collapsed. Just over a decade later, nuclear negotiations with North Korea were revived in a series of high-profile summits, which did not, however, make much progress toward denuclearization.

Like the Agreed Framework with North Korea, the 2015 nuclear deal with Iran started with hopes of achieving a signal triumph for the nonproliferation regime. At the time, Iran was making a major effort to enrich uranium for "peaceful" purposes. In a major oil exporter like Iran, this looked like an effort to develop nuclear weapons. It was especially troublesome because of the announced intention of the Iranian government to wipe out Israel, a sovereign member of the United Nations and an ally of the United States.

Under the terms of the 2015 agreement, Iran agreed to officially suspend and dismantle much of its nuclear program and to allow IAEA inspections for verification. In return, the United States agreed to provide relief from economic sanctions.[40] Both sides kept their commitments for the first three years of the agreement, but the United States pulled out in 2018 and reimposed sanctions. Iran responded by breaking its commitments, working on nonnuclear elements of nuclear weapons and increasing production of low-enriched uranium, an important step toward making bomb-ready material. European signatories have done their best to keep the agreement from collapsing altogether. (As this book goes to press, efforts are underway to restart negotiations.)

The issues debated in the United States in the run-up to the 2015 agreement displayed the full range of the disagreements between the two countries over geopolitics, mixed with scientific and technological uncertainty (real and imagined); risk aversion; and trust or mistrust in international institutions and national leaders. Defenders of each side of the argument seized on the technical arguments that supported their stance on the broader political issues. Was Iran really within a few months of having a bomb ready for testing? Could inspectors still detect possible undeclared nuclear activities after giving the suspected violators 24 days' notice that they were coming? Could a bunker-buster bomb destroy underground nuclear installations? None of these questions could be answered with absolute assurance.

In the end, the major issues at the time were political rather than technical: Was a poorer, sanctioned Iran with nuclear weapons preferable to a richer, sanction-free Iran without them? Could the IAEA be trusted to detect violations, even with enhanced inspection procedures? Would sanctions really be restored (the term of art was "snapped back") if Iran were to cheat on the agreement? These controversial questions concerned the trustworthiness of Iran and the IAEA, incorrectly assuming compliance on the part of all the *other* countries that negotiated the deal. Behind these questions was the broader geopolitical question of whether it was a good idea for the United States and Iran to have better bilateral relations, a subject of some disagreement between European and Middle Eastern governments. Ironically, no one at the time of the negotiations anticipated that the *United States* would be the one to pull out of the agreement.

The Second Nuclear Age: Civilization is Still in the Balance

The early years of the twenty-first century saw a change from the relatively stable, bipolar world of the Cold War into a much more volatile, multipolar world. The rise of China, the resurgence of the Russian military, the nuclearization of North Korea, the continuing tensions in South Asia and the Middle East, the deconstruction of the arms control regime, the deployment of anti-missile systems, and the continued doctrine of launch on warning all combine to create dangerous avenues to accidental nuclear war, on top of the long-standing risks of mutual assured destruction by nuclear superpowers and escalation of regional wars in Asia or the Middle East.

On the technical front, the near-term prospects of hypersonic missiles, autonomous weapons, low-yield atomic weapons, cyberattacks on nuclear command and control, and the application of artificial intelligence in strategic military decisions all increase the probability of accidental nuclear war still further and add new urgency to the need for both research and international discussions on arms limitations. All this at a time when existing regimes for arms control are being undermined and dismantled.

This "second nuclear age" provides numerous paths to unintentional or accidental nuclear war. For example, both Russia and China are flexing their muscles and expanding their influence in areas along their borders—Russia in the Caucasus and Ukraine and China in the South China Sea—evoking responses from the United States that could lead to the escalation of a regional conflict into all-out nuclear war. Moreover, the United States has deployed anti-missile systems, with the announced goal of defending against a possible limited attack from North Korea, but capable of being redirected toward China or Russia. This has led to an arms race in hypersonic missiles that could lessen decision times and lead to accidental nuclear war.

In the second nuclear age as in the first, nuclear weapons are the weapons of the weak, to be invoked when a power cannot prevail in a conflict using conventional weapons. This surprising conclusion was first reached in the 1950s by U.S.

President Dwight D. Eisenhower, who propounded the doctrine[41] of massive retaliation precisely because he did not want to try to match the manpower and conventional capabilities of the Soviet Union, instead declaring that he would respond with nuclear weapons to any possible invasion of Western Europe.[42]

Six decades later, Russian President Vladimir Putin propounded the mirror image of this doctrine. Realizing that he could not match superior conventional U.S. technology, Putin announced a doctrine of "de-escalatory first strike," also known as "escalate to de-escalate." In accordance with this doctrine, Russia would threaten a nuclear first strike with either strategic or tactical weapons, so as to deter a nuclear adversary from intervening in a conflict that would not be worth the risk of nuclear war.[43]

Imagine, if you will, that the Russian army is mobilizing for an invasion of Estonia, or that the Chinese navy is preparing an amphibious invasion of Taiwan, and that conventional means are unlikely to prevent them from succeeding in their plans. How would the prospective attack be deterred? As things now stand, the United States and its allies would face an unpleasant choice: either abandon an ally or else threaten full-scale strategic nuclear bombing of Russian or Chinese cities or military installations. Would a threat of massive retaliation be credible and hence an effective deterrent? Would the invaders pull back, or would they call our bluff? To threaten full-scale nuclear war over such a regional conflict would amount to the old teenage game of chicken, in which two cars are driven at each other at high speed and the "loser" is the first to turn aside to avoid a head-on collision.

In theory, one way out of this dilemma would be to deploy lower-yield nuclear weapons that could threaten damage big enough to deter the threatened invasion but not big enough to provoke strategic retaliation and full-scale war—although it has never been clear that it would be possible to prevent an escalating tit-for-tat exchange leading to full-scale catastrophe. This strategy of "bridging the escalation gap" would violate the fundamental taboo against the use of nuclear weapons, a logical consequence of which is that the United States and other nuclear powers would declare that they would never be the first to introduce nuclear weapons into a conflict—a no-first-use policy that is exactly the opposite of bridging the escalation gap.

Such a policy of no first use has repeatedly been advocated at high political levels in the United States and other countries. Nevertheless, no U.S. president from any part of the political spectrum—even those most sympathetic to the idea of no first use—has been willing to openly declare such a policy. Each has concluded that potential enemies can be deterred from attacking—and U.S. allies can be kept from panicking—only if they are convinced that the United States would retaliate with nuclear weapons if this were necessary to defend its allies from even a conventional attack (a commitment we earlier called extended deterrence). In other words, each president has decided that the United States should *not* declare a policy of no first use, even if that was the policy that he would actually follow if push came to shove. U.S. policy on first use has been kept deliberately ambiguous over the years, as has that of several other nations. In keeping with this ambiguity, when the issue of "escalate to de-escalate" came up in congressional hearings in 2017, the head of the U.S. Strategic Command, General John Hyren, assured the public that the U.S. arsenal was sufficient to provide "a variety of options to respond to any number of threats".

A war between secondary nuclear powers could result in a similarly accident-prone situation. Pakistan, for example, relies on its nuclear weapons to deter a possible invasion by superior Indian conventional forces. Even a "small" nuclear war between regional powers like India and Pakistan would be likely to have catastrophic global consequences. Beyond the immediate likelihood of many thousands of casualties from blast, heat, and radiation, the explosion of 50–100 nuclear weapons in such a war would set off fires that would fill the Earth's atmosphere with smoke, blocking the sun. These could remain in the atmosphere for decades, changing the weather and substantially reducing world agricultural output.[44]

Another possible avenue to nuclear war comes from the fact that Islamist terrorists have declared their desire to acquire or build a nuclear weapon, although this danger receded when ISIS lost nearly all the territory it controlled at its height. The long-standing interest of these terrorists in acquiring a ready-to-use weapon is well documented.[45] The threat of retaliation might mean little to fanatics bent on martyrdom. Some extremist Islamic theologians have even defended the potential acquisition and use of nuclear weapons by terrorists by arguing that the crimes of the West against Islam are so great as to justify even the killing of millions of innocents.[46]

To acquire nuclear weapons, terrorists would have to steal or buy an intact weapon—perhaps from North Korea, for whom weapons technology is a major export—or from a sympathetic future government of some other country. Alternatively, they might acquire weapons-grade nuclear materials, perhaps from an inadequately guarded installation in Russia or from reprocessed spent nuclear fuel, and fabricate a bomb of their own. There is considerable disagreement among experts over whether such an independent band of terrorists could design and fabricate a gun-type uranium weapon, with or without state assistance, once they become able to acquire a sufficient amount of weapons-grade material.[47] Even a gun-type weapon requires a considerable range of talents to fabricate and deliver: technological resources, financial backing, a secure facility, solid transportation, and an organized, armed, trained, and absolutely loyal team of operatives who can keep their mouths shut. In principle, terrorists might be able to assemble such a group, to provide them with the necessary facilities, to conceal them from detection, and to keep them from being infiltrated by "enemy agents," but this would certainly be a tall order.

All of this having been accomplished, the group would have to deliver a bomb to its target, probably by smuggling it into a port or coastal waterway, from which it could be transported to its intended target by truck. Old-fashioned intelligence and police work would be the best way to detect such a scheme, since a uranium bomb would give off little radiation and a plutonium bomb could be easily shielded so that neither could be detected by remote sensing. It might well take weeks to determine the source of the fissile material once it had been intercepted or after it had exploded, if, indeed, this could be done at all. This delay would create great public pressure to retaliate before the parties responsible had been definitely identified. One may hope that future terrorists will regard the risks of failure to be great enough that they would choose a different approach.

Lessons from Nuclear Experience

Experience with the arms control and nonproliferation regimes provides important lessons. First, international policy in these areas requires a detailed understanding and respect for the interactions of science, technology, and world affairs—the classic Survival Nexus. As with the climate change and stratospheric ozone regimes, these regimes have required national governments and international organizations to gain expertise in a complicated mix of technical and political issues and have required the scientific community to assume new advisory functions toward government and new educational functions toward the public. The active participation of the scientific community has also benefited the scientific community itself by increasing its sophistication in political matters and by stimulating the development of previously neglected scientific disciplines like seismology and oceanography.

The success of the regime also depends on public understanding and support for the idea of arms control. This support is especially important because the support of the nuclear science community for nonproliferation and arms control has not always been wholehearted. In the United States, for example, there was concern that these regimes would reduce support to the nuclear weapons laboratories, which are centers of research on nuclear science and technology and are also major employers in several congressional districts. This concern was alleviated by the creation and financing of such programs as that for nuclear stewardship, which became the price for congressional support for test bans and other extensions of the nonproliferation regime.[48]

Second, like the effort to control epidemic disease, the arms control and nonproliferation regimes require active tending and sustained international cooperation. Because nuclear technology, like epidemic disease, can spread across boundaries, it has given rise to regimes that oblige sovereign states to provide detailed information on domestic nuclear activities—a matter that would ordinarily be kept close—and to submit to intrusive inspections of domestic facilities in order to enhance global stability.

Third, governments are not the only actors capable of undermining the nonproliferation regime. The head of the Pakistani Atomic Energy Commission, Dr. A. Q. Khan, was discovered to have been making a career and a modest fortune as an enthusiastic and entrepreneurial proliferator. Starting from a base of stolen designs and documents from a Dutch manufacturer and working with an extensive clandestine network of unethical equipment suppliers and middlemen, Khan provided nuclear technology in support of weapons programs in North Korea, Iran, and Libya, and possibly other countries as well, making him a hero in his native Pakistan and a pariah in the rest of the world.[49]

Fourth, arms control is not necessarily a dominant theme of foreign policy in any country. It frequently gives way to other concerns when it conflicts with more pressing foreign policy objectives. In 2005, for example, the need for improved relations between the United States and India trumped concerns about India's nuclear explosions and eventually led to the de facto acceptance of India as a legitimate nuclear power.[50] This is but one of many examples in which short-term political or economic goals trump the need to manage serious long-term global dangers, a pattern we have already observed in the world's response to climate change and that we shall see again when we discuss pandemics in the next chapter.

Hypersonic Missiles: Even Shorter Times to React to a Possible Nuclear Attack

A *hypersonic missile* flies at lower altitudes than those monitored by existing radars and hence is undetectable by them. A hypersonic missile can also be steered while in flight, so that its target remains unknown to defenders until the final few minutes of its flight even if they have spotted it before then. This distinguishes it from a traditional ballistic missile, which also flies at hypersonic speeds but which follows a predetermined trajectory, so that both its origin and its target can be determined relatively straightforwardly well before it lands. (The term "hypersonic" refers to the ability of the new missile to fly at more than five times—in some cases, as many as twenty times—the speed of sound, but this is not its most important feature.

Hypersonic missiles can be designed to carry either a nuclear or a conventional warhead. Their military uses have yet to be precisely defined and, in fact, seem at first glance to be of doubtful value. They are most likely to be used in the first stages of a conflict. Conventionally armed hypersonic missiles, for example, could be used to assassinate key enemy leaders in a "decapitation" attack (not a good idea, since you'll need them alive to make peace at some point) or to destroy key military assets of an enemy—as for example, its anti-missile weaponry or the command-and-control systems of its nuclear weapons as a preliminary to a first strike intended to destroy the defenders' ability to retaliate (also not a good idea if you want the other side to be able to maintain control of its own forces and prevent unauthorized retaliation). Armed with nuclear warheads, hypersonic missiles could be used to evade enemy anti-missile defenses in a strategic first strike intended to wipe out the defender's ability to retaliate (also not a good idea if you don't want the preparatory missile build-up to start an arms race, as we discussed earlier). Nevertheless, half a dozen countries are racing to be the first and best at hypersonic missile technology, and several have stated their intention to deploy them at large scale as soon as they are ready. The resulting competition has all the earmarks of a classic arms race.

It is not at all clear that these weapons will make anyone safer. On the contrary, they seem more likely to make it easier for the world to blunder into a catastrophic, unintended nuclear war. Defenders are likely to design and launch an array of satellite-based sensors to detect this new kind of low-altitude, fast-moving missile while it is still in flight. Suppose, for example, that this satellite array, once developed and deployed, detected signals that it interpreted as meaning that a large number of hypersonic missiles are on their way to the homeland of a defender country. Defenders would then have perhaps five minutes to decide whether this was a real attack or a false alarm, and if it was real, to determine what the target of the attack might be.

What is more, the defenders would not know until the missiles had exploded whether they had carried conventional or nuclear warheads. They would therefore be likely to assume the worst. (Russia has already announced that it would assume that any incoming intercontinental missile would be nuclear.) This could lead them to launch an immediate nuclear counterattack while they were still able to do so—in other words, a launch on warning before the incoming missiles had landed. Since no political leader can be expected to make such an Earth-in-the-balance decision in five minutes, they would likely be forced to rely on the help of automated systems that evaluate the available evidence that an attack is really taking place and what its target and purpose are most likely to be.

This creates many opportunities for error. Any such system would have to use a form of artificial intelligence (AI) known as deep learning in order to evaluate information derived from a diverse range of sources, ranging from electronic sensors to political intelligence reports. The authors of the AI software could not possibly have anticipated all possible circumstances and considerations, nor could they detect errors and bugs under anything approaching realistic conditions. Besides, programs of this kind are inherently prone to errors and vulnerabilities that are nearly impossible to detect. Even more dangerous is the fact that it would be possible to give hypersonic missiles a degree of autonomy in choosing their targets without human intervention after they are launched. This would raise yet another set of deep moral and practical problems with huge implications for human survival, to which we now turn.

Fully Autonomous Weapons: Killing Without Human Involvement

Fully autonomous weapons systems can assess a situation, choose a target, and shoot it or bomb it without any human intervention in the decision-making process, all in a split second.[51] These systems can protect military personnel by removing them from the field, but they can also escalate a local conflict into a regional or global war that nobody wanted in the blink of an eye. Autonomous weapons have mainly been under the control of governments, though they could soon be available to individuals or even terrorists. As with the case of gene drivers and geoengineering, which we shall consider in chapter 12, autonomous weapons systems force us to ask ourselves: how do we—as individuals, experts, countries, and the global community—innovate responsibly?

Many situations involving autonomous weapons could require decisions much too fast for mere humans to respond. This could lead to catastrophic mistakes. For example, suppose that a large group of autonomous, armed drones (known to military planners as a wolf pack) crosses a border into a disputed area and begins to shoot at or bomb a large area. Defenders would have to make a quick decision as to whether this was the prelude to a major invasion or whether it was an accident or a mistake. Similarly, suppose that the commander of a naval ship in international waters detected signs of a swarming attack by dozens or hundreds of enemy missiles that were assumed to be autonomous and programmed to choose and engage targets within a designated area.

The same logic applies to cyberattacks. Suppose, for example, that there were signs of a strategic cyberattack on a country's critical infrastructure that could be countered only by launching an immediate hackback on the presumed perpetrator. This would be answered by a return hackback that would be answered by yet another return hackback, and so on, all within less than a second, resulting in a flash cyberwar that could have catastrophic results for the information and other infrastructure of both combatant countries. In all these hypothetical scenarios, our future would be in the "hands" of computers, for better or for worse.

Development of autonomous weapons is well underway in the United States, China, Russia, and at least three other countries.[52] The U.S. military is developing and testing an autonomous submarine hunter called the Sea Hunter for the Navy, robotic equipment transport and combat vehicles for the Army, and drones for the Air Force that are to be guided to a combat area by human pilots and then released to engage enemy radar or other targets on their own.[53] The U.S. Navy already deploys the Phalanx system, which engages enemy missiles at short range with a reaction time shorter than is possible for humans. Russia has tested (not very successfully, we are told) remotely guided tanks in Syria that can be converted to autonomous operation.[54]

Why Autonomous Weapons? Why Not?

Autonomous weapons are based on drone technology combined with deep learning, the technology that taught itself the games of chess and go starting from scratch and went on to beat world champions. Deep learning depends on so-called neural networks, which are designed to mimic the operation of the human brain. In military applications, a deep learning system trains itself to distinguish different kinds of objects (a cat from a dog, or a man holding a rifle from a man holding a pitchfork) by huge numbers of trials and errors, using a database supplied by the system designer. This mimics how a small child learns to make distinctions, say between an apple and a tomato. This basic technology is widely available through commercial channels, raising issues analogous to those raised by other do-it-yourself technologies like the gene drivers and geoengineering that we shall discuss in chapter 12.

Autonomous weapons are attractive to the military because they promise to be cheap, to greatly reduce the risk of human casualties (at least to the combatants who use them), to be able to operate in hostile and toxic environments, and never to get tired, stressed, or emotional. They can hover for days over hostile territory, searching for targets like artillery or antiaircraft radar. (Israel already has such a weapon, called the Harpy.) They can respond to an incoming artillery shell with an instantaneous counterattack on its point of origin. They would make it possible to defend against swarming missile attacks against expensive weapons systems like aircraft carriers, and also to carry out such attacks.

Artificial intelligence systems are subject to inherent flaws.[55] Neural networks may involve millions of connection points, called nodes, making their operation difficult or impossible to follow in detail. What is more, because deep learning programs teach themselves by trial and error rather than following logical rules set down by human programmers, it is often impossible to tell how they are making their decisions, making them hard to debug if things go awry. Once a deep learning system is deployed, operators may make mistakes that allow autonomous systems to run amok. Programmers cannot possibly foresee and provide instructions for all the wartime contingencies in which difficult decisions will have to be made in the heat of combat. Policies and attitudes may change between the time the program is written and its objectives are defined and the time a decision must be made in the field.

Even the best AI programs are incapable of recognizing anomalous situations or of understanding the broad context of a given situation—what the military calls "commander's intent"—so as to be able to adapt general guidelines to specific situations. The database on which the deep learning system is trained may have unsuspected biases, as, for example, when a civilian database of human faces includes disproportionately few women or people of color. Such a bias might, for example, affect a program's ability to distinguish a combatant from a civilian. Perhaps most important, AI programs cannot emulate the essentially human characteristics of empathy or moral sensibility that can be critical in split-second, life-or-death decisions in wartime.

Added to these considerations is the fact that military deep learning systems must be designed to operate in a hostile environment in which enemies will do their best to take advantage of and augment all these flaws. Like any other weapons, they can be captured by enemies and used against their developers. The basic technology is widely available and would be accessible to terrorists and other nihilists, who could use it for a bombing or shooting rampage against military targets or civilian populations. Deep learning programs are also vulnerable to subversion, as, for example, by camouflage (inserting small distortions that are undetectable to the human eye but that confuse AI algorithms), by poisoning (inserting items into the data that cause the AI system to make mistakes), and by hacking (exploiting flaws in the system on which the AI system is trained or in the design of the goals it is supposed to achieve).[56]

Proponents of autonomous weapons argue that these problems can be overcome with further research and development and with careful design, production, and training. They point out that the increased accuracy of autonomous weapons as compared with that of human weaponeers could minimize civilian casualties and that autonomous weapons could someday be proven to show better judgment than a stressed and fatigued soldier. On the other hand, it is almost impossible to avoid exploitable flaws in complex military systems that inevitably involve millions of lines of software code. These flaws may not be detected and repaired because of commercial or wartime pressures that speed the deployment of weapons, often before they are fully ready.

Arms Control for Autonomous Weapons

Whatever the advantages and vulnerabilities of autonomous weapons, there remains a fundamental ethical objection that giving a machine, any machine, unchecked authority to kill is just plain wrong.[57] Humans, it is widely felt, provide context, empathy, and a moral sense that are impossible to incorporate into a machine. Even if it were possible to program autonomous weapons to perform as well as humans in these areas, these safeguards could be turned off or ignored in practice. Far better, some argue, to forbid the development and deployment of such weapons from the beginning, lest they create opportunities for misuse on a grand scale. Popular protests have arisen in many countries demanding that autonomous weapons—dubbed "killer robots" by their critics—should be entirely banned by treaty. Thirty governments have called for a binding treaty forbidding the development and use of autonomous weapons systems.[58] Other proposals would allow development and deployment of such weapons but would place limits on their use.

These proposals raise a number of practical difficulties. First, the checkered history of the many previous efforts to ban specific weapons, from crossbows in the eleventh century to unrestricted submarine warfare in discussions before World War II, makes it likely that any such ban would be immediately violated in time of war.[59] Second, there are disagreements as to which levels of autonomy need to be regulated and how each of these levels might be defined. Third, even if we decided to ban or regulate only fully autonomous systems, it would be difficult to distinguish the forbidden software of a fully autonomous weapons system—one in which no human being was involved in the decision loop—from the permitted software in a semiautonomous system, for example, one in which a human operator sets limits on the overall activity of the weapon or approves the ultimate decision to fire (see note 52).

Finally, many of the possible misuses of autonomous weapons are already forbidden by the Geneva Conventions and other existing laws of war. These apply to any weapon and require that civilians and combatants be clearly distinguished, that reasonable efforts be made to minimize civilian casualties ("collateral damage," in the bloodless military argot), that the need for military action be sufficient to justify the predictable collateral damage, that military action not cause unnecessary suffering, and that no further harm come to combatants who have been wounded, captured, or otherwise out of action (hors de combat).[60] These rules are not always obeyed in practice, but it is hard to see what would be accomplished by a separate agreement that sets down similar rules for autonomous weapons.

From the point of view of practical geopolitics, the most powerful argument for a specific ban on autonomous weapons would be that their use can get out of control and lead to a war that none of the parties wanted—i.e., that they tend to be destabilizing. For this reason, a series of intergovernmental meetings have tried to fashion a practical approach to setting limits on the deployment and use of autonomous weapons. Even though substantial negotiations on autonomous weapons have repeatedly stalled at the United Nations, there are signs that countries may be able to reach agreement on broad general principles, of which the most promising is that there must be some level of meaningful human involvement in the decision to engage a target.[61]

An agreement of this sort could constitute a useful beginning. It could create a framework within which detailed rules could be developed through experience and that might evolve into informal codes of conduct or rules of the road. These might eventually be codified into a formal, binding international agreement that establishes criteria that are clear enough as to what constitutes meaningful human involvement, and hence what is and is not permitted, in any given situation.

There has been, to my knowledge, no comparable international effort to limit the development and deployment of hypersonic missiles. On the contrary, there is every sign of an arms race involving the United States, Russia, and China and several other countries, all of which aim to be the first to develop and deploy these weapons, either as a dependable force for retaliation or as a vehicle for an unprovoked first strike. In contrast with cyberweapons (chapter 9), which would be difficult to monitor, hypersonic missiles do lend themselves to the kind of arms limitation agreement that was effective for intermediate-range and intercontinental ballistic missiles, since, as mentioned earlier, a suitable satellite array could detect their test flights. It should therefore be possible to design an agreement to ban or limit the development or deployment of such weapons, should the major military powers wish to do so.

In sum, the world is racing to develop technology that could lead directly to a computer-directed nuclear exchange that no human had intended. The prospect of hypersonic missiles and autonomous weapons adds new urgency to the need for the resumption of international discussions on arms limitations, at a time when existing regimes for arms control are instead being undermined and dismantled.

Can We Handle the Nuclear Future?

Despite several narrow escapes, the legacy institutions of the Cold War have protected the world since the end of World War II. There has been, after all, no nuclear war, and the number of nuclear states is still far smaller than was once feared. Four countries have given up their nuclear weapons, and six more have abandoned programs that would in time have led to their development—despite political attacks on the IAEA, the withdrawal of the United States from key arms control agreements, the breakdown of agreements with Iran and North Korea, the fragility of the nuclear taboo, the counterintuitive nature of the doctrine of mutually assured destruction, and the inherent inequity and somewhat ragged edges of the nonproliferation regime. So far, so good.

Still, there is plenty to worry about. Two thousand nuclear weapons remain on hair-trigger alert in both the United States and Russia, ready for launch on warning—a doctrine that would allow a head of state perhaps ten or fifteen minutes to decide whether to begin a nuclear war in response to what might be a false alarm. The promised advent of hypersonic weapons would reduce this time still further. Combined with the launch-on-warning doctrine, these new weapons would require an automated response from computers programmed with deep learning, a form of AI that can neither be tested in advance under realistic conditions nor be reliably debugged.

There is also the continuing danger of nuclear war involving secondary nuclear powers, such as India, China, Pakistan, Israel, and North Korea. There are four new nuclear states, including a rogue state (North Korea) that does not accept the nonproliferation regime and threatens its neighbors with nuclear attack, and a nuclear state (Pakistan) that has a contiguous, hostile, nuclear-armed neighbor and besides is under constant threat of political instability. The very existence of another nuclear state (Israel) is unacceptable to at least one of its neighbors, which has resumed enriching and stockpiling uranium and apparently intends to develop nuclear weapons. Russia has abandoned its no-first-use doctrine, preferring to rely on tactical nuclear weapons to counter perceived but undefined threats to its interests.

Two mutually hostile nuclear states (India and Pakistan) share a common boundary, so any war between them would immediately threaten the vital interests of both. Both have adopted nuclear doctrines that increase the probability that a conventional conflict between them could escalate into nuclear war. Terrorist groups that say that they welcome martyrdom and hence cannot be deterred by the threat of retaliation have sought nuclear weapons, although it is not clear that they could obtain or develop them or could deliver them to a target.

The regime that has protected us from nuclear war since World War II has been being undermined and dismantled for some time without any serious effort to replace it, although this inaction may now be at an end with the renewal of the START agreement. The United States has withdrawn from key arms control agreements banning antiballistic and intermediate-range ballistic missiles, rather than update them to meet the challenges of new weapons systems and new geopolitical circumstances. The basic nuclear taboo is being eroded by military doctrines in several countries that dictate the first use of nuclear weapons, and by the deployment of low-yield tactical nuclear weapons that overlap in destructive power with that of conventional weapons. We are playing with fire and abolishing the fire department.

What is more, nuclear issues have receded from public attention. The bombing of Hiroshima and Nagasaki, the duck-and-cover drills, the testing of the hydrogen bomb on tropical atolls, the MAD doctrine, the evolution of the nonproliferation regime, and the constant background fear of nuclear annihilation that were once part of everyday experience and the subjects of extensive news coverage are all now subjects for history books and retrospective television documentaries. The erosion of public trust in institutions and expertise of all kinds has lessened public confidence in the policymakers who must deal with the complexity of nuclear issues, especially with their scientific and technological dimension.

Yet nuclear issues are still with us. What is the president to do if North Korea threatens Seoul or Guam or Honolulu with a nuclear attack? If there is a threat of a non-nuclear attack on a NATO ally? If a regional war between hostile nuclear neighbors evokes a nuclear explosion and threatens to escalate still further into a major nuclear exchange? If a terrorist group succeeds in smuggling a nuclear weapon into an American harbor and presents the president with a set of nonnegotiable demands? Most embarrassing to a U.S. reader and, indeed, to this U.S. writer, we need to add to this list a question that we have probably never asked ourselves: How would a foreign government respond if it was threatened with a nuclear attack by the United States? How would it react if the United States threatens to attack a third country for a reason it regards as insufficient?

We may someday confront one or more of these unthinkable questions. This may induce a certain uncomfortable nostalgia for the simpler, bipolar world of the Cold War. I can only hope that the reader has not already witnessed the need to face them and never does in the future.

Annex A
A Primer on Nuclear Science and Technology

A basic understanding of nuclear science and technology is critical to understanding the issues of arms control. I have therefore included in this annex concise answers to some of the technical questions that may be on the mind of the reader.[1]

Why are atomic explosions so much more powerful than chemical explosives?

Atomic weapons based on nuclear fission acquire their destructive power from the forces that hold together the 90-odd protons that are tightly crammed together, about half a trillionth of a centimeter apart, in the nucleus of the atoms of the heavy elements uranium and plutonium. The nucleus of a uranium atom contains 92 protons and is said to have an *atomic number* of 92; plutonium has 94 protons, so its atomic number is 94. These protons have positive electric charges and repel each other but are held together by the "strong" nuclear force, which is very powerful at these very short distances. Some of the enormous energy required to overcome the repulsion between the protons is released when the nuclei of these elements are split.

Chemical explosives like TNT, by contrast, get their explosive power from quantum mechanical forces on the electrons that hold their atoms together to form a molecule. These electrons form a cloud that takes up nearly all of the volume of each atom; the atomic nucleus is tiny by comparison. The electrons are a few hundred-millionths of a centimeter apart—still quite close together but about one hundred thousand times farther apart than are the protons of the uranium nucleus. They have negative electric charges and repel each other, but the binding energy that keeps the atoms that make up the molecule from flying apart is less than the binding energy needed to keep protons together in the nucleus. This means that the chemical energy available to be released when TNT explodes is much less than the nuclear energy that is released when uranium or plutonium atoms are split.

The Survival Nexus. Charles Weiss, Oxford University Press. © Charles Weiss 2022.
DOI: 10.1093/oso/9780190946265.003.0006

What's the difference between the isotopes of uranium? What's the difference between uranium and plutonium?

Uranium comes in several varieties, called isotopes, all of whose atomic nuclei have the same number of positively charged protons and negatively charged electrons (i.e., the same *atomic number*) but a different number of uncharged neutrons and hence a different *atomic weight.* Since the atomic number determines the chemical properties of an element, all isotopes of uranium have nearly identical chemical properties.

All atoms of uranium have 92 protons in their nuclei and hence have an atomic number of 92. There are two uranium isotopes of immediate interest. The first has 143 neutrons in its nucleus and hence has an *atomic weight* of 235, which is the sum of 92 + 143. Uranium-235 is written ^{235}U to save space. The second uranium isotope has 146 neutrons and has an atomic weight of 238, which is the sum of 92 + 146. It is written ^{238}U. Natural uranium contains about 0.7% of ^{235}U, which is the isotope used in nuclear energy and weapons. Natural uranium is usually artificially concentrated or "enriched" in ^{235}U before it can be used. A "lightly enriched" uranium mixture containing 3–5% of ^{235}U is useful for power plants, while anything above 90% is considered weapons grade. All isotopes of uranium are only very weakly radioactive. They emit alpha particles (helium atoms stripped of their electrons), which are easily shielded from biological and other sensitive materials, as well as from detection devices.

Uranium and plutonium are different elements with different atomic numbers and different chemical properties. Plutonium, the second element used in fission weapons, is an artificial element that does not appear in nature but is manufactured in nuclear reactors, including those used in nuclear power plants. Its atomic number is 94. Like uranium, plutonium is radioactive and emits alpha rays, so that a plutonium bomb can also easily be shielded from detection devices. The major isotope (^{239}Pu) of plutonium has atomic weight 239, but this is typically accompanied by a small amount of plutonium-240 (^{240}Pu), formed when ^{239}Pu absorbs a neutron.

What's a chain reaction? A critical mass?

When a neutron traveling at a certain speed strikes the nucleus of an atom of ^{235}U, the nucleus splits (*fissions*) into two, releasing a lot of energy and also three additional "secondary" neutrons. If any of these neutrons does not escape the confines of the chunk of matter of which the atom formed a part, that neutron can go on to split another ^{235}U nucleus. This again releases energy and three more neutrons and so on. This process is known as a *chain reaction*. If, on the other hand, more than two-thirds of the secondary neutrons escape, the process stops.

A piece of matter that is just big enough to allow secondary neutrons to encounter enough ^{235}U nuclei to keep this process going is called a *critical mass.* (If it's bigger than that, it's a *supercritical* mass; if smaller, it's *subcritical.*) The size of a critical mass depends on the total amount of ^{235}U and on its concentration, i.e., on how many atoms of ^{235}U there are in a given volume of material. Under normal conditions, a critical mass of pure ^{235}U is a little less than seven inches in diameter—somewhat smaller than a volleyball—and weighs about 114 pounds. A critical mass of pure plutonium weighs about 22 pounds and is about the size of a softball or a small melon.

What's the difference between the chain reaction in an atomic bomb and the one in a nuclear power plant?

In a nuclear power plant, just enough secondary neutrons are generated to keep the chain reaction going at a steady state so that useful energy can be extracted in a smooth, sustained, carefully controlled process. A reactor uses a moderator—typically ordinary water, heavy water (deuterium oxide), or graphite—to slow down the neutrons so that they can be captured more efficiently and keep the chain going. Because of this higher efficiency of neutron capture, power reactors can use natural or low-enriched uranium as fuel. The nuclear reactions that produce energy also produce plutonium, which can be separated chemically ("reprocessed") from the residual fuel and other byproducts. This means that any nuclear reactor—even a small research reactor—can be a source of plutonium that can be used in a nuclear weapon.[2]

In contrast to a nuclear power plant, an atomic explosion requires that the secondary neutrons give rise to a larger number of tertiary neutrons, the tertiary into a larger number of quaternary, and so on, for about eighty more generations—all compressed into less than a millionth of a second. If this is to result in a full-scale explosion, the piece of matter containing the fissile material must not blow apart before the process is well underway. If it blows apart too soon, the result is a "fizzle."

What's the difference between an atomic bomb and a hydrogen bomb?

In a hydrogen or fusion bomb, energy is released by the fusion of hydrogen nuclei into helium nuclei, the same nuclear reaction that powers the sun.[3] Fusion bombs can be thousands of times more powerful than atomic (fission) bombs.

A hydrogen bomb explodes in two stages. The first stage consists of a fission bomb "trigger," which provides the energy needed to ignite the second stage. The second stage consists of lithium deuteride, a compound of the light metal lithium with deuterium, an isotope of hydrogen. A "spark plug" of fissile material is also included in this second stage. Deuterium has both a proton and a neutron in its nucleus, whereas ordinary hydrogen gas only a proton (^{1}H). Both have only one electron. Deuterium is used instead of ordinary hydrogen because a helium nucleus has two neutrons, and it wouldn't get them from the fusion of two hydrogen nuclei, which don't have neutrons. Lithium deuteride is used instead of straight gaseous deuterium because it is a solid and does not explode under normal conditions.

In the first stage of the explosion of a hydrogen bomb, the heat and X-rays from the explosion of the fission bomb trigger compress the lithium deuteride to a small fraction of its original volume and heat it to temperatures comparable to those at the core of the sun. Under these conditions, the nuclei of two deuterium atoms (^{2}H, sometimes written ^{2}D) fuse to form nuclei of helium (^{4}He). The fusion of the deuterium nuclei not only produces prodigious amounts of energy but also produces neutrons that speed the fission reactions in the trigger in the first stage and in the fissile spark plug in the second stage. As a result, half of the overall energy of the explosion of a hydrogen bomb comes from fission and half from fusion.

What's the difference between the bomb exploded at Hiroshima and the one exploded at Nagasaki?

There are two basic designs of an atomic bomb, each of which creates a critical mass of fissile material in a different way. In a "gun-type" bomb, a bullet consisting of a subcritical mass of weapons-grade uranium is fired down the barrel of an artillery piece into a target that also consists of a subcritical mass of uranium, creating a supercritical mass that stays together just long enough to produce an explosion. (If the critical mass flies apart too soon, the result is a fizzle—a failed explosion that can still produce a kiloton or so worth of damage.) A bomb of this design was exploded over Hiroshima at the end of World War II, more than seventy-five years ago.

A gun-type bomb is relatively simple to design and build, and, in fact, the Hiroshima bomb design was not even explosively tested before it was used in warfare. For this reason, the major obstacle for a government that wants to develop a gun-type bomb is thought to be the acquisition of weapons-grade fissile material. Aside from poorly guarded stockpiles of intact weapons, the most readily accessible source is nuclear power plants: either diverting material from the fuel cycle or reprocessing the spent fuel to recover the plutonium it produces.

Plutonium bombs like the one exploded over Nagasaki in 1945 operate on a different principle, that of implosion. Simultaneous chemical explosions compress a sphere consisting of a near-critical mass of plutonium to about two-thirds of its original volume, forcing its atoms closer together. This creates a critical mass, enabling sufficient neutrons to encounter secondary nuclei and generate secondary, tertiary, quaternary, and subsequent neutrons to give rise to an explosive chain reaction. To ensure uniform compression, the electronics that control these chemical explosions must be timed to coincide within a few nanoseconds, a more sophisticated and difficult task than the technology involved in the design and fabrication of a gun-type bomb.[4]

What causes the damage from an atomic bomb?

Both atomic and hydrogen bombs give rise to three sources of destruction: heat, blast, and radiation.[5] Radiation comes in three forms: alpha particles, which are helium atoms stripped of their electrons; beta particles, which are electrons, and gamma rays, which are light rays of very short wavelength and thus very high energy per light particle (photon). Gamma rays can wreak havoc with the chemical bonds of DNA and thus cause cancer or possibly lethal mutations. It is easy to shield against alpha and beta rays, but shielding an object that emits gamma rays requires surrounding it with a lot of mass. This is most easily achieved with high-density materials like lead or uranium from which ^{235}U has been removed (depleted uranium).

The heat, blast, and some forms of the radiation damage caused by nuclear explosions are virtually instantaneous, while the DNA damage caused by radiation may take days to manifest itself as radiation sickness or years or even decades to manifest itself as leukemia or other forms of cancer.[6] A nuclear explosion close to the ground also produces a huge cloud of radioactive fallout—soil and water contaminated by the radioactive materials produced in the explosion—that gradually falls to Earth downwind and can contaminate a large area.

How do natural, low-enriched, and weapons-grade uranium differ?

Natural uranium is a mixture of two isotopes of uranium: 99.3% ^{238}U, which is not used in bombs, and 0.7% ^{235}U, which is. Low-enriched or reactor-grade uranium has between 3 and 5% ^{235}U and is useful as fuel for nuclear power plants.

A nuclear weapon can be made of uranium consisting of 20% of ^{235}U, so that 20% enrichment is considered the boundary between low-enriched uranium (LEU) and highly enriched uranium (HEU). Weapons-grade uranium has at least 90% ^{235}U.[7]

How is uranium enriched?

^{238}U and ^{235}U have the same atomic number (the number of protons, which is equal to the number of electrons) and, hence, have almost identical chemical properties. They can be separated from each other only by taking advantage of their slight difference in atomic weight (number of protons plus number of neutrons) and, hence, in their density (weight per volume). The two most common ways to do this are by centrifugation and by gaseous diffusion.[8] Both begin by purifying natural uranium ore and converting it into a mixture of uranium oxides, a solid called "yellowcake," and thence into the highly toxic gas uranium hexafluoride, UF_6.

The UF_6 produced in this way is a mixture of $^{235}UF_6$ and $^{238}UF_6$—i.e., uranium hexafluoride made from ^{235}U and uranium hexafluoride made from ^{238}U, respectively. These have the same chemical properties but differ in molecular weight by 1.14%. In the centrifuge method, these are separated by compressing the gas and feeding it into centrifuges made of special maraging steel, which spin so fast that their outer edges reach 400–500 meters per second, faster than the speed of sound. The lighter molecules of $^{235}UF_6$ are forced outward a little faster than the heavier $^{238}UF_6$, in a sort of high-tech version of a salad spinner[9] The UF_6 in the layer of gas near the outer edge of the centrifuge chamber thus has a slightly higher proportion of ^{235}U compared with that in the inner parts and is fed into a series or "cascade" of thousands of centrifuges until the desired level of enrichment is achieved. The principle of the gaseous diffusion method is similar, except that the UF_6 is forced under pressure through small holes in a series of special membranes, through which the lighter $^{235}UF_6$ diffuses slightly faster than the heavier $^{238}UF_6$.

The centrifuge method is well suited to proliferation because it does not require huge quantities of space or energy and is relatively easy to hide. It is therefore important to prevent maraging steel from falling into the hands of potential proliferators, which is why this strategically important material is carefully controlled by the London Suppliers Group. Nevertheless, the Pakistani master proliferator A. Q. Khan was able to obtain maraging steel and other needed materials and equipment from unscrupulous suppliers and middlemen for use in the Pakistani and Iranian nuclear programs.

Why is it harder to enrich natural uranium than it is to enrich 3% or 20% ^{235}U?

For every thousand atoms of natural uranium, there are seven atoms of ^{235}U that are useful for bombs and 993 atoms of ^{238}U that are useless for bombs. To produce low-enriched or reactor-grade uranium, which has between 3 and 5% ^{235}U, we need to keep the seven atoms of ^{235}U that are useful and remove 860 of the 993 atoms of ^{238}U that aren't, so that there remain 133 atoms of ^{238}U to accompany our original seven atoms of ^{235}U for 5% ^{235}U. For 3% ^{235}U, we would need to remove 769 atoms of ^{238}U from the original 993 so that there would be 224 left.

From this point on, enrichment gets easier. To achieve 20% low-enriched ^{235}U from a 5% mixture, we must get rid of 105 additional atoms of ^{238}U, so that there are seven atoms of ^{235}U to 28 of ^{238}U. Weapons-grade uranium with 90% ^{235}U would have nine atoms of ^{235}U for every atom of ^{238}U, so to produce it from 20% ^{235}U, we have to get rid of only 19 atoms of ^{238}U from our 20% mixture. That's why it's easier to go from low-enriched uranium (20% ^{235}U) to weapons-grade uranium (90% ^{235}U) than to go from reactor-grade uranium (3% ^{235}U) to low-enriched uranium (20% ^{235}U). It's easier to get rid of 19 atoms of ^{238}U than to get rid of 700 800.

What is reprocessing and why is it a proliferation risk?

The term "reprocessing" refers to the technology for extracting purified plutonium-239 (^{239}Pu) from the spent fuel of a nuclear reactor. Reprocessing facilitates proliferation because metallic plutonium is a compact material that is much easier to steal and conceal than the spent fuel from which it is extracted. What is more, a major reprocessing program would produce so much plutonium that it would be difficult to detect a theft that was relatively small compared with overall production but was enough to produce a bomb.

Plutonium is produced in a nuclear reactor because the neutrons that split uranium atoms of ^{235}U are also absorbed by ^{238}U to produce a different isotope of uranium, ^{239}U. (The extra neutron increases the atomic weight from 238 to 239 but doesn't change the atomic number or the name of the element.) One of the neutrons in the nucleus of ^{239}U then breaks up into a negatively charged electron and a positively charged proton to produce a new artificial element called neptunium-239 (^{239}Np) in a process called beta decay. ^{239}Np then undergoes a similar beta decay to form ^{239}Pu, a fissile material well suited to making bombs.[10]

When nuclear energy was a relatively new technology, it was commonly believed that uranium was a rare and expensive strategic material. It was therefore thought important to reprocess nuclear fuel, or else to develop breeder reactors that produced plutonium in large quantities so as to assure an adequate supply of fissile material for fuel and weapons. This was thought to be worthwhile even though extracting plutonium from spent fuel is a difficult, environmentally hazardous, and expensive chemical process that for technical reasons must be carried out a large scale, so that spent fuel must be transported to a large central reprocessing plant, creating possibilities for theft and sabotage.

It turned out that uranium is not a scarce resource at all but is, in fact, a common mineral. It was thought to be scarce because no one had bothered to look for it. Most countries have realized that the world would be better off with less plutonium, so reprocessing of nuclear fuel should be discouraged. Nevertheless, some countries still plan to build reprocessing facilities.

Why is a graphite reactor a good source of plutonium?

Graphite-moderated reactors of the kind used in North Korea before the Framework Agreement are well suited to producing plutonium because they are fueled by natural uranium, which has a higher proportion of ^{238}U than enriched uranium. ^{238}U doesn't sustain a chain reaction the way ^{235}U does, so it can't be used to make an atomic bomb, but it can absorb a neutron to become plutonium. So a graphite-moderated reactor is more efficient than other nuclear reactors for producing plutonium and is well suited to proliferation, as are heavy-water reactors.

6
Global Health
Security and Inequality

A cholera outbreak in 1991 forced the government of Peru to make an awkward choice between economics and public health. The minister of health argued for a policy of total transparency. He proposed to appear on television to warn people not to eat ceviche, the national delicacy of raw fish marinated in lemon juice, because it might carry the germ that causes cholera. On the other side was the powerful fishing industry, the source of a third of Peru's export revenues, which wanted news of the outbreak to be suppressed. For President Alberto Fujimori, the decision was easy. He fired the minister of health and ate ceviche on national television, using fish harvested far offshore. To show his loyalty, the Peruvian minister of fisheries followed suit, using fish purchased at a local market. The unfortunate minister was shortly thereafter hospitalized with "influenza."[1]

Fast forward to 2020 and the COVID-19 pandemic in the United States, the United Kingdom, Italy, Russia, and Brazil—countries that, unlike the Peru of 1991, have world-class science and good to excellent systems of public health. (The China story is more complicated, and we will discuss it later in this chapter.) With variations from country to country, political leaders suppressed evidence of virus spread, downplayed the importance of the epidemic and the need to observe measures to protect the health of the population, ignored the opinions of local experts, and publicized bogus "cures"—all to avoid economic damage and public panic, but equally importantly to consolidate political power and to show themselves as strong leaders who were firmly in control. The result was the squandering of critical weeks or months that could have been spent in preparation, the loss of hundreds of thousands of lives that could have been prevented, and trillions of dollars of economic damage due to the shutdown of much of the world economy.

The story of the global response to the COVID-19 pandemic encapsulates the elements of the Survival Nexus. The response involved politics, economics, business, law, culture and psychology. Epidemiologists,[2] immunologists, ecologists, zoologists, economists, lawyers, political and business leaders, finance ministers, biotechnology firms, and international organizations all played important roles. Despite ample warnings long in advance that a catastrophic pandemic was inevitable sooner or later, investments in necessary preparations had long suffered

The Survival Nexus. Charles Weiss, Oxford University Press. © Charles Weiss 2022.
DOI: 10.1093/oso/9780190946265.003.0007

from crisis-to-crisis funding and low priority relative to more pressing needs.[3] In the United States, simple protective measures—wash your hands, wear a mask, maintain social distance, avoid indoor crowds—became sources of political controversy.

Efforts to trace the origin and spread of the virus became embroiled in top-level, big-power politics between the United States and China, resulting in a major threat to the integrity and the financing of the World Health Organization (WHO), the United Nations technical organization with responsibility for global health and for coordinating the global response to the pandemic. International collaboration and competition among scientific researchers and biotechnologists unraveled the details of the structure and functions of the virus with extraordinary speed. More than a hundred different vaccines were developed in parallel in the United Kingdom, United States, Germany, China, and other countries. To avoid delays in scale up of production facilities, the U.S. government in Operation Warp Speed took on the multibillion-dollar risk of guaranteeing in advance the purchase of hundreds of millions of doses of each of several vaccines deemed to be promising, even if they ultimately failed their final trials. (In contrast, the government had been widely criticized for taking on a somewhat similar risk in the case of the solar energy company Solyndra.)

The pandemic also raised ethical issues. It fell disproportionally on low-income countries and communities and on historically marginalized groups all over the world—Native Americans, Latinos, and African Americans in the United Sates and migratory workers in India, for example—in part because their living and working conditions made them particularly susceptible to exposure and in part because they did not have access to adequate healthcare services. The COVID-19 pandemic also raised the question of whether the citizens of countries who develop a vaccine or a cure should have first claim on its benefits (vaccine nationalism), or whether there is a moral obligation to make it available to everyone, whether or not they could pay for it, giving priority to those at greatest risk regardless of where they live, given that infectious disease spreads rapidly all over the world and variant viruses can arise anywhere, so that no one is safe until everyone is safe. (Nevertheless, early vaccines went almost exclusively to high-income or scientifically advanced countries. COVAX, a consortium of governments and foundations, has been formed in the hopes of redressing this balance.)

Finally, the story of the COVID-19 pandemic shows the intimate connections among the natural world, human health, and the global economy. It probably spread from a large, open-air market in Wuhan, a city of eleven million people in Central China, where pangolins are sold as a delicacy and a source of traditional medicine. (Pangolins are nocturnal mammals known for their scaly exterior and their super-long tongues, which they use to catch ants.) Horseshoe bats that live in caves near the city had infected some of these pangolins with one of the many viruses of which bats are known to be carriers. The viruses jumped to humans, creating the disease we now call COVID-19. That makes COVID-19 a zoonosis—a disease that "spills over" to humans from nonhuman animals. From bat caves to a global economic shutdown in just a few months. Welcome to the Survival Nexus.

Influenza: The Annual Pandemic Zoonosis

Influenza is also a zoonosis. We all need to get a new flu vaccine every year because of ducks and pigs in China. For thousands of years, Chinese farmers have been raising pigs, ducks, and fish together in an integrated and ecologically balanced system. Duck cages are placed over pigsties, which are, in turn, placed over a fishpond. The ducks eat pests and weeds, the pigs eat duck droppings, and the pig manure fertilizes the aquatic weeds and microscopic plants on which the fish live. The farmers can use less fertilizer and fewer pesticides because of this system, but it also gives rise to a new strain of influenza virus every year, and hence to the need for annual vaccinations.

Here's how it works. Domestic and wild ducks can carry influenza virus in their intestines without showing symptoms, so pigs can contract influenza by inhaling or ingesting the droppings of seemingly healthy ducks. In pig stomachs, different strains of influenza virus can exchange DNA and create new strains of the virus. Migrating wild ducks can become infected with the new strains of the virus while dabbling in a nearby fishpond or while pecking for food on the ground and can carry these new strains all over the world. We, as humans, have limited or no immunity to these new strains, which is why we need to be vaccinated against a new strain of the influenza virus every year.[4] (We explain how vaccines work in Annex B.)

Although an annual flu vaccine may be an inconvenience, behind it lurks a major danger. Right now, most strains of influenza are not particularly lethal and are not passed that easily from person to person. However, if a future lethal strain of influenza were to evolve the capacity to be passed much more easily from person to person, the result could be yet another major pandemic. After all, the influenza pandemic of 1918 killed twenty to fifty million people in about nine months. The coronavirus pandemic, although less lethal than the 1918 flu, has shut down much of the world economy, showing the effect of today's much more globalized world and greatly increased international travel that spreads infection much more quickly. (We explain the lethality of the 1918 influenza virus and of the new coronavirus in Annex B.)

Influenza and COVID-19 are by no means the only infectious diseases that pose global threats. Malaria, yellow fever, Zika, dengue, and chikungunya are all on the rise and are predicted to increase further as climate change expands the range of the mosquitos that spread these diseases to humans.[5] The most serious of these diseases is malaria, from which two hundred million people suffer and half a million people—two-thirds of them children under five—die each year. Another two hundred million suffer from schistosomiasis, a debilitating disease spread by snails in tropical waters.[6]

Neither disease has a proven vaccine, although a promising one for malaria is in advanced trials. This is partly because it is hard to develop vaccines for parasitic diseases, as we explain in Annex B. But mostly, it's a matter of market economics. Most people with malaria and schistosomiasis lack significant money or political power. Even though four hundred million people suffer from the two diseases combined, there is not enough of a commercial market or other incentive for investment in research and development. For a pharmaceutical company, it is much more profitable to invest in developing a pill that affluent people will take every day for the rest of their lives for some chronic condition than to invest in a one-time cure or vaccine for less affluent people. To a government, there is little political payoff for public investment in cures or vaccines for diseases that afflict people in far-off countries. Besides, from the commercial point of view, making vaccines is a risky business. Vaccines are tested for safety and efficacy on at most tens of thousands of volunteers and are then administered to millions of healthy people. If anything goes wrong in this much larger population, the manufacturer may face expensive liability suits.[7]

The long delay in developing a malaria vaccine is one sign of a more general problem that affects every aspect of global health. Some 80% of the world's burden of disease is borne by low-income people, who benefit from only about 20% of the funds the world expends on health problems.[8] Perhaps 1–2% of the money spent on health research is on the diseases of low- and middle-income countries—and the diseases that do get attention tend to be those that have captured the attention of the higher-income countries.[9]

We're All in this Together

The COVID-19 pandemic showed how infectious diseases can bring far-off troubles home to us very quickly. That's one reason why we talk about "global health," a term that implies that disease knows no national boundaries, so that we're all in it together and can learn from each other's experiences, rather than "international health," an older term that suggests that *we* in the higher-income countries are providing aid, expertise, and technology to lower-income countries to help *them* to deal with *their* problems.[10] One might even use the term "planetary health" to call attention to the fact that climate, food, people, environment, and disease all continuously interact with each other and with global health—a prime example of the close relation of the many disciplines of science and technology with that of global health.

As we have already seen with pangolins and COVID-19, public health can create global interest in politically sensitive sociological and environmental issues that would otherwise be of purely domestic concern. Such "intermestic" issues challenge traditional notions of national sovereignty, which hold that such matters are the business of no one but the country's own government and people.[11] In the best-known example on the sociological side, HIV/AIDS, a pandemic that has affected hundreds of millions of people, is often spread through unprotected sex and unsanitary drug use. This has led public health professionals in one country to be concerned with patterns of sexual behavior and drug use in other countries, and with the stigma, education, and support surrounding these behaviors.

In another example, the Russian prison system has become a topic of global health concern because it is a major focal point for drug-resistant tuberculosis. (Annex B explains the origin of antibiotic resistance in bacteria.) Prisoners who are held for long periods in overcrowded prisons can be exposed to tuberculosis, but once they are infected, they cannot access a full course of antibiotic treatment because of unreliable supply. As a result, infected prisoners may receive an antibiotic long enough for the pathogen to become resistant to that specific antibiotic, but not long enough for the patient to be cured. They are then started on a new round of antibiotics. Once the infected prisoners are released from prison, they act as "epidemiological pumps" or superspreaders who can pass the antibiotic-resistant strain on to anyone, but in particular to immunocompromised people or to people living on the streets or in crowded apartments. If infected people travel abroad, they can infect people in the countries they visit.[12] An epidemiologically analogous situation arose in the Philippines and Vietnam, where sex workers are provided with regular doses of antibiotics, to which the gonococci that cause gonorrhea develop resistance. Traveling clients then spread these resistant organisms around the world.[13]

On the environmental side, as human settlements expand into wilderness or forest, people come into closer contact with wild animals and are more likely to contract the diseases that these animals carry. Zoonoses include some of the deadliest pandemics in recent years, including HIV/AIDS, MERS, SARS, Ebola, and most recently, COVID-19. The spillover event, when the disease passes from animal to human, can occur in many different ways but often includes eating infected meat from wildlife, as is common in many countries (much as North Americans eat wild venison and Europeans eat wild boar). People probably first contracted HIV/AIDS by consuming meat from infected chimpanzees or monkeys, and it is likely the 2014–2016 Ebola outbreak in Western Africa began when a young boy played with or ate infected bats.[14] Once these diseases are in the human population, they can be transferred through intimate human interactions, like burial rituals (Ebola) and sexual exchanges (HIV).[15]

The study of the origin and spread of these zoonoses has led to the concept of One Health—the idea that human health is intertwined with both animal health and environmental health. The One Health framework is multidisciplinary. Its research teams include not only epidemiologists, doctors, and nurses, but also veterinarians, animal and plant ecologists, environmental and agricultural scientists, anthropologists, statisticians, zoologists, microbiologists, and molecular biologists.

Vertical and Horizontal Health Services

Advances in science and technology in wealthy countries, made in close collaboration with scientists in low-income countries, have made possible many of the great triumphs of modern medicine: techniques that made possible the eradication of smallpox (see Annex B) and major progress against polio; DDT and other pesticides that control mosquito-borne diseases; antiretroviral therapy that transformed HIV from a death sentence to a manageable chronic disease; oral rehydration solution and a rotavirus vaccine that have saved millions of children from dying of cholera and other diarrheal diseases; and Ivermectin, the anti-roundworm medicine that is close to wiping out river blindness, a fly-borne scourge of the African Sahel.[16]

These technology-based medical triumphs reflect the historic emphasis of the public health profession on tackling infectious diseases through dramatic "vertical" campaigns, using technology tailored to the requirements of campaigns against individual diseases. In general, official development assistance agencies and private foundations have tended to favor such vertical programs because they promise direct results and are easier to manage and evaluate.

Since the late 1970s, however, there have been efforts to effect badly needed improvements in "horizontal" primary healthcare and the general healthcare system. These would require a quite different approach to the management and application of technology—developing a tiered system of health delivery institutions, beginning with rural or neighborhood clinics with relatively inexpensive equipment, building to regional centers for more demanding cases, and topped off with apex institutions with the most sophisticated and expensive equipment the country can afford, with provisions for sending a small number of especially difficult cases abroad. Such a system is especially important for addressing non-communicable diseases like cancer, heart disease, depression, and dementia, as well as the broader context of public health: poverty, nutrition, housing, pollution, access to clean water and sanitation, and promoting avoidance of life-threatening behavior like smoking, alcohol and drug abuse, poor diet, and unsafe sex.

The choice between vertical and horizontal systems involves the full panoply of political, economic, legal, and institutional issues that interact with science and technology to constitute the Survival Nexus. It raises the philosophical question of whether money spent to improve the health of a population is an economic investment or the fulfillment of a human right, as well as the political question of who has the right to decide about the priorities that determine how that money should be spent. The history of global health since World War II is in large part the history of how these issues have played out.

The World Health Organization and the Politics of International Cooperation

The world has sought global solutions to global health problems, especially those of infectious disease, for at least 170 years. In 1851, European countries participating in the Paris International Sanitary Conference agreed to give up some of the traditional aspects of their national sovereignty in order to harmonize national regulations governing the quarantine of infected travelers arriving from foreign ports. The agreement over quarantine regulations was intended to facilitate world trade and was one of the earliest examples of functionalism, which political scientist Ernst Haas defined as international cooperation for a technical purpose that could not be achieved by unilateral action. Functionalists, he wrote, "believe in the possibility of specifying technical and 'non-controversial' aspects of governmental conduct, and of weaving an ever-spreading web of international institutional relationships on the basis on meeting such needs."[17] In practice, countries do cooperate to manage global health—but national self-interest and power dynamics are never far below the surface.[18]

Fifty-six years after the Paris Conference, in 1907, an International Office of Public Hygiene was established. This evolved into the health organization that was one of the most successful aspects of the League of Nations, and eventually into the WHO. The WHO is headquartered in Geneva and has regional offices all over the world. It was one of several intergovernmental organizations established during the same period to manage technical services like the international exchange of weather information and the management of international postal service. These evolved into today's World Meteorological Organization (WMO) and Universal Postal Union, respectively.

The stated goal of the WHO, the only international authority with the expertise and prestige to declare pandemics and to rally and coordinate the world's resources to fight public health emergencies, is "to build a better, healthier future for people all over the world." A major, though largely unsung, virtue is that it can call the world's attention to neglected problems of public health that may not be regarded as part of medical science. For example, it was one of the first international organizations to highlight the public health implications of traffic safety.[19] However, it has limited organizational flexibility and freedom of action, and its decentralized and collaborative management structure, while important for earning the trust of its diverse membership, creates obstacles to quick response to emergencies.

The commercial interests of the wealthier countries, especially those of the pharmaceutical industry, have long had a substantial influence over WHO policies and have sometimes forced its professional staff to tread carefully.[20] For example, in the early 1990s, the United States withdrew its contribution (about a quarter of the core WHO budget) in protest of the WHO decision to issue a model list of essential drugs suitable for local manufacture, on the grounds that this would discourage imports.[21] The need for careful, deferential diplomacy vis-à-vis China led to controversial concessions by WHO regarding the conduct of research into the origins of the coronavirus epidemic, and it led to a conflict between the U.S. and Chinese governments that played a part in the U.S. presidential election campaign of 2020 and to the withdrawal of the United States from WHO on the grounds that the organization was "Chinese-centered."[22]

In another example, the WHO was forced to stand largely on the sidelines during the international negotiations over the Agreement on Trade-Related Aspects of Intellectual Property (TRIPS) during the early 1990s, in which the United States successfully pressed low- and middle-income countries to adopt stricter rules governing intellectual property (IP) to the advantage of U.S. pharmaceutical exporters.[23] Stronger IP rules might well benefit more technologically advanced low- and middle-income countries that could attract foreign investors with state-of-the-art technologies and that might wish to nurture their own technologically innovative industries. To the lowest-income countries, however, they meant much higher costs for essential pharmaceuticals. The WHO later did try to make up for this inaction by advising low- and middle-income countries on how best to cope with the new IP regime.

To its credit, the WHO promotes breastfeeding against the "ferocious" opposition of manufacturers of infant formula[24] and has taken a strong stand against the international tobacco industry. More recently, the WHO has called attention to the increasingly critical global problem of antibiotic resistance, which threatens to reverse the conquest of infectious disease that is one of the signal triumphs of medical science (see Annex B). Here it must deal not only with the overprescribing of antibiotics caused by the expectations of patients that doctors will prescribe them even when they are not needed, but also with the economic interests of the agricultural industry, which uses antibiotics to promote livestock growth and is the major user of antibiotics worldwide.

"Tropical Health": The Colonial and Postcolonial Background

By the end of the nineteenth century, nearly all equatorial countries were colonies. The colonial health agenda was driven by the priorities of European governments. It emphasized health services to people essential to colonial government and industry, as well as programs targeted against individual infectious diseases, rather than support to broad-based healthcare systems to meet the needs of most of the population. Until the 1930s, investment in colonial health services was limited, at least as compared with the analogous investments being made in the mother country. In the early days of colonialism, this investment was made to support the military expansion of colonial rule. Later, it was intended to legitimize colonial rule as part of the presumed "civilizing burden" of the colonial power, to reinforce the stability of colonial rule, and to safeguard the health of colonial officials, their families, and local workers in extractive industries.[25]

Colonial powers like Great Britain and the Netherlands established schools of "tropical medicine," which developed moderately effective (although extremely toxic) remedies for some tropical diseases, especially leishmaniasis and the early stages of African sleeping sickness (trypanosomiasis).[26] For researchers in these institutions, colonial rule presented an opportunity to carry out experiments free of regulatory or ethical limits regarding the safety of human research subjects.

The biggest overall health impact of colonial rule came from the human and environmental exploitation that accompanied the African slave trade, which from the sixteenth until the nineteenth centuries enslaved and killed millions of people. In addition, African men were forced to do dangerous work, and the crowded camps in which they were required to live facilitated the spread of infectious disease. Widespread plantation agriculture opened new grassland habitat for the tsetse fly and led to serious epidemics of sleeping sickness, which were contained partly by the new medicine and partly by quarantines.[27]

Colonial investments in health services increased in the 1930s in anticipation of the need for healthy colonial troops to fight in World War II. These investments continued to increase after the war in an effort to improve living conditions so as to counter growing movements for independence. These increased resources were still only enough to support vertical interventions limited to urban areas and targeted against specific diseases, which were relatively inexpensive and easy to manage. On the other hand, few resources were given to horizontal support to general health services for the entire population, including the rural majority.[28] After World War II, the WHO continued to emphasize major vertical campaigns, including a partly successful campaign against malaria and a heroic and dramatically successful campaign from 1967 to 1980 that eradicated smallpox—one of the great triumphs of the public health profession (see Annex B).[29]

When independence came, most former colonies were left with limited funds, skeletal civil services, and few trained health professionals. Their health policies and levels of investment varied from country to country. Many newly independent countries committed themselves to a policy based on equity and did their best to provide basic health services to their entire population through regional health centers and smaller rural health posts. Others continued the emphasis on urban-oriented services inherited from colonial rule, with a focus on the new African elites, sometimes spending their limited resources on a few sophisticated facilities in major cities. Still others could not manage significant investment in any kind of health service delivery.[30] In some cases, development assistance from the United States and former colonial powers attempted to bridge the funding gap, though it was never sufficient to provide healthcare to all.

The 1970s: Is Health a Human Right or an Economic Investment?

After World War II, the international community declared health to be a fundamental human right. In 1946, the WHO Constitution declared that "[t]he enjoyment of the highest attainable standard of health is one of the fundamental rights of every human being,"[31] a principle later reaffirmed by a variety of United Nations resolutions[32] and decisions of national courts that blocked laws or projects that they deemed exceptionally detrimental to public health.[33]

Until about 1980, however, the meager development assistance for public health belied this principle. Both the development assistance community and the governments of recipient countries regarded funding health initiatives in the developing world to be less economically productive than funding in agriculture, industry, and infrastructure. Health was afforded relatively low priority. By the late 1970s, the divide was clear: some decision makers saw health as a human right and wanted to reach as many people as possible, while others argued that health for all was neither attainable nor a high priority.

The difference between these points of view played out in a series of confrontations over the relative importance of primary healthcare and technology-oriented, disease-specific campaigns. At a major conference in 1978 in Alma Ata, Kazakhstan (then part of the Soviet Union), the world community signed a global commitment to locally based, universally accessible, horizontal systems of primary healthcare, under the slogan "health for all by 2000."[34] This Alma Ata Declaration was the culmination of a decade of effort by Dr. Halfden Mahler, the WHO Director-General, and WHO staff, backed by the voting bloc of newly independent African countries in the World Health Assembly, the WHO governing body. The Declaration went far beyond the delivery of health services by doctors, nurses, and hospitals, to include virtually all aspects of development that affect health: education, clean water, basic sanitation, maternal and child healthcare, including family planning, and much more. It was, in effect, a statement of the Survival Nexus as it affects global health. As a to-do list for a low-income country with limited government capacity and serious health problems, however, it constituted an immense policy, administrative, and political challenge.

Although the Alma Ata Declaration was an internationally recognized commitment, it was quickly undermined by a meeting the following year, convened by the U.S.-based Rockefeller Foundation and attended by high officials of leading development assistance agencies. The meeting took its name from its location at the Foundation's gloriously beautiful study center at Bellagio, overlooking Lake Como in northern Italy. Unlike the Alma Ata conference, where participants had been representatives from almost every country in the world, the Bellagio meeting was attended by a select few foundations, development organizations, and government policymakers.

Contradicting Alma Ata, the Bellagio meeting agreed that the objective of universal healthcare was "unattainable" and that the scarcity of financial and human resources devoted to global health required the international community to identify and put into practice specific, cost-effective, vertical interventions using relatively simple "appropriate" technology. It advocated campaigns against a limited number of diseases that had the greatest mortality or morbidity (sickness) and for which interventions of proven efficacy existed. Selective primary healthcare was said to be an "interim strategy," the best that could be put into practice at the time.[35] The vertical approach came to dominate international health programs, many of whose alumni ascended to senior positions in the global health establishment.

The United Nations Children's Fund (UNICEF) built on the high-level endorsement of vertical programs at Bellagio to launch the GOBI program, which promoted four interventions. Each of these interventions centered on a low-cost technology, promised tangible and measurable results, and was represented by a letter in the GOBI acronym: (1) *G*rowth measurement to detect infant malnutrition, complemented by advice to weaning mothers on proper nutrition; (2) *O*ral rehydration to cure dehydration due to diarrhea, a major cause of death in childhood (see Annex B);[36] (3) *B*reastfeeding instead of less nutritious and less hygienic infant formula, which at the time was being heavily promoted by food companies; and (4) *I*mmunization, building on the Expanded Program of Immunization (EPI) that had previously been launched by the WHO. Later, three more interventions were added: (5) *F*ood supplementation to at-risk pregnant mothers; (6) *F*amily spacing; and (7) *F*emale literacy, all of which have been shown to reduce the risk of infant death, making the complete acronym GOBI-FFF.[37]

The conflict between Bellagio and Alma Ata highlighted two critical, intersecting issues regarding global health: what should be the relative priority of specific, disease-oriented interventions as compared with the building of general health systems, and who should exert the decisive influence over a major policy issue concerning global health?

Vertical programs of the kind favored by the Bellagio meeting have a long and distinguished place in the history of global health: vaccination campaigns for polio and other infectious diseases; mosquito control campaigns for malaria and yellow fever; the eradication of smallpox; and the control of epidemics of HIV, SARS, and Ebola. Vertical programs date at least from the early twentieth century; the construction of the Panama Canal was made possible through vector control programs, leading to the formation of the International Office of Public Hygiene mentioned earlier. At around the same time, the Rockefeller Foundation provided shoes and improved sanitation to protect poor people in the Southern United States and many other countries from feces- and soil-borne hookworms.[38]

From the point of view of a foreign donor, vertical systems have the advantage of specific and measurable goals by which their performance can be evaluated and justified to a parliament or to the board of directors of a nongovernmental organization (NGO) or private foundation. They make it easier to address the specific characteristics and intervention points associated with a particular disease. Vertical systems also enable a foreign donor to avoid what is often the complex and difficult task of building a system of primary healthcare in a low-income country. To patients and recipient governments, on the other hand, they have the major *dis*advantage that they address priorities and objectives chosen by the donor that may or may not coincide with those of the supposed beneficiary.

This discussion begs an important question: why are the health needs of a low-income country for donors to decide? Isn't this the role of the governments of the countries concerned, hopefully in consultation and with the consent of the governed, and assisted by the WHO and by bilateral, multilateral, and private donors? We will take up these questions once again after finishing our historical overview.

The 1980s: Tough Times for Global Health

The increase in oil prices and simultaneous decline in commodity prices of the 1980s hit African economies hard, especially in the many newly independent countries that were suffering from economic mismanagement. To avoid economic collapse, these countries were forced to seek relief through loans from the World Bank and the International Monetary Fund, a condition of which was an urgent reduction in public spending and government deficits. Governments of low-income countries, both democratic and nondemocratic, found it easier to make drastic cuts in social expenditures like education and health delivery systems than to cut politically essential expenditures affecting well-organized constituencies, like military budgets and fuel subsidies.[39] These "structural adjustment" policies were not usually accompanied by "safety net" measures that could have compensated for their impoverishing effects and consequent harm to human health, especially among the most vulnerable populations.[40]

In the health sector, the World Bank pressed for the imposition of "user fees"—small charges for access for doctor and hospital visits—to ease the pressure on government budgets. In practice, however, these copayments discouraged poor people from seeing a doctor when they were sick. The effect on African health systems, whether equity based or elite oriented, was immediate and drastic. Personnel fled, maintenance was neglected, and equipment deteriorated. The WHO, which was staffed mainly by medical doctors and experts in public health and lacked expertise in economics and other social sciences, was ill equipped to take on the challenge of opposing these policies.

The effectiveness of the WHO suffered another series of blows in the 1980s. In the United States and elsewhere, governments that were distrustful of multilateral organizations came into power. They froze the size of the WHO budget and often delayed or reduced their "required" annual national contributions. The WHO also suffered from poor leadership that led to the departure of many highly qualified professional staff. As a result, donors often created new organizations to deal with global health challenges as they arose, rather than entrust additional responsibilities to WHO. The resulting institutional rivalries further undermined the WHO's authority as the primary international organization concerned with global health.

The upshot of these changes was a major reduction in the budget directly managed by WHO leadership and staff. For some time, this "unrestricted" budget covers only about a quarter of the WHO's overall program.[41] This leaves the WHO staff to raise the remainder from grants for individual projects from national governments and private foundations in wealthier countries, so that in practice, these donors set most of the programmatic priorities of the organization. This reflects the lack of confidence on the part of donor countries in the ability of the WHO to do its job but also makes it difficult for the organization to reform itself. It is part of the more general trend of governments limiting the freedom of action of multilateral entities not under their direct control.

The 1990s: HIV/AIDS and the Resurgence of Global Health

The new millennium saw revived interest in global health and a flood of new money stimulated by four new developments. First, the U.S. intelligence community defined the HIV/AIDS pandemic as a prime *security* threat, justifying substantially increased U.S. government expenditure.[42] For the first time, a global health issue appeared on the "big-time" geopolitical agenda.[43]

Second, the new discipline of health economics strengthened the case for investments in public health by demonstrating that such investments contribute to economic growth, not just to human welfare. Studies showed that healthier people are more productive and that the direct financial costs of poor health in the form of lost wages, lost productivity, and medical bills constitute a significant drag on economic growth. Cost/benefit analyses based on this insight concluded that expenditures on health services and infrastructure were economically productive and hence "legitimate" investments. These studies helped bridge the gap between economics and the ethical conception of human rights.[44]

Third, the work of Paul Farmer and his colleagues in Boston, Haiti, Peru, Rwanda, and elsewhere showed that it was possible to deal effectively with diseases of poverty like multi-drug-resistant tuberculosis and HIV/AIDS even in remote areas of some of the poorest and most corrupt countries in the world—*if* enough money was made available for the purpose.[45] This shattered the somewhat condescending consensus that there was no good way to spend large sums on public health in poor countries. Farmer and his colleagues argued that low-income Haitians have the right to diagnoses and cures, even expensive ones, just like everyone in high-income countries, and successfully campaigned for major increases in assistance to the Haitian public health budget. They also helped to counter the cynicism that had developed surrounding the HIV epidemic, which had incorrectly assumed that low-income, poorly educated patients would be unable to follow the demanding multi-drug, multi-pill daily regimens that were changing HIV from a fatal to a chronic disease in high-income countries.

Fourth, in 2000, billionaire Bill Gates launched the Bill and Melinda Gates Foundation. With a $40 billion–odd endowment, the foundation quickly became one of the major players in global health, greatly increasing the funding available to fight the diseases of low-income countries. The Gates Foundation was the second largest donor to WHO in 2018.[46]

These four landmark events prepared the way for the injection of new money into what had been a sorely underfunded field. In the United States, the Republican administration of President George W. Bush launched the Global Fund to Fight AIDS, Tuberculosis and Malaria in 2002 and the President's Emergency Plan for AIDS Relief (PEPFAR) in 2003, which together tripled the total budget for responding to HIV.[47] Working in almost fifty countries, PEPFAR has provided free antiretroviral medicines, HIV counseling, support to orphans and vulnerable children, and—since 2013—training for local health workers. Its achievements reinforced Farmer's lesson that it was possible to control HIV/AIDS even in the lowest-income countries of the world.

The impact of these programs has not been exclusively positive, however. PEPFAR has undermined local health delivery in many countries by creating a parallel institutional structure that bleeds local hospitals and clinics of their most qualified staff by offering higher salaries and better management and working conditions. In addition, PEPFAR as originally conceived included prohibitions of contraceptives and abortion, and the spending of more than a billion dollars on "abstinence-only and faithfulness" campaigns that have had no measurable effect on risky sexual practices.[48] These strictures responded to the concerns of politically powerful religious groups within the U.S. Republican Party but limited the value of PEPFAR to recipient countries.[49]

Luck also intervened to infuse new funds into U.S.-based global health programs. The newly established Gates Foundation found itself with a huge sum of money burning a metaphoric hole in its pocket, which it was obligated to give away in order to preserve the Foundation's tax-exempt status under U.S. law. The resulting $750 million grant to the Global Alliance for Vaccines and Immunization (GAVI), a public–private partnership that was a successor to UNICEF's immunization campaign, helped immunize nearly half a billion children, saving an estimated eight million lives.[50] GAVI also funds field tests of vaccines for malaria and cures for African sleeping sickness, trachoma, and other tropical diseases.

On the research front, new tools for investigating the scientifically fascinating ways that parasites have evolved to evade or manipulate the human immune system led to a resurgence in interest on the part of the world scientific research community in long-neglected tropical diseases like malaria, leishmaniasis, African sleeping sickness, elephantiasis, and Chagas' disease (see Annex B). The National Institutes of Health (NIH), the major government funder of health-related research in the United States, and the Gates Foundation became major funders of research on these and other vector-borne diseases. Other public–private partnerships between consortia of research institutions and private companies collaborated with researchers in low- and middle-income countries, many of whom had been trained by WHO, to develop cures and vaccines for kala-azar (visceral leishmaniasis), dengue (breakbone) fever, trachoma (a major cause of blindness), and other tropical diseases.

The 2000s: Improved Response to Pandemics

In February 2003, Carlo Urbani, an Italian doctor working at the French Hospital in Hanoi, Vietnam, recognized that his patient, an American businessman who had just arrived from Hong Kong, was suffering from a new and highly contagious form of pneumonia. He christened the new illness "severe acute respiratory syndrome" (SARS). Dr. Urbani had the hospital quarantined to prevent further spread of the disease, even though it put healthcare workers at risk of contracting the illness. He and many members of the hospital staff died from SARS but thanks to their heroic self-sacrifice, the epidemic did not spread further in Vietnam.[51]

The SARS story was quite different in China.[52] By the time Dr. Urbani had made the first public report of the disease to the appropriate international authorities, there were already some three hundred cases in the southern Chinese province of Guangdong. Nevertheless, local officials suppressed information of the SARS outbreak for six critical months, in accordance with Chinese traditions of bureaucratic secrecy, the traditional view that domestic health problems were China's own sovereign concern, and the desire of the Chinese government to "uphold social stability" (i.e., to maintain its hold on political power). During that time, the disease spread as far as Canada and Germany, and the number of cases grew to the point where the epidemic became much more difficult to manage. The WHO had to make an unannounced visit to the provincial capital of Guangzhou to gather information on SARS that contradicted official denials. Fortunately, SARS turned out to be less contagious than had at first been feared, and the epidemic was brought under control.

The experience with SARS led the Chinese government to strengthen its ability to respond to epidemic disease and to become much more forthcoming with its epidemiological data, as its officials realized the damage that this experience had wrought on their reputation for credibility. Despite these reforms, a somewhat similar sequence of events unfolded sixteen years later when the COVID-19 virus was discovered in Wuhan, China, with much more serious consequences.

The SARS experience led to an important change in WHO policy. Before the SARS epidemic, the WHO, as a technical agency of the United Nations, had waited for governments to submit official statistics regarding infectious disease, a process that took months or years and sometimes did not take place at all. The WHO then accepted these statistics without question, even when its own experts had strong reason to believe that they were in error. This policy was no longer tenable after the experience with SARS. The WHO revised the International Health Regulations, which are legally binding, to expand the list of diseases that governments are obligated to report to international authorities and to require member countries to report "events that may constitute a public health emergency" within 24 hours and to develop "core capacities" for surveillance, assessment, reporting, and response.[53]

The WHO also created a worldwide network of technical institutions, called the Epidemic Intelligence Open Source System, to give early warning alerts of possible outbreaks of epidemic disease. The System uses data from nongovernmental sources that are insufficient to diagnose a specific illness in an individual patient, but that in the aggregate may indicate the early stages of an outbreak of disease, a technique known as syndromic surveillance. Such data may include purchases of over-the-counter medicines, Web searches, worker absenteeism, ambulance log sheets, visits to hospital emergency rooms, articles in local newspapers, and even calls to the offices of veterinarians, since sick pets may be early signs or carriers of impending human disease. (The technology needed to track this information incidentally benefits from and contributes to that used for military intelligence.)

Despite this improvement, the outbreak of Ebola in 2014 showed that the early warning systems to detect epidemic diseases were still inadequate, especially in vulnerable African countries with weak public health infrastructure.[54] In an event that foreshadowed the course of the COVID-19 pandemic, the WHO delayed declaring Ebola to be a "Public Health Emergency of International Concern," a finding that requires a coordinated international response, in order to ensure continued cooperation with the affected West African countries. Since then, the WHO and the Atlanta-based Centers for Disease Control and Prevention (CDC), both of which are critically important leaders in global efforts to prevent, detect, and respond to epidemic disease, have invested substantial sums in a "health security" program that helps low-income countries fulfill their obligations for surveillance and reporting under the International Health Regulations.

The Human Right to Health as a Workable Principle

GAVI, PEPFAR, and the Global Fund are all vertical programs, in the sense that they are directed against specific diseases or conditions and have specific, measurable goals. On the other hand, they address priorities chosen by the donor for their worldwide importance. If HIV, malaria, or tuberculosis is not the major public health problem in a country participating in a PEPFAR-funded program or if none of the three is the disease affecting a particular patient seeking help from a PEPFAR-financed clinic, too bad. (Maybe the patient has a broken leg!)

Some countries have dealt with this issue pragmatically, by adding their own horizontal investments to complement the support from vertical programs like PEPFAR, giving rise to a diagonal approach. Ethiopia, for example, used PEPFAR money to recruit health workers from rural communities to provide immunization and maternal and child health services, resulting in improvements both in the care of HIV and in overall health indicators.[55]

More generally, the deficiencies of the vertical approach have led to a resurgence of support for the horizontal approach to primary healthcare as an embodiment of the concept of health as a human right. This concept has figured prominently in the advocacy of universal health coverage by the newer leadership of the WHO and in the seventeen Sustainable Development Goals adopted by the United Nations in 2015, at least three of which ("zero hunger," "good health and well-being," and "clean water and sanitation") are directly related to health.[56]

The most important initiatives to give practical meaning to the concept of health as a human right have built on the seminal work of Julio Frenk and his colleagues in Mexico. By the time Frenk became minister of health in the year 2000, life expectancy in Mexico had increased to seventy-five, and low-income communities suffered not only the diseases of poverty but also from the non-communicable diseases typical of an aging population. The out-of-pocket costs of this double burden of disease were financially catastrophic for many low-income families, most of whom were excluded from the health insurance programs then available in the country.

Frenk established universal access to basic health services, independent of income or employment status, as a core ethical principle of Mexican national health policy.[57] He won strong support from the new, reform-minded President Vicente Fox, from his minister of finance, and from the parliament, by showing that public health could be a centerpiece of the country's economic development: creating jobs, improving productivity, making education more effective, protecting savings, and creating a fairer distribution of wealth.[58] Here he benefited from the strong Mexican economy, which was buoyed at the time by high prices for its oil exports.

Based on epidemiological and economic evidence, Frenk and his colleagues identified the health problems of highest priority in Mexico and designed a package of institutions, policies, and health services that could be delivered to the whole population within the limited resources of the country.[59] This package consisted of relatively inexpensive interventions of proven effectiveness for commonly occurring conditions, plus a smaller number of more costly interventions needed to prevent catastrophic financial losses to patients and their families. The package was accompanied by comprehensive measures to raise the quality of the services provided, and to evaluate their effectiveness at regular intervals by measuring improvements in child mortality, nutritional status, and the frequency of a number of communicable diseases.[60] Frenk and his counterparts in other Latin American countries also put in place policies to address the "social determinants" of health, for example, by placing taxes on tobacco, sugary beverages, and unhealthy processed foods.[61] Sadly, many of Frenk's innovations have been cancelled by subsequent governments.

Many other middle-income countries have followed the Mexican example, showing that they can marshal enough of their own resources to provide their people with basic health services without dependence on foreign donors.[62] Improved health delivery systems in these countries have also made them more effective participants in the identification, surveillance, and control of infectious diseases that could spread beyond their borders and develop into regional or global pandemics.

In the long run, the only sustainable solution for the health problems of low-income countries is for them to develop their own capacity to diagnose their own needs and to develop policies and programs to improve the health of their people, at the same time as they participate in global efforts to spot and contain incipient epidemics. Given the broad range of issues that affect health and the many political pressures from donor countries, the global trade regime, and multinational companies—not to mention the health effects of climate change—this is no small order.

COVID-19 Brings the Survival Nexus into Stark Relief

The COVID-19 pandemic has provided an intensive, real-time crash course on the workings of the Survival Nexus. It illustrates how scientific issues can become intertwined with domestic and big-power politics. It demonstrates the importance of effective political leadership, clear public messaging, trust in public institutions, willingness to sacrifice for the common good, international collaboration among scientists, and incentives to private industry to invest in innovation for the public good. It showcases the tension between the requirements of politics and those of public health and the value and the limitations of scientific models. It illustrates the difficulties in making policy and informing the public while scientific understanding is still incomplete, in updating policies and explaining them to the public as the crisis unfolds and knowledge improves, and in controlling the spread of conspiracy theories and misinformation in the age of social media. It also raises familiar international issues concerning inequality among and within countries, international trade, intellectual property, the role of multilateral institutions, and the conflict between traditional bureaucratic secrecy and the need for transparency in decision-making.

The coronavirus pandemic is much more serious than seasonal influenza. The virus passes from human to human even before the appearance of symptoms, carried by the aerosols exhaled during the course of ordinary breathing, speaking, and singing, not to mention coughing and sneezing.[63] Most of those infected suffer only relatively mild symptoms or even no symptoms at all, but the disease can be fatal, especially among the elderly, the immunocompromised, and people with underlying conditions such as diabetes or high blood pressure. It also has serious effects on the long-term health of many who recover from the acute phase of the disease.

Scientists and biotechnologists responded to the discovery of the new virus with a worldwide, all-hands-on-deck collaboration. Within a relatively few weeks, they had unraveled the details of the submicroscopic structure of the virus and the mechanism by which it invades the cells of the human respiratory tract, building on the results of years of research on other coronaviruses, such as those that cause SARS and Middle Eastern respiratory syndrome (MERS). In this process, they identified numerous targets for possible vaccines and cures. Working at breakneck speed, immunologists all over the world pursued over a hundred different possible anti-coronavirus vaccines. Several vaccines were pronounced ready for widespread distribution in less than a year—about a tenth of the time traditionally needed for vaccine development. Two of the first vaccines to prove successful used a novel technology involving messenger RNA (mRNA) to prime the immune system, a technology explained in Annex B.

Early epidemiological models, using assumptions based on previous coronavirus epidemics and preliminary data from the first countries to experience the COVID-19 epidemic, gave optimistic predictions that the pandemic could be contained without major changes in public behavior, so that the major objective was to prevent hospitals from being overwhelmed by a flood of critically ill patients. It took some weeks for scientists to reach a consensus, and for the WHO to announce, that COVID-19 can be transmitted person to person, and by people who do not show symptoms. They were then forced to change their advice and recommend that almost everyone practice social distancing, i.e., wearing a mask and maintaining a six-foot distance between people, a practice that applied even to those that seemed perfectly healthy. This meant shutting down much of the world economy at enormous economic, social, and political cost. This, in turn, created great political pressure in practically every country to reopen the economy as soon as possible, and in a few countries to speed the development and release of vaccines and cures even at the risk that they might prove ineffective or unsafe. A few months into the pandemic, researchers had more realistic data on which to base their models, although they still could not predict when the government of any individual country would adopt their recommendations, or to what extent the people of that country would comply with them. Different epidemiological models made different predictions regarding the rate of spread of the virus, allowing political leaders flexibility in deciding when to reopen the economy.

The keys to the effective response to a pandemic are that the problem be acknowledged and taken seriously, that people trust their institutions, and that they understand that they are all in it together and that the disease isn't "someone else's problem"—i.e., that no one is safe until everyone is safe. Once an epidemic disease is identified, controlling it requires respect for public health measures (in the case of COVID-19, wearing masks, avoiding crowds and indoor gatherings, social distancing, and frequent handwashing); limiting the size of public gatherings and closing or restricting gathering spots like theaters, churches, bars, restaurants, sporting events, and gymnasiums; widespread testing to identify the infected; and isolating those who test positive and tracing and quarantining their recent contacts, with special attention to superspreader events at which many people congregate, especially if they are unmasked and closely spaced. If the virus is not yet established in an isolated country, region, or island, these measures can be augmented by closing borders and restricting travel.

All this requires sustained national and international leadership, respect for expert recommendations, transparency in decision-making, and truthfulness and consistency in public messaging, as well as investment in large-scale testing and in hiring and training contact tracers. Despite these demanding requirements, countries at many different levels of per capita income were able to contain the initial phases of the COVID-19 epidemic in the spring of 2020. In many other countries, however, political leaders delayed, disregarded, or pooh-poohed these essentials, so infections quickly became too numerous for containment to be practical. Some downplayed and concealed the extent of the epidemic; sidelined, muzzled, and undermined qualified experts and technical organizations; suppressed telltale statistics, hawked unproven remedies, and amplified conspiracy theories regarding the origin of the virus. Those who followed this pattern typically took too long to shut down their economy and reopened it too soon, causing hundreds of thousands of unnecessary casualties. What is more, many countries, including many that had successfully navigated the early stages of the epidemic, opened up their economies during the summer or fall of 2020 and fell victim to a second wave of infections in the fall and winter of 2020–21.

Early Chinese and U.S. Responses to the COVID-19 Pandemic

The tensions between public health and high-level domestic and geopolitics, and between the conflicting traditions of scientific openness and bureaucratic secrecy, came into full display in the early Chinese response to the COVID-19 pandemic.[64] To summarize a rather complicated story, Chinese doctors recognized the danger posed by the new virus early on and alerted their counterparts all over the country to the possibility of human-to-human transmission. However, they were forbidden to alert the public or their professional counterparts in foreign countries without clearance from political authorities, which they knew would be only reluctantly given. The high-level decision to undertake the drastic measures needed to control the pandemic came about as a result of leaks of information to the Internet, revelations by investigative journalists, posts of scientific data on international research websites, quiet international conversations among scientists, and finally, an intervention by influential Chinese doctors to gain the attention of high-level officials. In the end, the Chinese government put in place draconian measures to contain the epidemic, but only after a delay of more than three weeks that helped the virus to begin its worldwide spread.[65]

The story is worth recounting in detail. Doctors in Wuhan, China, first identified the disease that came to be known as COVID-19 in early December 2019. Within a few weeks, they had recognized its similarity to the SARS and MERS viruses, warned public health officials all over China about the dangerous new virus, issued guidelines on the proper precautions to be taken in treating the new disease, sent informal warnings to the United States and the WHO, published the sequence of the nucleotides in the genetic material of the virus in an international scientific database, developed diagnostic tests, distributed samples of the virus to Chinese genomics companies, and joined the worldwide scientific race for a vaccine. Local authorities quickly shut down the Wuhan Seafood Wholesale Market where pangolins and other "exotic" animals were sold, because it was thought to have been an important source of contagion. So far, so good.

From the beginning, however, Chinese authorities mismanaged public communications regarding the pandemic.[66] The Chinese public and the WHO first learned of the new virus not from an official pronouncement, but from the fact that the initial confidential warning to Chinese public health officials quickly leaked to the domestic Internet and spread rapidly via chat groups and an international, nongovernmental medical news service. A few days after the first warnings, China's long-standing tradition of government secrecy reasserted itself. At almost the same time as the Wuhan market was being shuttered, Li Wenliang, a Wuhan doctor who had warned the public of the new disease, received an intimidating visit from the police, who told him to stop spreading "rumors" about a virus "hoax." National authorities ordered local laboratories to turn over or to destroy all their samples of the virus and one laboratory was shut down. Government-controlled communications downplayed the seriousness of the virus. Local health officials did not file the usual reports to Beijing, nor was the official notification sent to the WHO, as is required under the binding International Health Regulations.

Fearing to use official channels, Chinese doctors and public health officials resorted to a variety of workarounds to get their message to the attention of the WHO and the authorities in Beijing. There were quiet conversations between Chinese scientists and their overseas colleagues, and between the directors of the Chinese and U.S. Centers for Disease Control. Genetic and epidemiological data were posted on publicly accessible scientific databases. Finally, one of China's most influential doctors, the 84-year-old Zhong Nanshan, responded to the urgings of his colleagues and led a special delegation, first to Wuhan for a confrontation with local political leaders and then to Beijing to brief top officials, whom he convinced to take prompt action to lock down Wuhan and neighboring cities, a measure that had eventually to be extended to half of China's population.

In sum, while the Chinese response to COVID-19 was a major improvement over its earlier experience with SARS and eventually resulted in effective control over the epidemic, it reflected both bureaucratic secrecy and the increased concentration of political power in Beijing. Along the way, investigative journalists exposed lies and missteps in official handling of the epidemic, revelations that were followed by an explosive public reaction to the death from COVID-19 of Li Wenliang, the Wuhan doctor who had received the first warnings from the police. Facing a political crisis, the Chinese government tightened its control over communications and scientific collaborations.[67] An army of paid trolls flooded Chinese social media with favorable accounts of the government's management of the epidemic. Censors blocked all terms associated with the new virus from Chinese social media. The government refused to allow an impartial investigation of the origins and early spread of the virus, even though such an investigation could lead to useful lessons for the control of this and future pandemics and could have eliminated the suspicions that had fallen on its own scientists.[68]

The delays in the United States and many other countries in responding to the pandemic were much more prolonged and more serious in their impact than those in China. As part of its trade war with China, the United States had withdrawn its team of more than a dozen public health experts from Beijing a few months before the outbreak, disrupting a decade of close collaboration that had helped control a dangerous new strain of influenza three years before. This cut off an important source of information regarding public health developments in China. While U.S. public health officials did their best to convey to the public the seriousness of the pandemic and to emphasize the need for testing and social distancing, political leaders and their allies at home and abroad undermined both of these messages and even encouraged the idea that efforts to mitigate the disease were violations of personal liberty. In the United States, refusal to wear a mask and practice social distancing became a symbol both of proud individualism and of loyalty to President Donald Trump and to the Republican Party.

The WHO became caught up in the crossfire between the United States and China. Early in 2020, both the United States and WHO had observed diplomatic niceties in their communications with China: the United States because it was engaged in delicate trade negotiations and the WHO because it needed China's cooperation in gathering information about COVID-19. Later in the year, as hostility to China became a theme of the President's reelection campaign, the United States announced and carried through its plans to withdraw from the WHO and to rescind its required financial contribution, citing the WHO's praise of the Chinese leadership. Coming at the midst of the pandemic, U.S. withdrawal was a body blow to the efforts of the WHO to coordinate the fight against the pandemic, although collaboration with U.S. scientists continued. (President Joe Biden, Trump's successor, rejoined WHO shortly after his inauguration.)

Newly Developed Vaccines Raise Issues of Equity

Scientists, biotechnologists, private industry, and foundations, especially the Gates Foundation, mobilized to develop and manufacture multiple safe and effective vaccines at record-breaking speed.[69] Despite logistical and political hiccups, most wealthier countries succeeded in vaccinating substantial parts of their population by the spring of 2021. Even so, there are still concerns that refusal to accept vaccination may endanger efforts to create community immunity (also known as herd immunity)—a population in which a high enough proportion are immune to the virus that it no longer has enough susceptible people left to sustain the epidemic.[70] Once a population has attained community immunity, even if one or a few people contract the disease, there are too few susceptible people for the disease to spread exponentially and become epidemic. This means that the remaining 10–20% are protected even though they are not themselves immune. Any delay in achieving community immunity also gives the virus more opportunity to evolve dangerous new variants, some of which may escape existing vaccines and require periodic revaccinations like those for influenza.

The COVID-19 pandemic made it impossible to ignore the inequities in the social determinants of health and in investments in health services, in both high- and low-income countries. In higher-income countries, rates of infection and death were much higher among low-income people and historically marginalized communities, who had less access to health services and tended to have jobs and living arrangements that increased their risk of exposure. International inequities in vaccine distribution are also likely. These inequities come into even sharper focus in the next chapter, in which we turn to inequalities in the global burden of disease that long preceded the coronavirus pandemic and that will be with us long after it has receded.

Annex B
Science and Global Health

Note: The answers to questions regarding COVID-19 are provisional, as the pandemic is still unfolding and information concerning the virus and the pandemic is still developing as I write.

How do vaccines work?

Vaccines against diseases prime the immune system for the possible future appearance of a particular pathogen—i.e., a particular agent of disease. If the actual pathogen appears after vaccination, the immune system recognizes it and produces antibodies specific to the pathogen, along with other chemicals that overwhelm and destroy it before it can cause disease. In general, any substance that the immune system recognizes as foreign and that produces antibodies to fight it is called an *antigen*.

To accomplish its objective, a vaccine must be innocuous and non-infective, and it must be similar enough to the actual pathogen to be identified as such by the various cells of the immune system whose job it is to recognize pathogens that must be destroyed. At the same time, the vaccine must pass through "checkpoints," or filters, that double-check whatever the immune system has identified as a foreign intrusion. These checkpoints protect the body from an overreaction that could bring on an autoimmune disease analogous to rheumatoid arthritis, in which the immune system attacks the body's own tissues.

Historically, early vaccines consisted of killed or altered (attenuated) forms of the original pathogen, like the Sabin polio vaccine, or closely related but non-infective microbes or viruses, like the cowpox virus of Edward Jenner's eighteenth-century vaccine against smallpox. These vaccines are relatively difficult to manufacture, often must be kept cold until administered, and can in rare cases revert to an infectious form that actually causes disease. As the technology advanced, vaccine developers increasingly developed vaccines based on protein subunits of pathogens, which cannot cause disease and typically do not require refrigeration. These are typically produced by recombinant bacteria (bacteria produced by genetic engineering), and the vaccines are called recombinant protein vaccines. The hope is that these subunits will induce the same immune reaction as the whole pathogen. Subunit vaccines typically require helpers,

The Survival Nexus. Charles Weiss, Oxford University Press. © Charles Weiss 2022.
DOI: 10.1093/oso/9780190946265.003.0008

called adjuvants, both to ensure that the cells of the immune system recognize the subunits as parts of pathogens that need to be attacked, and to ensure that the checkpoint filters do not eliminate the subunits as unavoidable noise in the system that do not justify an immunological response.

What are mRNA and adenovirus vaccines and how do they work?

A new approach to vaccines, the use of messenger RNA (mRNA), came to public attention when it led to the first vaccines against the COVID-19 virus to gain regulatory approval. In brief, rather than inject the desired antigen into your body, the vaccinator injects mRNA that teaches your body to manufacture the antigen by itself in quantities sufficient to prime the immune system. In adenovirus vaccines, vaccine makers disable a common virus that normally causes sore throats and other infections and use it to transfer the genetic blueprint for the antigen directly into the nuclei of your cells, so that they can manufacture the antigen themselves.

First, a bit of background. Proteins consist of a necklace-like string of small molecules called amino acids, linked up in a specific sequence and shaped in a way that allows them to fulfill their functions. DNA in the nucleus of every cell in your body stores the blueprints for the sequence of amino acids for all of the proteins that your body can manufacture. These are your genes. Protein manufacture in the cell does not take place in the nucleus but, rather, in a separate protein factory called the ribosome. mRNA takes the information from the nucleus to the ribosome. It transcribes the information from the DNA in the nucleus, travels to the ribosome, and guides it to string the amino acids together in the proper order so as to make the desired protein, which then folds into the shape it needs to perform its function.

To make an mRNA vaccine, vaccine makers identify a virus protein subunit that can act as an antigen and can prime the immune system to identify and fight the real virus in case of a future attack. In the case of COVID-19, the antigen chosen is the spike protein on the surface of the virus that is essential for the initial infection of a cell. Biotechnological methods are used to manufacture mRNA corresponding to this protein. When this artificial mRNA is injected, it enables the body's ribosomes to manufacture the virus protein subunit. This provides an ample supply of antigen with which to prime the immune system and convey immunity. The first dose of an mRNA vaccine primes most of the immune system to produce antibodies against the spike protein. The second dose brings antibody production to its maximum level by stimulating cells of the immune system known as memory B cells, which received insufficient stimulus from the first dose.

Adenovirus vaccines also enable the recipient to manufacture the desired antigen, but by a different mechanism that uses DNA rather than RNA. Adenoviruses are DNA viruses. Vaccine makers make the virus unable to cause disease by disabling its ability to reproduce itself within an infected cell. They then insert into its DNA the genes that provide the blueprint for the desired antigen, which in a COVID-19 vaccine would be the spike protein. When the vaccine is injected, the disabled virus transfers the new genes into the DNA of the nucleus of the cells of the vaccine recipient. The cells then make mRNA from the newly transferred genes and send it to their ribosomes to manufacture the antigen, which primes the immune system and creates immunity.

Why was the influenza pandemic of 1918 so deadly?

The 1918 influenza virus was especially deadly because it was more likely than other strains of the virus to cause a deadly pneumonia, especially in young, healthy adults. Here's why.

The influenza virus cannot penetrate a person's cells to cause infection until the virus is split (cleaved) by an enzyme called a protease. This frees a part of the virus, the so-called hemagglutinin, so that it can fuse with the membrane surrounding the cell it seeks to infect. This enables the genetic material of the virus to enter the cell and infect it. The virus does not make its own protease but, instead, relies on a protease that occurs naturally in the cell to be infected. Not every protease will do. Most strains of influenza virus require the specific protease that is produced by the cells of the upper respiratory tract and hence do not infect the lungs. They can make you miserable, but they do not give you pneumonia.

The 1918 virus, by contrast, could be cleaved by a protease that is found in the normal human lung, so it could directly infect the lungs, making pneumonia a much more frequent complication of the infection. The 1918 virus sometimes stimulated the immune system of the infected person to go into overdrive and produce a so-called cytokine storm. This caused large quantities of fluid and blood cells to seep through the walls of small blood vessels, blocking the patients' airways and, in effect, drowning them. Cytokine storms happened most often with patients with strong immune systems, which was why scientists think the 1918 flu was especially deadly in healthy young people.

Why is COVID-19 so dangerous?

The virus that caused the COVID-19 pandemic, known formally as severe acute respiratory syndrome coronavirus-2 (SARS-Cov-2), has extraordinary properties, the details of which are still being elucidated as this book goes to press[1]. First of all, and unlike influenza, COVID-19 can spread from human to human even before symptoms appear. In the first few days after a person is exposed to the disease, the concentration of virus in the nose and respiratory tract rises rapidly even before symptoms arise. Despite this high virus load, symptoms do not immediately appear because the virus suppresses the production of interferons and other proteins that would have helped stop the virus from multiplying, and that also provide the signals by which the body summons the immune system to fight off infection. These signals would have brought on the coughs and sneezes that are the usual symptoms of a respiratory infection. The result is that there are plenty of viruses to be transmitted from one person to another, but no symptoms to alert either of them to their presence. An infected person can therefore transmit the virus to someone else before they show symptoms, or even if they never have any symptoms at all.

Second, the virus blocks the action of proteins called cytokines, whose job is to recognize it as foreign, as well as the antibodies that try to neutralize and destroy it. It also reduces the number of T cells of the immune system, whose job is not only to attack and destroy viruses but also to summon other cells of the immune system to join the fray. Third, COVID-19 infection attracts white blood cells called neutrophiles, which produce netlike structures called neutrophil extracellular traps. These normally serve the useful function of trapping virus particles. In serious COVID-19 infections, however, so many of these traps are produced that they can trigger abnormal clotting that blocks small blood vessels all over the body and can cause strokes, heart attacks, and long-term neurological problems and damage to the lungs, kidneys, and other organs.[2]

Finally, and most important, although in the great majority of cases the COVID-19 virus cannot overcome the body's immune system, in a small percentage of cases it causes the immune system to turn on itself and to go into a life-threatening overdrive that fills the lungs with fluid. The mechanism by which this occurs is rather complicated and illustrates some of the intricate workings of the immune system. We have already mentioned that COVID-19 suppresses interferon production at early stages of infection. When interferon production resumes, the body overreacts and sends a flood of different kinds of white blood cells (specifically, monocytes, which, in turn, differentiate into macrophages and dendritic cells).

These cells infiltrate the lungs and produce proteins called cytokines and bradykinin, which cause leaks in the endothelial cells that line blood vessels all over the body and contribute to the long-term damage that the virus may cause in many organs. These leaks flood the lungs with fluid, where it is absorbed and retained by molecular sponges of a chemical called hyaluronic acid. The result for the patient can be pneumonia, respiratory failure, and death. (The cytokine storm caused by the 1918 influenza produced a similar result by a somewhat different series of events.[3])

How do researchers search for a cure for COVID-19?

Researchers first identify the virus's vulnerabilities by determining its structure, by unraveling the details of how the virus infects a cell, and by figuring out how serious or prolonged COVID-19 infections can lead to blood clotting, cytokine storms, and other pathological effects. Researchers then look for a way to interfere with each step without creating undesirable side effects. Given the urgency of the COVID-19 pandemic, they look especially carefully for drugs that are already in use against other diseases and whose safety has therefore already been demonstrated to the satisfaction of regulators so as to avoid delays.

Under the electron microscope, the COVID-19 virus looks like a ball surrounded by spikes, something like an old-fashioned naval mine. (It looks like a crown in two-dimensional photographs, whence the name coronavirus, from the Latin word for "crown"). The genetic information of the virus is coded in the form of RNA. In brief, the COVID-19 virus enters a cell, hijacks its reproductive system, and forces it to make many copies of the virus. Then it bursts the cell and releases these new viruses so that they can infect other cells.

This is how the COVID-19 virus infects a cell. The proteins that make up the spikes that surround the virus recognize and fuse with a protein called ACE-2, which is located on the surface of the target cell and normally helps to regulate blood pressure. The virus then uses the ACE-2 protein to force its way into the cell. The ACE-2 protein is found on the surface of cells in many organs throughout the body and can serve as the entry point (receptor) for the virus, a fact that may account for some of the effects of the virus outside the respiratory tract. The virus binds to the AC-2 more strongly than other coronaviruses, a fact that may explain why it is so contagious. Mutations in the RNA can produce variant strains of the virus, some of which can be transmitted from person to person more easily than others. The stronger the binding of the spike protein to AC-2, the more contagious the variant.

Once the virus is inside it, the cell delivers the viral genetic material to its ribosome protein factories, where the genetic material becomes an integral part of the instructions from which the cell manufactures proteins.[4] The hijacked ribosomes then obey the new instructions and produce hundreds of copies of two sets of viral proteins; one of these holds the virus together and the other performs its biochemical functions. Meanwhile, the virus creates a separate enzymatic mini-factory (called an RNA polymerase) that turns out hundreds of copies of viral RNA. The newly synthesized proteins and RNA then assemble into hundreds of new viruses. The cell now splits apart (lyses) and releases the new viruses to infect other cells or other human or animal victims. Each of the steps in this process is a potential target for a treatment. Researchers look first for a treatment that has already been approved for some other application, so that it need not wait for regulatory approval. Failing that, they try to design a drug that specifically blocks some essential virus function, a process that may take several years.

The COVID-19 virus comes equipped with an efficient proofreading system that catches and corrects mistakes and minimizes mutations, a feature unusual among RNA viruses. Nevertheless, the virus has infected so many people, some of them for many months, has reproduced itself so many times, and hence has had so many opportunities to mutate, that several important new varieties of virus have appeared, some of which are more contagious or cause more serious disease than the one that first appeared in Wuhan. Some of these new varieties may be resistant to vaccines developed for older varieties, raising the possibility that COVID-19 may evolve into a continuous epidemic or a recurrent one like influenza, requiring regular revaccinations. Alternatively, the virus might exhaust its potential for "useful" mutation, or researchers may develop a universal vaccine that is effective against all varieties of virus.

Why did early modelers have difficulty predicting the course of the COVID-19 pandemic?

Epidemiologists use two broad types of models to predict the course of a pandemic. The first kind of model simulates human behavior. Model builders divide a population into groups by age, gender, and other demographic characteristics and estimate the rate at which the members of each group are likely to encounter an infected person for long enough for the infection to have a chance of being transmitted. This requires a lot of data on the behavior of the population to be studied, which is typically not available in many places at early stages of the pandemic. Model builders then estimate how likely that each encounter is to transmit the infection, and how likely the new patient is to recover, to be hospitalized, or to die from the disease. These estimates may be based on empirical data from the current epidemic—which is not reliable at early stages—or on experience with previous epidemics with the same or similar disease agents, which may or may not be applicable to the current situation. All of these assumptions are likely to change as public policies and behavior evolve.

The second kind of model does not try to simulate human behavior but, rather, uses mathematical patterns that describe how this or similar diseases have spread in the past. It assumes that the rate of incidence of sickness or death from the disease will follow a predetermined mathematical form, whose parameters for the peak, the inflection point in the rate of increase of infection, and the final total number of infections can be derived from empirical data from past experience in different places where the epidemic had already hit. In modeling COVID-19, modelers did their best to correct for differences in the point at which policies were established to encourage or require social distancing but did not take into account the fact that people in different countries would respond to these measures in different ways.[5] Moreover, with few exceptions, early epidemiological models of the pandemic did not estimate the economic costs of a pandemic or of the measures needed to address it. They therefore did not provide policymakers with estimates of the economic costs that would ensue if a "second (or third) wave" of COVID-19 infection were to result from premature relaxation of social distancing rules.

How do public health experts prepare for a potential pandemic?

The key to pandemic control is advance preparation, taking into account the fact that all epidemic diseases are different. At the national and local levels, this involves investments in surge capacity: essential supplies; medicines; equipment; infrastructure; spare beds; and capacity for public messaging, case identification, and contact tracing that is not needed day-to-day but would be required in case of a sudden increase in caseload, such as would happen in an epidemic. These investments require public and political support and must compete with other more pressing budgetary needs, leading to a cycle of crisis-oriented, feast-or-famine funding.

This preparation also involves planning the response to hypothetical future epidemics and conducting rehearsal exercises involving all potential participants, including first responders, public health officials, physicians' organizations, hospitals, transportation coordinators, and business and political leaders. These allow the various participants to get to know one another and to become familiar with each other's capabilities and limitations. They also identify flaws in the system, such as incompatible information or communications systems, shortages of essential equipment or supplies, and regulations or procedures that could get in the way of quick action. They also facilitate preparing a strategy for providing clear information to the public, and combating misinformation and possible disinformation. Preparation for a possible pandemic affecting many countries involves analogous planning and exercises at the international level, involving top levels of government, business, and finance, as well as international organizations like the World Health Organization (WHO), the International Monetary Fund, and the World Bank.

Ironically, faced with the COVID-19 pandemic, political leaders and policymakers in many countries disregarded elaborate planning exercises that had previously been conducted in their own countries and relied instead on their own intuition and judgment. Moreover, these exercises had often assumed that personal protective equipment like masks and gowns for hospital workers and other first responders would be available "just in time," not taking into account the possibility that supply chains could be affected by the very pandemic they were simulating, as, indeed, happened with COVID-19.

Why is it difficult to develop a vaccine against HIV/AIDS?

Antiretroviral therapy has changed HIV/AIDS from a fatal to a chronic disease and has greatly reduced transmission of the disease from mother to newborn. Nevertheless, despite billions of dollars in support of research, there is still no effective cure or vaccine for the disease.

Most vaccines act by teaching the immune system to recognize the coat proteins on the outside of the disease-causing pathogen, which are relatively easy for antibodies to lock onto and disable. The HIV/AIDS virus disguises itself so as to render this approach difficult or impossible. Like the malaria parasite, which we will discuss in response to a later question, it changes its coat protein faster than the immune system can recognize it as being "non-self"—i.e., not a human protein. It does this differently than the malaria parasite does. HIV is a retrovirus, which means that the genes that specify the structure of their proteins are encoded in ribonucleic acid (RNA) rather than in deoxyribonucleic acid (DNA). Most RNA viruses mutate more rapidly than DNA viruses (COVID-19 is an exception) because they make mistakes when their genetic material is copied as the virus multiplies inside the host, and lack DNA viruses' careful proofreading system that catches and corrects these mistakes. Most of the resulting mutations lead to proteins that are useless or actively harmful to the virus, but a small number of them allow the virus to evade the body's immune system.

These mutations make it possible for the HIV/AIDS virus to assume a series of disguises, each of which changes faster than the immune system can keep up. The proteins on the virus' coat are most accessible to antibodies and are hence most vulnerable to their attack, but the genes that specify these proteins are also the ones that mutate the most rapidly. This means that the immune system has only a relatively short time to recognize these coat proteins and attack them before they are replaced by a mutated version. (In contrast, the malaria parasite is *programmed* to turn the coat protein genes on and off; they don't mutate randomly.) The immune system then has to retool to recognize the new coat protein, by which time it may have mutated yet again. This sequence of events enabled the virus to defeat several early attempts at an HIV vaccine.

Researchers have therefore turned their attention to designing vaccines that would stimulate the production of "broadly neutralizing antibodies" against proteins in the interior of the virus, which are essential to its functioning and tend to mutate much more slowly than do coat proteins. Here the problem for vaccine designers is that these proteins are deep inside the structure of the virus and are therefore hard for antibodies to reach.

How was smallpox eradicated?

Smallpox was a highly contagious, disfiguring disease with a 30% fatality rate. It was the first disease to have been entirely eradicated—a major triumph by the public health profession. The last case was identified in 1977, and eradication was officially certified by the WHO three years later. Routine vaccination against smallpox had been discontinued in the United States in 1971.

Smallpox was a disease uniquely suited to eradication. The smallpox vaccine has been known at least since 1796. It is safe and confers lifelong immunity from a single dose. It was easy to train community health workers in the technique of vaccination. Smallpox infected only humans, so there is no animal reservoir to sustain the disease once there were no more infected humans. Its symptoms were very visible, easily recognizable, and appeared soon after infection, so there was little chance that sufferers could spread the disease before they had become aware of it. Survivors of smallpox were left with lifetime immunity and could not be reinfected.

The strategy behind the eradication of smallpox was based on three elements: focused surveillance, ring vaccination, and the bifurcated needle. In focused surveillance, vaccination campaigns aggressively sought out smallpox patients, often by showing neighbors pictures of what smallpox looks like and asking whether they knew anyone with a similar rash. Once identified, smallpox patients were isolated from other people, and everyone with whom they could have been in contact was vaccinated (ring vaccination). Vaccinators had to overcome obstacles posed by floods, wars, famines, and refugee camps, not to mention national and international bureaucracies. But they did the job.

This process was facilitated by the invention of the bifurcated needle. This was a thin, disposable steel rod with two prongs at one end, designed to hold exactly one dose of smallpox vaccine, to puncture the skin of the recipients, and to deliver the vaccine to the arm in just the right amount and at just the right depth. It greatly speeded and simplified the vaccination process and cut the amount of vaccine needed for each recipient by 75%.

Once smallpox was eradicated, there ensued a lively debate over whether to destroy the remaining stock of smallpox virus that was stored in freezers in the Center for Disease Control and Prevention in Atlanta and in the Institute for Viral Preparations in Moscow. The argument became moot when scientists synthesized the virus from scratch from its nucleic acid and protein components. The stocks of stored virus were retained on the chance that they might be needed in the future to facilitate the production of new supplies of vaccine.

In 2011, rinderpest, a cattle disease related to human measles, became the second disease to be officially eradicated. A task force of the Carter Center in Atlanta has identified polio, mumps, Guinea worm, pork tapeworm, rubella (German measles), and elephantiasis (lymphatic filariasis) as human diseases that are possible future candidates for eradication. None of these diseases, however, has all of the characteristics that facilitated the eradication of smallpox. For example, although polio has no animal reservoir, only a few percent of its sufferers show obvious symptoms, making it hard to identify all infected people and apply the ring vaccination strategy.

Why is malaria so tough to eradicate?

Malaria is much harder to eradicate than is smallpox.[6] It's hard to detect cases and thereby to track the spread of the disease because sufferers can be sick but display no obvious symptoms. Even laboratory methods of diagnosis are not reliable. There is no vaccine. Survivors of infection are left with only partial immunity and can be reinfected, especially if migrants from malarial regions reintroduce the disease into areas from which it has been previously cleared and in which even partial immunity has declined.

Malaria is spread by the bite of a female mosquito from any of several species, most of which do their biting at night. Individuals can protect themselves by sleeping under fine-mesh bed nets treated with chemical insecticides. Several programs distribute these bed nets at subsidized prices, but using them is uncomfortable in hot climates. Eradicating the disease with currently available technology would also require a comprehensive program of repeated spraying of houses in malarial areas with chemical insecticides, complemented by systematic surveillance of populations to identify people suffering from the disease, and giving them curative drugs.

The WHO-sponsored program of malaria eradication from 1955 to 1969 did, in fact, use this approach and succeeded in eradicating the disease in two dozen countries.[7] However, the program ran into numerous difficulties in sub-Saharan Africa and was finally abandoned without achieving its goal of total eradication. The mosquito vectors developed resistance to the insecticides, especially since they were being used at the same time in even greater quantities for agricultural purposes. On the pharmacological side, the parasite that causes malaria has developed resistance to a series of drugs: first to quinine, then to chloroquine, and most recently to artemisinin, leading to a race to develop new medicines faster than the parasite can evolve resistance. (The standard cure uses a cocktail of medicines, to all of which an infecting parasite must be immune if it is to survive.) Finally, mosquitos reproduce quickly even in very unfavorable conditions. In dry areas, larvae can develop into adults in a few days from eggs laid in tiny amounts of water, like those remaining in a worn-out tire left outside during a brief rainstorm.

The WHO eradication program also faced logistical and social issues. Spraying had to be carried out just right in ways that could vary from one house to another. Poor transport infrastructure made it difficult to transport chemicals and equipment. Spraying inside houses didn't protect people who sleep outside on nights. Repeated house visits, which were often necessary to eliminate the mosquitos, were difficult or impossible under wartime conditions. Even people who might accept the need to clear out their homes once in preparation for spraying, and who are willing to tolerate the smell and possible toxicity of the insecticide, might well object to doing this several times.

Why is it difficult to develop a vaccine against malaria?

Malaria is caused by a one-celled parasite called a *Plasmodium* and is spread by mosquitos. Over millions of years of evolution (malaria has existed in humanoid and nonhuman animals for at least that long), the parasite has developed a number of clever ways to outwit the defenses of both its human and its mosquito hosts. For this reason, people have succeeded in building only partial natural immunity to malaria even after repeated exposure to the parasite, giving the vaccine developer an inadequate foundation of natural immunity on which to build.

The complicated life cycle of the mosquito gives the vaccine developer many targets, but each presents difficulties. Starting with the time that an infected mosquito injects an asexual form of the parasite into a person's bloodstream, the parasite has perhaps twenty minutes to hide before the immune system recognizes and destroys it. To evade the immune cells in the blood, the parasite changes its "outer garments"—the coat proteins on its surface—every few minutes, faster than the immune cells can recognize the coat proteins as belonging to a foreign interloper that they should destroy. To do this, the parasite carries genes for several different coat proteins. It disguises itself by turning off the genes that correspond to the old coat protein and turning on (expressing) the genes that correspond to the new one.

The parasite then takes advantage of the fact that the liver and red blood cells are relatively safe havens where it can hide from the immune system and reproduce itself. As soon as it can, the parasite finds a hiding place in a red blood cell or in the lining of a blood vessel, where it's hard for the immune system to reach it. From there it makes its way to the liver, where it is also relatively out of reach of the immune system. There it turns itself into a sexual form and reproduces rapidly. After about a week, the progeny leave the liver and flood the bloodstream with new parasites, producing the periodic high fever and chills that are the hallmarks of acute malarial infection. The end result is that the malaria parasite spends so little time in places in the human body where the immune system can get at it that people develop only partial immunity. This hinders efforts to develop a vaccine that would enhance this natural immunity.

Why and how do germs become resistant to antibiotics?

Antibiotics target and disable biochemical processes that bacteria have and that humans don't—making a cell wall, for example—or processes that bacteria need to make people sick, like getting into your lungs or through the lining of your intestines. Different species of bacteria are susceptible to different antibiotics because each species of bacteria has its own biochemistry and each type of antibiotic targets specific biochemical processes.

Even if almost all bacteria of a certain species in your body are killed off by a particular antibiotic, there may be a relatively few that have one or more genetic mutations that make them resistant. These few bacteria have a big advantage over the susceptible ones when they are exposed to that antibiotic and may even be the only ones that can survive and reproduce under these conditions. In the technical term, they have a huge "selective advantage." After a few generations of bacteria have been exposed to antibiotic treatment—sometimes in only a few hours—all or most of the bacteria of this species that remain in your body will be resistant to that antibiotic. They will be the ones that multiply in your body from then on.

If you have a contagious infection with antibiotic-resistant bacteria and you give it to someone else, the bacteria infecting that person will be resistant to that antibiotic. If the antibiotic in question is given to many patients infected with this kind of bacteria, this process will be repeated until the antibiotic is more or less useless against that disease in your neighborhood, and eventually in your state, your region, or even worldwide. What is more, the resistant bacteria may be able to pass their genes for resistance on to bacteria in your body of a different species that causes a different disease.

Humans promote resistance to antibiotics by overusing them. First, many human patients insist that their doctors prescribe antibiotics for colds or other virus infections even though they are ineffective against such infections. Second and even more important, farmers who raise livestock typically treat their animals with large doses of antibiotics, both to protect them from infection (disease spreads fast among animals penned close together in factory farms) and to make them grow faster. This creates additional places where there is selective pressure on bacteria and hence more sources of resistant bacteria, which then infect farm workers, who spread them to everyone else.

Bacteria become resistant to an antibiotic in any of a number of ways, of which the simplest is to expel the antibiotic from the bacterial cell or to wall it off inside the cell ("sequester" it) so that it can't hurt them. Another way for a germ to develop resistance to an antibiotic is to develop a different method of accomplishing the task that that antibiotic is blocking—say by using a different set of enzymes to achieve the same biochemical function. Still another is to swamp the antibiotic by "overexpressing" (making and turning on many copies of) the gene that controls the process with which the antibiotic is interfering. This speeds up whatever the antibiotic is trying to stop and, in this way, counteracts its effect.

How does oral rehydration therapy for cholera and infant diarrhea work?

Cholera and diarrhea kill by dehydration. Oral rehydration therapy (ORT) is a simple cure that has saved millions of lives by preventing this dehydration, maintaining needed concentrations of sodium and potassium ions in body fluids, providing essential food energy, and in so doing giving the body time to cure the disease.

ORT is a solution of sugar (dextrose) together with sodium and potassium salts. The caregiver can make this solution by adding the contents of a sachet distributed by the WHO to a liter of water, which they can measure using an ordinary soda bottle. If no such sachets are available, they can use a bottle cap to measure the necessary amounts of salt and sugar. The biggest problem in introducing ORT was to convince caregivers to keep feeding their children and loved ones a water solution that was obviously coming right out again in the form of watery diarrhea, and to continue doing so until the patient recovered.

How is the global burden of disease calculated?

The simplest way to measure the global burden of disease (GBD) is to gather data on the number of people who die (*mortality*) or who become ill (*morbidity*) from a given cause in a given year. About 70% of the deaths in a given year worldwide (40 million out of 74 million total) are due to non-communicable diseases like cancer, heart disease, and malnutrition. Half the deaths in low- and middle-income countries are due to causes that are much less common in high-income countries: communicable diseases, conditions arising during pregnancy and childbirth, and nutritional deficiencies.[8]

Mortality and morbidity data have serious limitations. First of all, they may not tell whether the people affected were young when they became sick or died, or whether they had already lived long lives and were statistically likely to have died soon of something else. They also don't tell you whether they were healthy throughout their lives or suffered from some disability. They further need to be broken down by income, age, geographic region, and the level of per capita income of the country in which they live.

These more nuanced requirements led to the development and application of a battery of new epidemiological concepts and indices. Of these, the most important is the disability-adjusted life year (DALY), which measures the number of years of *healthy* life lost to a population because of a particular disease or condition, either through death or disability. (A year lived with disability is counted as a fraction of a year of life lost, depending on the severity of the disability.) Measuring DALYs allows policymakers to put all people and all sources of death or disability on an equal footing.

Three complementary indices, years of life lost (YLL), years lost to disability (YLD), and healthy life expectancy (HLE), complete the statistical description of the global burden of disease. YLL measures the total number of years that people in a given population would have lived, statistically speaking, if they had not succumbed to a given cause of death. This highlights the causes of death that cost the most years of life in any population. YLL statistics give extra weight to causes of death in infancy and childhood, since these young people would have had the most years to live if they had survived.

YLD measures the number of years that people *live* while suffering from a loss of quality of life due to disability. This measures the human and economic cost of illness and disability *short of death*. Back and neck diseases, loss of sight and hearing, and depression are the major causes of loss of years of quality of life due to disability in all parts of the world—even in China, India, and sub-Saharan Africa. There are regional anomalies, as other causes of mortality or morbidity are predominant in certain regions: iron-deficiency anemia in South Asia, HIV/AIDS in sub-Saharan Africa, and violence in North Africa and Latin America. HLE is defined as life expectancy minus years lost to disability.

How do demographers know the life expectancy of a population while it is still alive?

Life expectancy is defined as the average additional years a person of a given demographic can be expected to live. To understand how life expectancy is calculated, imagine a person, born today, whose chances of dying before age five are the same as the mortality rate of the under-five cohort today, as reported in the latest available statistics for the population to which he or she belongs. Similarly, their chances of dying between the ages of five and ten are imagined to be the same as the mortality rate of the five-to-ten-year-old cohort of the same population today. Then make the same calculation for each of the span of ages between ten and fifteen, between fifteen and twenty, and so on up to 105–110 or whatever you consider to be the maximum human life span. Combine all of these numbers, and you have the life expectancy at birth for this hypothetical person. For persons who have already attained some age, start with their present age and perform the same calculation to find their expected remaining years of life. Expected values of DALYs, and YLDs, are calculated in an analogous manner. Fertility rates are calculated by imagining a woman whose chances of giving birth between the ages of fifteen and twenty are those of the present cohort aged fifteen to twenty, and so on up to age forty-five, as fifteen to forty-five are assumed to be the reproductive years.

7

Globalization and the Burden of Disease

First, the good news. People all over the world are living longer and healthier lives. The conquest of infectious disease, coupled with the rise in living standards, has led to a dramatic increase in worldwide life expectancy from 46.8 years in 1950–1955 to 71.4 years in 2015. Even so, there are important differences within and among nations, with low-income countries like Mali and Congo experiencing almost negligible improvement over the past several decades.

Life expectancy worldwide is expected to reach 83 years by the year 2100. (We explain how life expectancy is calculated in Annex B). Mass vaccination campaigns and reduced rates of premature birth, diarrhea, and pneumonia have lowered the rate of infant and child mortality in low-income countries so much that experts project that it will decline in a few decades to the level of the highest-income countries.[1] Many fewer mothers are dying in childbirth. Fewer years of life are being lost to malaria, HIV, and diarrhea, and to lesser scourges like tetanus, measles, meningitis, and African sleeping sickness. (We do not in this chapter consider the impact of the COVID-19 pandemic.)

Overall patterns of death and disease in low- and middle-income countries are becoming more and more like those of wealthier countries—a process known as *convergence*. In this *epidemiological transition*, infectious and parasitic diseases, while far from eradicated, become less important than non-communicable diseases in most of the low- and middle-income world, whereas non-communicable diseases like heart attack, cancer, diabetes, depression, and dementia become the major sources of illness and death. The underlying causes of these diseases are linked to environment and lifestyle via pollution, smoking, a fatty and sugary diet, and reduced physical activity.

The major exceptions to these trends are South Asia and sub-Saharan Africa, where HIV and malaria still loom large, many children still die young, and the diseases of poverty still predominate. These parts of the world suffer from a *quadruple burden* of disease that combines the diseases of affluence and those of poverty: infectious disease, non-communicable disease, malnutrition, and maternal mortality. To top it off, climate change and the COVID-19 pandemic are worsening these problems.

The Survival Nexus. Charles Weiss, Oxford University Press. © Charles Weiss 2022.
DOI: 10.1093/oso/9780190946265.003.0009

Demographic Transitions and the Quadruple Burden of Disease

The epidemiological transition is the first of the three great transitions in global health. In the second or *demographic transition*, the world's population is expected to level off at about 11 billion by the year 2100, up from 7.3 billion in 2015, representing an overall slowing of the rate of population increase around the world.[2] This projection is linked to worldwide urbanization. We can understand this better from the example of a couple who leaves a rural area in a low-income country and migrates to the city. Here they are likely to want a smaller family rather than a large one. Their children are less likely to die of diarrheal or infectious disease and are more likely to survive to adulthood, so they no longer need as large a family to assure themselves of support in their old age. Not only do they no longer need a large family to help on the farm, but also in the city they will have to pay school fees—a major burden if they have many children to support. In the future, their children are also more likely to also choose to have fewer children, as will their grandchildren.[3]

In the third or *nutritional transition*, low- and middle-income countries are undergoing a transformation from not having enough to eat to eating too much or eating unhealthy (fatty, sugary, or salty) foods. Such under- and over-nutrition may appear in the same region (overfed rich, underfed poor), in the same family (overfed father, underfed children), or occasionally in the same individual (eating too much of the wrong foods and not enough of the nutritious ones)!

The early data that enabled these transitions to be identified have been confirmed and greatly augmented by an enormous effort, involving many hundreds of physicians and epidemiologists all over the world, to gather detailed statistics on the "global burden of disease" (GBD).[4] GBD data bring into sharp relief the worldwide neglect of important sources of death and disability like non-communicable disease, mental illness (including depression, which shows up in health statistics as suicide and violence), dental disease, traffic accidents, and disabling conditions like cataracts, fistula, and cleft palate that are easily cured by surgery.[5] They call attention to poverty, poor housing, dangerous traffic, and lack of access to clean water and sanitation. They spotlight the fact that the resources that the world devotes to research, development, and the application of technology on health are not focused on its dominant health problems.

GBD statistics also highlight the many external factors that influence the state of public health in a country: natural disasters, war, poverty, malnutrition, ignorance, workplace safety, access to clean water and sanitation, pollution, natural resource degradation, and transportation safety. To this long list, we must add a variety of sociological and cultural issues: family structure, gender relations (including gender-based violence and infanticide, genital mutilation, and discrimination against women and girls in healthcare and nutrition), cultural attitudes toward mental illness, the presence or absence of community support structures, and deleterious lifestyle choices like smoking, unsafe sex, substance abuse, obesity, and lack of exercise. The World Health Organization (WHO) has a somewhat similar list, which it calls the "social determinants of health."[6]

These problems may reinforce each other in a syndrome called syndemics, defined as "clusters of disadvantage." Think of a low-income, indebted, heavily stressed single mother anywhere in the world. She may have little education, suffer from diabetes, live in a rickety house in a violent area far from the nearest well-stocked grocery store, and be unable to maintain a proper diet or see her physician because she is holding down two jobs in order to support herself and her children. Her situation will be still more complicated if she suffers from diabetes-related depression or other mental illness.[7]

From all these considerations, we can identify technologies relevant to health, starting with closely related technologies like those for water supply and sanitation, pollution, industrial safety, and natural disaster preparedness and response, but expanding to technologies for transportation, education, and mass media, and eventually including science and technology related to food, agriculture, housing design, and even economic growth and job creation. These technologies would be insufficient, however, without the policies and regulations needed to ensure that they are put into practice in a way that benefits the entire population. This is the Survival Nexus in spades!

Health, Globalization, and Trade

In 1966, the government of Thailand sought to control an epidemic of lung cancer and other smoking-related diseases by banning the import of foreign-made cigarettes. The potential spread of this ban directly threatened the global strategy of major tobacco companies, which had targeted the expanding markets in low- and middle-income countries when their markets in high-income countries threatened to shrink because of aggressive (and to my mind, laudable) anti-tobacco campaigns. Acting on their behalf, the United States brought suit to overturn this ban through the dispute settlement procedures of the old General Agreement on Trade and Tariffs (GATT, later folded into the World Trade Organization), arguing that tobacco was a product within the meaning of the treaty and that to restrict its import without banning its domestic equivalents constituted a forbidden restraint on international trade.

The Thai tobacco case put into sharp relief the influence of globalization on non-communicable "lifestyle" diseases like heart disease, cancer, diabetes, and chronic lung disease, which are largely caused by behaviors like smoking and unhealthy diet. While these behaviors are usually voluntary, they are heavily influenced by commercial advertising, much of it from abroad—so much so that diseases related to tobacco and unhealthy foods have been called "communicat*ed*" diseases. Such diseases claim forty million lives each year.[8]

According to the international rules of free trade, any regulations that a government applies to imported products must be applied equally to their domestic equivalents. Based on this doctrine, the U.S. brief to the GATT tribunal pointed out that Thailand placed no restriction on domestic cigarettes sold by the government-owned tobacco company and argued that the prohibition of cigarette imports wasn't really a public health measure but was, rather, a device to protect the domestic cigarette industry. The Thai government replied that tobacco was harmful and that if the Thai market for cigarettes were to be opened to foreign competition, ads from American cigarette makers would overwhelm anything the government might do to discourage smoking and would addict many to nicotine. The United States won the case. American cigarettes remain free to enter the Thai market and thereby to increase the incidence there of the many diseases related to smoking.[9]

The controversy over tobacco imports to low- and middle-income countries eventually led to the Framework Convention on Tobacco Control, the first treaty ever sponsored by the WHO.[10] This treaty commits the 181 countries that have become parties to educating the public regarding the dangers of smoking, restricting advertising and promotion of tobacco products and regulating their packaging, setting indoor air controls, controlling tobacco smuggling, and instituting other measures intended to limit demand for tobacco products. Its practical effect has been to increase the bargaining power of low- and middle-income countries in their efforts to control multinational tobacco exporters.

The controversy over tobacco exports is a classic example of how international trade agreements enable higher-income countries to impose their economic and political power on lower-income ones to the detriment of the health of their people. Like tobacco companies, multinational producers of sugar-sweetened beverages and processed foods with excess fat, salt, and sugar have taken advantage of free trade rules to expand their sales in the emerging markets of Latin America, South and Southeast Asia, and sub-Saharan Africa, thereby contributing significantly to the increase in obesity, diabetes, and heart disease in these regions.

There are two sides to this story. Tobacco and processed foods are commercial products for which there is high demand everywhere, regardless of their unhealthy characteristics. Exporters of cigarettes and unhealthy foods are in business to sell legal products that people want to buy. The same is true for exporters of alcoholic beverages and small arms, both of which also are items of large-scale, legal international trade but have major implications for public health.

Domestic tobacco, food, and beverage companies in lower-income countries would meet these demands even in the absence of foreign imports. It would be difficult for governments to limit the sale of these products in the name of public health because it would be hard to overcome the political pressure from consumers who like good food, drink, and tobacco, and from businesses that profit from selling them. But multinational food and beverage companies are more effective marketers and distributors and have greatly accelerated this unhealthy part of the nutritional transition. All of these commodities illustrate the complex connections between non-communicable diseases and the Survival Nexus of science, technology, economics, commerce, law, communications, psychology, and trade.

At the same time that higher-income countries are in effect *ex*porting obesity, heart disease, alcoholism, violence, and lung cancer to low- and middle-income countries, they are *im*porting the badly needed doctors, nurses, and other health workers whom the lower-income countries have invested scarce resources to train. This movement of capable and trained health workers to wealthier nations is known as "brain drain." Again, this is not a nefarious scheme. Like other migrants, most low- and middle-income-country doctors and nurses would prefer not to leave their home countries, despite the higher salaries and undoubted attractions of life in more affluent countries. They typically emigrate to the United States and Europe because their opportunities to practice their chosen professions and their career prospects in their home countries are discouragingly poor, not only due to the countries' lack of resources and perhaps instability, but also because of the frequent underfunding, corruption, and mismanagement in their domestic health systems.[11] Many of them continue to collaborate with colleagues and to contribute to the economy in their home countries after they have emigrated.

A notorious example of the conflict between the rules of free trade and global health came when the U.S. and European pharmaceutical industries attempted to take advantage of WTO rules in order to maintain the profit margins on their exports of antiretroviral therapy for HIV/AIDS to lower-income countries. Led by Nelson Mandela, the South African government vehemently protested U.S. policy, calling HIV/AIDS a public health emergency and arguing that the country desperately needed to import low-cost generic (off-brand) medicines at a time when the disease infected a quarter of its population.[12] The South African government eventually won a prolonged legal battle, but only after the industry was shamed by worldwide public pressure into withdrawing court suits in several African countries.[13]

Bioprospecting and Clinical Testing in Low-income Countries

More complicated issues are raised by bioprospecting—studying the pharmaceutical properties of traditional medicines and other local plants and animals—and by drug tests of new or experimental medicines on people in low-income countries. Both are legitimate activities of multinational pharmaceutical corporations, but both should be carried out with the consent and collaboration of local people and in a way that shares benefits and builds local capacity. This has not always been the case.

In the case of bioprospecting, a bit of background may be useful. Many traditional medicines have become part of the standard toolkit of Western medicine. Aspirin, quinine, reserpine, digitalis, curare, and the malaria medicine artemisinin are prominent examples.[14] Multinational drug companies therefore explore the possibility of deriving valuable new medicines from the knowledge of traditional healers, and if they find a promising candidate, they may invest up to hundreds of millions of dollars in identifying the chemical or chemicals responsible for its efficacy—the "active principle"—and in improving them, for example, by increasing their potency (i.e., reducing the effective dose) and hence their toxic side effects.[15]

There is, however, an unfortunate history of drug researchers smuggling samples of traditional herbs out of low-income countries, or of hiring local biologists or healers for a pittance to gather samples, and then themselves identifying, patenting, and commercializing their active ingredient back in their home country. Researchers have been known to do this without so much as informing the local people of the outcome, let alone rewarding traditional healers for their knowledge, involving local scientists in the research, or sharing the profits with the country in which the biodiversity resource was located. The archetype of such "biopiracy" was the 1876 export of a shipload of rubber seeds from Brazil to Kew Gardens, the British botanical garden, so that rubber plantations could be established in British colonies all over the world.[16] This ruined the Brazilian wild rubber industry and inspired enduring resentment, as did the patenting of the active principle of neem oil, a traditional Indian remedy with a wide range of uses (although I understand that that patent has never actually been invoked).

The controversy over bioprospecting and biopiracy entered the diplomatic arena during the negotiations in the early 1990s over the Convention on Biodiversity (CBD). Low- and middle-income-country negotiators seized the opportunity to define traditional medicines and other biodiversity resources as natural resources subject to national sovereignty in the same sense as petroleum or rubies.[17] Here the challenge is to ensure that countries of origin, and especially their indigenous peoples, share in any commercial benefits from their biodiversity resources and indigenous knowledge, while still preserving the freedom of scientists to carry out essential conservation-oriented research on biodiversity.[18] From the scientific point of view, the solution would be for low-income countries to develop their own expertise to commercialize biodiversity-based products.[19] India and China, for example, have substantial programs of scientific research on their systems of traditional medicine.[20]

The conduct of trials of experimental medicines in low-income countries raises somewhat similar issues. Many clinical researchers in Africa welcome these opportunities as a source of research funding and as a supplement to their often meager salaries. On the other hand, these projects divert time and attention that could have been spent on research of direct local benefit—the more so since the medicines being tested, if successful, may end up being more expensive than patients in a low-income country can afford. Indeed, the Indonesian government once refused to send more samples of a local strain of influenza virus to international researchers on the grounds that they would probably not be able to afford any vaccine or cure that might result from their research.[21] (They eventually relented.)

The most notorious case in which the needs of a powerful country have taken precedence over the health needs of a less powerful one is the widely reported false polio vaccination campaign said to have been undertaken by the CIA in Abbottabad, Pakistan, in order to confirm the identity of the family of the Islamic terrorist Osama bin Laden in preparation for his assassination. This ruse undermined trust in the integrity of the genuine efforts of WHO, UNICEF, and Rotary International to wipe out polio in Pakistan, one of the last countries in which it is still endemic.[22]

The power of wealthy countries over global health can take subtler and less tangible forms, especially in mental health, a field that suffers from extreme underinvestment and shortage of trained personnel everywhere, especially in the lowest-income countries. The definition of normal and aberrant behaviors varies based on cultural and socioeconomic contexts, as does how people with mental illness are expected to behave. Moreover, different cultures have different ideas concerning the origin of mental illness, and these ideas influence the stigma attached to mental illness and the type of care that is traditionally provided to patients. These factors can lead to misdiagnoses and wrong treatments when Western-trained physicians treat mental illness in cultures different from their own.[23]

Health and Military Security

Infectious disease is a major threat to social and political stability and is thus also a threat to security. After all, if we were to face a military or geopolitical threat that shut down our entire economy and that threatened the lives of many millions of people, as the COVID-19 pandemic does, we would surely consider it an issue of national security. Infectious disease is also a direct threat to the military. Until World War II, more wartime soldiers died from disease than in combat. In more recent history, the ability of then-Soviet Russia to raise an army in its later years was seriously limited by the poor health of its potential draftees, many of whom were debilitated by drug use, alcoholism, and general neglect.[24]

Infectious diseases have affected the military for millennia.[25] The Bible recalls the sudden departure of apparently invincible Assyrian invaders, possibly due to an attack of disease.[26] Nearly a millennium later, an epidemic of bubonic plague thwarted the ambitious plans of the sixth century Byzantine Emperor Justinian to restore the glory of the Roman Empire.[27] In the seventeenth century, the European settlement of North America was aided immeasurably by the fact that smallpox, measles, and the other diseases that had spread from Spanish colonies farther south had reduced the population of Native Americans by as much as 90%.[28] In more recent examples, inhabitants of Caribbean islands and the northern coast of South America were able to fight off nineteenth century invasions by large French and Spanish armies, which lacked immunity to yellow fever and other vector-borne diseases.[29] During World War II, malaria devastated the armies of both the Allies and the Japanese in the South Pacific theater, causing at least as many casualties as did actual combat.[30]

Warriors have long done their best to help these natural phenomena along by acts of what today would be called biological warfare. The fourteenth-century Mongol hordes were reputed to catapult the corpses of plague victims into cities they were besieging.[31] The British general Jeffery Amherst notoriously proposed to provide his Native American enemies with blankets that had been used by victims of smallpox.[32] The Soviet Union launched similar efforts on an industrial scale. A twenty-year program, costing billions of rubles and involving over sixty thousand scientists and engineers, developed antibiotic-resistant strains of deadly diseases like anthrax and tularemia (rabbit fever) and "weaponized" them by putting them in a form that would infect an enemy and developing means to deliver them.[33] North Korea also has the intention and the capability to produce biological agents, although it is not clear that it has actually done so.[34] According to South Korean and U.S. intelligence, North Korea began a program of research on biological warfare parallel to that of the Soviets in 1961, received "strains of the causative agents of anthrax, plague, and cholera" from a Japanese source in 1968, and obtained samples of the smallpox virus from the Soviet Union in the 1980s.[35] The United States continues to pursue extensive biodefense research, conducted principally at the United States Army Medical Research Institute of Infections Disease (USAMRIID).

From the strictly military point of view, germs make undesirable weapons. They are tricky to produce and keep alive in dispersible form, and difficult to transport and to deliver to one's enemy. Worse, if bioweaponeers succeed in causing an epidemic, there is no way to assure that it does not "blow back" on their own troops or their civilian population in the absence of an easily detectible program of mass immunization. In part for this reason, biological weapons were among the first subjects of an arms control agreement. As early as 1925, an international agreement prohibited its parties from using "bacteriological methods of warfare"—although the United States did not become a party until 1975.[36] A stronger agreement, the Biological Weapons Convention, which also entered into force in 1975, forbade its parties to "develop, produce, stockpile, . . . acquire or retain . . . biological agents or toxins . . . that have no justification for . . . peaceful purposes; [or] . . . weapons, equipment or means of delivery designed to use [them] for hostile purposes or in armed conflict,"[37] a prohibition to which the Soviet Union agreed but which it flouted on a grand scale. In less flagrant situations, however, the line between peaceful and hostile purposes, or that between offensive and defensive purposes, can be hard to define, so biological warfare will forever be a possibility for which the world must prepare.

On a happier note, military laboratories have been important pioneers in research on tropical diseases like malaria and yellow fever, which, as we have seen, are sources of numerous casualties in military operations. Military personnel from many countries have rendered assistance to victims of natural disasters like the Haiti earthquake and the Indonesian tsunami, and in epidemics like that of Ebola in West Africa. Health can also be an important tool of "soft power," defined the ability of a country to persuade others to do as it wishes without military or economic coercion,[38] as when the Cuban government of Fidel Castro made friends around the world (and earned foreign exchange as well) by exporting some of its many doctors and other health professionals to friendly countries.

Who Makes the Decisions Regarding Global Health?

The transformation of a single-minded, Seattle-based entrepreneur into a passionate philanthropist is one of the most consequential events in the recent history of global health. In 1993, Bill Gates, then the richest man in history, was awakened to the severity of poverty in low-income countries while on safari, where he witnessed African women carrying water long distances in jugs on their heads.[39] This realization was reinforced by advice from his father: "Son, it's time for you to begin thinking about what you can give back."[40] In a 2013 interview, Gates recalled how reading an overview by WHO and the World Bank had made him realize the potential contributions of technological innovation, led by private business, to global health.[41]

Until that time, Gates had been entirely focused on the running of his company, Microsoft, owner of the immensely profitable Windows operating system that powered desktop computers, then the dominant information technology. Like most high-tech companies of the time, Microsoft paid little attention to broader policy issues. But in the latter years of the Clinton Administration, the company faced a strong probability of being broken up for the alleged abuse of its monopoly power. Microsoft fought off the antitrust challenge, but the experience forced Gates to adopt a broader perspective on his business activities and ultimately led to his epiphany concerning the importance of global health and his potential contribution to it.

The contributions of the Gates Foundation to the Global Alliance for Vaccines and Immunization (GAVI), to the UNICEF-sponsored GOBI-FFF program described in chapter 6, and to WHO, are comparable to those of the U.S. government. Its support to these programs illustrates at the same time the immensely valuable contributions of private philanthropy to global health, and the difficult problems of governance created when key decisions regarding what diseases should be addressed, what approaches and technologies should be adopted, and what research should be supported, are taken far from the countries that suffer from the problems.

The Gates Foundation benefits both from its relatively immense resources and from a flexibility that is denied to governments and to intergovernmental organizations like the WHO. These advantages enabled the Foundation to respond quickly to the need for major funding to support critical research on COVID-19. They also enable it to wield outsized power and influence over critical policy and strategic decisions concerning the direction of global health, eclipsing the WHO and most of the world's governments. It makes these decisions with accountability only to its own donors and directors, and not to the people it seeks to benefit.[42] Indeed, the very size and importance of its resources discourage criticism from experts who are dependent on its financing.[43]

The Gates Foundation, for example, has succeeded in its efforts to establish as an international objective the complete eradication of malaria, rather than control—something that many global health leaders perceive to be overly optimistic and expensive.[44] This issue requires some background. There is controversy among experts as to whether the world should aim to eradicate malaria completely or to control it indefinitely at some manageable level. Eradication would have the virtue of being a permanent solution, but it is not at all clear that it is feasible, since previous attempts have been expensive and have achieved only limited success (see Annex B). On the other hand, a strategy based on control would also be expensive and difficult and would have to be sustained indefinitely.

The point here is not which side is correct but, rather, that the Gates Foundation exercised great influence over this major strategic choice. The

President's Emergency Plan for AIDS Relief (PEPFAR) program that we described in the previous chapter exhibits a similar pattern, in that it is accountable only the U.S. government and specifically to the U.S. Congress, and not to the people it is intended to benefit. We have also seen in the previous chapter how the Bellagio meeting and the Structural Adjustment Policies of the World Bank and International Monetary Fund undercut efforts by newly independent ex-colonies to provide equity-based programs of primary healthcare. Those who pay the bills set the priorities. But they also provide energy and flexibility to an otherwise cumbersome legacy system.

The influence of major donors extends to the use of the term "global health security." As originally defined by WHO, this concept referred to infectious diseases that threaten to spread beyond national boundaries.[45] As we have observed, these can be threats to political and social stability, and hence to national security as it is usually defined. In principle, addressing these problems should go hand in hand with efforts to improve national health services in low-income countries, since this would at the same time improve national surveillance capabilities and hence their ability to participate in WHO and CDC programs of global health security. The problem stems from the fact that when resources are tight, as they almost always are, the needs of health security—i.e., the concerns of higher-income and more powerful countries—are likely to take priority over those of domestic health services. This leads to some resistance to the term "health security" on the part of low-income countries.[46]

Constructive Tension: Health Security and "Health for All"

Global health presents a mixed picture. The good news is that people all over the world are living longer and healthier lives than ever before in human history. Healthcare in low-income countries is improving. Researchers are solving the mysteries of long-neglected vector-borne diseases, and a trickle of promising vaccines and cures are in the testing pipeline. But there is lots of bad news, too. Millions of people die each year from diseases that could be cured if they had access to proper healthcare.

The public health profession has understandably focused on the control of infectious diseases, which have killed tens of millions in the past and could do so again. The world's capability to respond to potential pandemics has substantially improved over the past decade, but there is also clearly still much room for improvement. The WHO, the titular leader of global epidemic response, is the only available organization with both the authority and the technical capacity to carry out this critical global function. It needs to be redeemed from its present downward spiral of decreased effectiveness, frozen funding, and increased politicization.

More generally, as we have seen, there is a glaring disconnect between the causes of death and disability revealed by the GBD statistics, on the one hand, and the level of resources that are being brought to bear on these problems worldwide, on the other. Research on poor people's diseases like malaria and schistosomiasis, which between them affect more than a billion people worldwide, is grossly underfunded relative to their importance. The same is true of research on mental illness, traffic injuries, and natural disasters, all of which affect rich and poor alike.

As people grow older all over the world, they contract more noncommunicable, lifestyle diseases like cancer, heart disease, and diabetes and acquire more disabilities like mental illness, blindness, and back problems. For the poor, these add the burdens of chronic disease to older problems like infectious and parasitic disease and maternal and child health. These diseases require a health strategy based on universal access to health services and focused on prevention, follow-up care, information, education, and behavioral change. Public health authorities also need to deal with the long list of social determinants of health—housing, access to clean water and sanitation, and on and on—and with the health effects of climate change and environmental pollution. Finally, they need to face up to the health effects of international commerce in such items as liquor, tobacco, unhealthy foods, and small arms, even when these conflict with strong commercial and political interests.

From the broader perspective of science, technology, and world affairs, global health shows striking parallels with both international agricultural research and nuclear proliferation. In chapter 10, we shall see how "vertical" research on specific crops has taken priority over the building of a "horizontal" capacity for research and extension in individual countries, even though the latter is key to improving farm productivity and farmers' incomes, and how research on fruits and vegetables—key elements of proper nutrition—has lagged because of political tensions between Taiwan and China.

This is a formidable list of issues and involves an equally formidable list of players in business, scientific and technological research laboratories, national governments, and international organizations and private foundations in different countries, each with its own role and each with its own strengths, limitations, and motivations. As a practical matter, the preferences of those who pay the bills will continue to reflect their political and commercial interests and their administrative and managerial convenience, in addition to their humanitarian interest in public health. Fortunately, there is a good deal of overlap between the universal need for "health security" in the sense of preparation for global pandemics, and the needs of low-income countries for "health for all" via improved access to health services. The task is to turn the difference between these two sets of motivation into a constructive tension.

8

The Internet and Social Media

Euphoria and Repression

The educated but underemployed fruit vendor who set himself on fire in Tunis in 2011 lit a match, both literally and figuratively, that set off political firestorms not only in his native Tunisia, but also in Egypt, Syria, and Libya. In all three countries, cell phones and the Internet facilitated demonstrations protesting the economic stagnation and political repression of decades-long authoritarian and kleptocratic rule. During the Arab Spring, long-standing dictators in all three countries were ousted—in Egypt and Tunisia by popular demonstrations and protests, in Libya by civil war.

Urban protesters in Egypt shared news and launched organized demonstrations by way of Facebook and Twitter, which were then relatively new to the country. Here they were assisted by the *ancien regime*'s ham-handed attempts to control the new media—at first ignoring it and then infuriating the business establishment by belatedly pulling the plug.[1] Ten years earlier in the Philippines, "smart mobs" assembled by cell phone had outwitted the police and sparked demonstrations that forced out President Joseph Estrada.[2] These events seemed to confirm the predictions of prominent techno-optimists that the new Internet and social media technology would destroy "the power of local governments to assert control over online behavior."[3]

Almost all of these early successes were short-lived, however. The Arab Spring produced long-lasting democracy only in Tunisia. Determined governments found ways to suppress what they regarded as subversive or undesirable communication among their citizens, at least to the extent needed to maintain political control. In Egypt, the democratically elected government that succeeded the dictator Hosni Mubarak lasted only twelve months, and the military dictatorship that overthrew it established a pervasive surveillance network, blocking websites and hacking cellphones of suspected activists and other dissidents in order to track their movements and communications, much of it using European and American technology.[4]

Other countries learned from the Egyptian example. The Burmese military used cell phones and social media in a campaign of ethnic cleansing and genocide against the Rohingya, a Muslim minority. In Syria and Bahrain, Internet-fueled protests were firmly and brutally suppressed. Even when its use was

The Survival Nexus. Charles Weiss, Oxford University Press. © Charles Weiss 2022.
DOI: 10.1093/oso/9780190946265.003.0010

not controlled, the Internet turned out to be less of a force for democracy than techno-optimists had hoped. For example, once people in Eastern Europe and the Soviet Union were able to gain access to the Internet without restrictions, they preferred discussing music, sports, and entertainment over news and politics.[5]

Convenience and Repression

The counterattack against liberal ideas of free expression and access to information through the Internet takes its fullest form in China, where people enjoy the best and suffer the worst effects of the Information Revolution. This Jekyll-and-Hyde approach to technology, writ large, is at the heart of the high-stakes economic and geopolitical rivalry now underway between Chinese techno-authoritarianism and the liberal post–World War II world order.

Some Chinese people experience astounding opportunities afforded by information technology. They can look into the facial recognition camera on the counter of their favorite Beijing fast-food restaurant and have the price of their meal deducted automatically from their account on Alipay, the Chinese online payment platform. They can build their car from choices displayed on their cell phone and have it delivered to them for a fixed price from a five-story vending machine.

In sharp contrast, the Uyghurs, a Muslim minority in Western China, experience the worst of the Chinese Internet. Their every communication is monitored by an army of official eavesdroppers, and their every move is tracked by a pervasive surveillance system of checkpoints and closed-circuit television cameras, all backed by the latest artificial intelligence technology for facial recognition. Their DNA and iris scans are collected at mandatory fake physical examinations and verified at police checkpoints, spyware is installed on their mobile phones, and all Uyghur-owned cars must install the Chinese equivalent of GPS so that their movements can be tracked.[6] In effect, the Uygers live in cyberspace, a model that, as we shall see in chapter 9, may be extended to the rest of China and exported to authoritarian countries around the world.

The revolution in information and communication technology has been driven in part by its own momentum—a process known as technological determinism. These innovations have reshaped international commerce, finance, and diplomacy; given rise to new dimensions of warfare; facilitated globalization of commerce and finance; provided new tools for both democratic activists and authoritarian governments (not to mention terrorists and conspiracy theorists); and created an ever-changing menu of strategy, policy, and governance issues for business, governments, and international organizations—all in just half a lifetime![7]

But the direction of many of these innovations has been determined not so much by any technological imperative as by human preferences and choices, as expressed either through commercial markets or government policies. For example, the scientists and computer visionaries that created the Internet did not foresee the rise of social media, online banking, online gambling, or online ride hailing, and the balance between privacy and security struck by each country and company has been based more on social, political, and business choices than on technical limits to encryption.[8]

Throughout its early development, the Internet has been an American project, and its underlying technology has been American. The United States has therefore been able to ensure that the governance and technical structure of the Net embodies liberal values of freedom of expression and bottom-up innovation, although this has come at a cost to privacy and security. Control of the Net has also given U.S. intelligence access to global communications, most of which pass through the United States at some stage. Only recently have countries whose governments do not share these values been able to mount effective challenges to this control. These challenges will grow stronger with time.

The 1970s and 1980s: Early Euphoria Plants the Seeds of Later Issues

Many of the issues raised by the revolution in Information and Communications technology (ICT) have their roots in decisions made early in its history. Some of these were conscious choices, made in awareness of their possible long-term consequences. Others were the products of the libertarian spirit shared by many of the Internet pioneers, later reinforced by the determination of U.S. and other Western governments to keep the Internet free of censorship and open to innovation.[9]

The early Internet was designed to meet three disparate requirements: a national telecommunications network that could survive a strategic nuclear attack; convenient links between individual computer networks (called intranets), many of which used different hardware and operating systems; and a communications system that enabled scientific researchers to exchange their latest findings quickly and easily.

This last requirement gave rise to a new concept: an information space called the World Wide Web, in which anyone could be an author or an innovator, and onto which anyone could post a document, a video, or an application, without getting the approval of an editor, a publisher, or any other intermediary. Invented as a vehicle for exchanging preliminary results of scientific research by Tim Berners-Lee of the European Center for Nuclear Research (CERN), an international laboratory in Geneva, Switzerland, for basic research in physics, the Web quickly merged with the Internet so that the two terms have become virtually interchangeable in common usage. (Technically, the Web is an application atop the Net.)

The full realization of Berners-Lee's concept required four further inventions: hypertext, which enables a user to pass seamlessly from one document to another without ascending one hierarchy and descending another; the universal document locater (URL), which assigns a unique name to every document on the Web; the browser, which finds documents and other information and displays them in a form readable by humans; and the search engine, which enables a user to search for information within the Internet using ordinary language.

Underlying the entire system is the technology of packet switching, by which a message to be transmitted is broken up into small pieces, each of which is sent separately by way of the most convenient available path through the network and assembled in the appropriate order at its destination. In this way, a message can get through to its destination even if the network is severely degraded, such as might happen after a strategic nuclear attack. As an added bonus, packet switching eliminates the need to maintain a dedicated open circuit for each connection between individual computers or computer networks, greatly improving network efficiency.

The earliest direct ancestor of the Internet was the ARPANET, an experimental system pioneered by John Licklider of the Advanced Projects Research Agency (ARPA), a branch of the U.S. Defense Department charged with facilitating the development of promising advanced technologies and promoting their application within the defense establishment.[10] A colleague of Licklider's, Robert Taylor, was frustrated by the need to maintain a separate computer and monitor screen for each of the two dozen or so networks with whom he maintained links.[11] In response, Licklider commissioned a team of graduate students, chiefly at MIT, Stanford, and the University of California, to develop software that would link these individual networks seamlessly in a single combined network. The students' freewheeling attitudes, together with the more general libertarian and anti-regulation attitudes of the Silicon Valley, came to define the culture of the nascent Internet. This laid the groundwork for later conflicts with the more traditional outlooks of the defense and intelligence communities, international organizations, and many foreign governments.

The informality of the early Internet extended to its management and governance structure. Jon Postel, a universally admired and trusted computer scientist, oversaw the development and definition of the technical specifications of hardware and software. He was also responsible for maintaining the critical heart of the Net, the directory that assigned a unique number, called an Internet Protocol

(IP) address, to each computer on the ARPANET.[12] Working with the Internet Engineering Task Force (IETF), an informal group of volunteers, Postel also established the administrative system of top-level domain names like .com, .org, and .edu and was placed in charge of managing it.

The users of the ARPANET were researchers and educators at elite institutions who trusted each other and could physically exclude potential intruders from using their computers. A later version of the Net, the NSFNET, was run by the National Science Foundation (NSF), the major U.S. government funder of basic scientific research. It linked universities and researchers (and later some civilian firms) who also knew and trusted each other. Computers and computer memory were expensive at the time, and even the military at first saved money by storing classified and unclassified material on the same network without protecting it from being hacked! This meant that seemingly unnecessary features like network security could be postponed for later development. (The military later established a more secure system, the MILNET, for classified information.)

In the beginning, there was no assurance that the newfangled ARPANET would work, and therefore every reason to keep its design simple until its technical feasibility had been demonstrated. (The Wright Brothers, it was said, did not test their first airplane in a hurricane, nor did they have comfortable seats and steward service!) The warnings that free access to the Net would make it vulnerable went back as far as 1967 but were largely ignored. The chief scientist of ARPA, for example, argued that the information technology that underlay the new system was so far ahead of that of any other possible competitor that the United States would not have to be concerned with destructive hackers, commercial and military spies, or asymmetric warfare for decades to come.[13] Decades have passed since then without a fundamental reworking of the Net infrastructure, and here we are.

Nor was the information carried by the nascent Internet protected from malicious interference. On the contrary, to assure the free flow of information, the network was kept "dumb," in the sense that it was not designed to filter or censor the information that flowed through it. Network security was defined as keeping *outsiders* off the computers that were connected to the network. Those *inside* the network were free to post and to download whatever they wanted. There were no secrets, and no way to keep insiders from posting misinformation, *disinformation* (intentionally misleading information, including postings intended to influence democratic elections), unwanted information (like pornography or Nazi propaganda), or links to malicious software (malware). These five features of the early Internet—informal organization, free flow of information, lack of foreign-based technological competition, simplicity, and lack of security—were to have important consequences for its subsequent development.[14]

The 1990s: "The Babe That Roared"

By the early 1990s, the NSF realized that the NSFNET had become too big for it to manage. After all, the NSF was a basic research funding agency not well suited to managing large infrastructure projects. The NSF therefore turned the management of NSFNET over to private companies called Internet service providers (ISPs), opening the Internet for the first time to private businesses. At the same time, the U.S. Congress encouraged the development of national Internet infrastructure for the "information superhighway" through the High Performance Computing Act of 1991. This legislation was promoted by then-Senator Al Gore. (And no, he never said that he had "invented" the Internet, only that he had helped make it possible.[15])

Neither the U.S. Congress nor the NSF specified whether these ISPs were to be regulated like broadcast radio or television stations, which are responsible for the information they convey and can pick and choose what they broadcast, or were to be considered common carriers like telephone companies, which cannot favor one customer over another and bear no responsibility or liability for what their customers say over their network or how they use it. This ambiguous status ultimately fed the continuing controversy over whether social media platforms like Facebook should be responsible for the content that they convey, and in particular, whether they should filter political material that comes from foreign sources or that contains obvious misinformation. This issue was further complicated by Facebook algorithms that encouraged the formation of online communities, including those devoted to sharing disinformation and conspiracy theories.

The privatization of the Internet resulted in an anarchic explosion of new applications and new users—the number of users in the United States more than quadrupled in the five years after privatization[16]—and the disruption of a seemingly endless and extraordinarily diverse group of industries: banks, retailers, publishers, music producers, newspapers, mapmakers, auction houses, taxicab companies, real estate brokers, and universities, to name but a few examples. All this in a technological revolution that was still very much in its infancy!

The early period of the new, public Internet was a time of great optimism and exhilaration. Enthusiasts predicted that the new technology would free people all over the world to speak their minds and learn what others were saying and thinking, defying or evading government attempts at censorship and control.[17] Markets would be free and impossible to regulate, both domestically and internationally. Young people would form global communities independent of geography and national borders. All these expectations were thought to be inherent in the technology; there were few predictions of effective backlash by people or countries that did not share the values of free expression and innovation.[18]

To be sure, there were caveats. Critics warned that the new technology could be hacked by malicious intruders or could be misdirected in the commercial interest of the companies that controlled it. But these concerns were for the most part waved aside in the excitement over the power of the new technology to open new paths of innovation and to disrupt long-stagnant industries. As a result, key decisions during that period both shaped the immense capabilities of information technology and created deficiencies in cybersecurity and privacy that spawned the issues that we face today.

The dot-com boom of the 1990s, culminating in the dotcom bust in 2000–2001, left an indelible imprint on the Net. As the boom gathered momentum, supposedly hardheaded venture capitalists rushed to finance any business plan, no matter how shoddy, as long as it was labeled "Internet," resulting in the classic "irrational exuberance" of a technology-inspired bubble. When the inevitable bust occurred, this wildly excessive investment left behind a huge overcapacity in fiber-optic cable and other infrastructure, financed by the losses of now-bankrupt investors. This made it possible for their successors to offer customers very cheap connectivity and opened the way to a huge variety of Internet applications. It also left a legacy of applications with less attention to security.

Four new U.S. technological behemoths, the so-called GAFA, emerged from this frenzied expansion to join the original digital behemoth, Microsoft. These new giants were *G*oogle, near-monopolist of the search engine; *A*pple, pioneer of the laptop computer and its successor, the smartphone; *F*acebook, dominant leader of the new social media; and *A*mazon, master of online retail. Each in its own way benefited from network economies: the fact that its existing customers benefit when others buy its product. (For example, my phone becomes more valuable when you buy a phone and I have one more person to talk to.) Each solved the problem that had bedeviled most of the early pioneering start-up firms: how to make money from an information product that costs a great deal to develop and practically nothing to produce once it had been developed, perfected, disseminated, and become easy for competitors to copy. Google, Facebook, and many other Internet-based companies did this through an implicit bargain with their customers. They provide a nearly free service, and they profit from selling the information their customers provide—or access to customers on the basis of the data they provide—in exchange for the convenience they receive.

In retrospect, the unruly characteristics of the early days of the ICT revolution were typical of the early stages of other technological revolutions. For example, there are remarkable parallels with the early days of the exploration of the New World and with the early days of radio and television.[19] Each followed a somewhat predictable path. First came a time of innovation, "the stage of tinkerers and inventors," during which the new technology was largely ignored by businesses and governments. Next came the second stage of commercialization, when the technology moved onto the frontier, where pioneers saw the potential for profit—and also where pirates exploited that potential. Then came the third stage of "creative anarchy," in which great fortunes were won and lost, dominant players established themselves, and governments struggled to catch up with and control an initially ungovernable situation. Once the winners were well established, they collaborated with governments to establish rules and laws that protected their interests in a fourth stage of consolidation.

Amid the hundreds of applications that emerged during the 2000s, two pairs stand out for their international impact: first, mobile phones and social media, which together made possible both the Arab Spring and the later repression of most of the democratic "color revolutions," and second, the combination of data analytics (big data) with deep learning, a branch of artificial intelligence that made possible such technologies as accurate recognition of faces and lethal autonomous weapons, which we discussed in chapter 5.

2000 and Beyond: The Free Internet Faces Competition

By 1995, the Internet was very much an international resource. Participation had spread to almost all countries, though still at very low rates. Over the next decade, the new Internet expanded rapidly, especially in high-income countries, so that by 2004, half the population of the countries that are members of the Organization for Economic Cooperation and Development (the OECD, the "rich countries' club") was online.[20] As the power and the importance of the new infrastructure became increasingly apparent, concern grew over the prospect of a "digital divide" that would keep low-income countries permanently behind.[21]

The growth and development of the Internet and social media in China were even more dramatic. From 2000 to 2010, over 430 million Chinese users joined the Internet, making it the largest Internet market in the world.[22] At first, Chinese companies imitated U.S. business models, using largely imported technology. They soon discovered opportunities to "leapfrog," skipping over technologies based on laptop computers so as to move directly to mobile applications, and discovering opportunities to integrate diverse services that in the West were provided by separate, competing, turf-conscious companies, making possible the seamless retail experience we described earlier. China's enormous population, combined with its relative unconcern with personal privacy, provided an enormous market for information-based services and became the source of a huge database, a resource for major advances in artificial intelligence and big data analytics. The database was augmented by hacking personnel records from the United States and elsewhere.[23]

In China as in the United States, the market for Internet-related services is dominated by huge companies. Four of the most prominent of these are

WeChat, Alibaba, Xiaomai, and Baidu, the Chinese companies corresponding to Facebook, Apple, Google, and a combination of Amazon and PayPal, respectively. (The Chinese conglomerate TenCent is even larger than these four.) Their U.S. competitors have little or no market share in China—in part because of obstacles on foreign firms placed by the Chinese government, in part because of the U.S. companies' own failure to adjust their business models to the special characteristics of the Chinese market, and in some cases because of their laudable resistance to government demands for censorship, spying, or the sharing of intellectual property.[24]

As in the United States, the business model of Chinese Internet based firms rests on an implicit bargain with their customers: in the Chinese case, convenient, cost-free service in exchange for acceptance of government surveillance of all aspects of life. (U.S. companies offer similar services in exchange for acceptance of "capitalist surveillance" for commercial purposes.[25]) This bargain finds an expression in the concept of social credit, a score that was intended eventually to be awarded to each Chinese citizen and to companies doing business in China. Social credit scores for individuals are largely based on signs of financial reliability analogous to those that determine a U.S. credit rating but are also increasingly derived from other indications of "trustworthiness," such as legal records, online activity, and potentially, lack of participation in political demonstrations. An unfavorable social credit rating not only puts a Chinese citizen at a disadvantage for bank loans and mortgages but may also get them banned from buying a ticket on railroads and airplanes.[26]

A variety of pilot social credit programs are being tested in several Chinese cities and by major Chinese high-technology companies, with plans to expand some variation of the system nationwide. A full implementation would gather data from a wide variety of public and private sources, from almost all government agencies to major e-commerce giants like Alibaba, with the potential to draw even more information from cell phone and automobile GPS data and from the huge existing network of urban closed-circuit television cameras, augmented by facial recognition technology. China is also considering the launch of a non-anonymous, government-sponsored e-currency—a digital renminbi issued by the central bank—legal tender that could provide the government with detailed data on economic transactions that are currently conducted in cash.[27]

The Chinese Government Masters the Internet

Of the predictions of the techno-optimists, one of the most important was that authoritarian governments would find it technically impossible to control the flow of information from the outside world. To cite the classic words of one enthusiastic futurist, "trying to ward off the virus of liberty by erecting guard posts at the frontiers of cyberspace . . . will not work in a world that will soon be blanketed in bit-bearing media."[28] After all, the Internet was vastly more powerful than the mere cassette recorders that had carried the smuggled sermons that had roused popular support for the overthrow of the Shah of Iran in 1979.

These hopes were especially strong regarding China. The assumption was that Chinese residents would have to be allowed access to the whole of the Internet if they were to obtain the information on technology and markets needed to maintain China's high level of exports and economic growth. This hope was further fed by the clumsiness of the communist regime's initial efforts in the early 2000s to block access to forbidden websites and to intimidate Web users, Internet cafe owners, and service providers. These efforts slowed the very connectivity that Chinese authorities were anxious to encourage and, besides, were easily evaded by Chinese surfers with the aid of foreign computer hobbyists, who set up a kaleidoscope of "mirror" sites faster than the authorities could blacklist them.

Within a few years, however, the Chinese government's control measures became more sophisticated and effective.[29] The government established a huge internal network, encompassing all Chinese users and surrounded by firewalls and filters that became known as "the Great Firewall of China." All international Internet connections were directed through a small number of easily controlled nodes. Internet cafes and service providers were required to keep records of their customers, and literally to look over the shoulders of their customers to make sure they were not visiting "objectionable" websites. An army of censors numbering tens of thousands cut off access to websites, using criteria that were made deliberately vague, so that Web publishers and service providers would exercise wary self-censorship. A "50-cent army" of trolls, said to number in the millions, flooded social media with pro-government propaganda. To complete the intimidation, a few high-profile "offenders"—most prominently, a software engineer named Lin Hai, who had turned over a list of thirty thousand Chinese e-mail addresses to a pro-democracy magazine based in the United States—were arrested and sentenced to prison. In two incidents that were particularly important to global health, Chinese censors filtered information about the SARS and COVID-19 epidemics from chat room traffic, contributing to the delay in the worldwide response to both of these epidemics (see chapter 6).

The goals of Chinese Internet policy have been to guide public opinion, to block collective action or political mobilization, and to stop ideas and information, whether from foreign or domestic sources, that could undermine the regime.[30] Information about forbidden topics like Taiwanese and Hong Kong independence, the Tiananmen Square protests, Falun Gong, and the situations in Tibet and Xinjiang, as well as criticism of the national government and the rule of the Communist Party, was systematically suppressed. Chinese Internet and social media were left free enough to provide entertainment and limited freedom of expression to Chinese citizens, as long as they do not threaten the political control of the Communist Party. From the point of view of central government authorities, social media provide a useful and easily monitored safety valve for popular discontent, as well as an early source of information regarding local scandals, corruption, and environmental problems—issues that, if allowed to fester, could threaten "social harmony," i.e., party control. As added benefits, the Great Firewall helped Chinese Internet-based companies that conform to government censorship requirements to flourish in the absence of effective foreign competition and might make it possible for a protected Chinese intranet to function domestically even if it were to be cut off from the outside world during a crisis.

When Xi Jingping assumed power in 2012, measures to ensure ideological conformity were raised to the highest levels of the Chinese government and tightened and expanded into an elaborate, high-priority system, reflecting Xi's conviction that Western ideas of freedom of information are anathema to Communist Party control. Information control is enforced by an enormous bureaucracy and encompasses the full spectrum of controls on all the platforms of communication available to Chinese netizens.[31] Their objective is now far-reaching: to silence dissenters, to mobilize party members in support of party values, to spread propaganda, and to keep out news, information, and ideas from foreign sources—even at the risk of limiting economic growth, hampering scientific research and technological innovation, and calling into question the credibility of the government.[32]

The scope of these efforts is quite breathtaking. At least two million people are engaged in monitoring the Internet, social media, messaging platforms, chat rooms, mobile apps, and Zoom meetings. Chinese information control measures extend even to websites that are hosted abroad and are blocked from Chinese viewers, as well as to nonpolitical Chinese dating and video-sharing platforms, since these can also be used for communication. Much of this work is outsourced to private "censorship factories" that work closely with government, creating substantial opportunities for profit. Locally hosted websites and Internet companies are expected to monitor content and to remove banned material. Their work is guided by thousands of detailed official directives, many of them kept opaque to ordinary citizens in order to maintain flexibility and encourage cautious self-censorship. Back doors and specially developed technology enable Chinese police to extract and analyze information from older versions of smartphones in the field (although as a practical matter, people are likely to give the police their passwords). "Offenders" are subject to a variety of penalties, including content removal, loss of social media accounts, and even arrest and jail for individuals, and Internet shutdowns for localities or even whole provinces. Foreign newspapers and magazines that do not echo the party line are blocked, as are Wikipedia and similar sources. Workarounds like virtual private networks (VPNs), which had previously enabled knowledgeable Chinese netizens to evade government censorship, have been severely restricted, although the cat-and-mouse games between censors and determined Chinese netizens still continue. As we discussed in chapter 6, the full spectrum of police controls was put into effect during the COVID-19 pandemic in order to limit discussion of the virus and its effects on public health and the economy, as well as reports of police use of excessive force in enforcing public health restrictions.[33]

The size of the Chinese market places foreign multinational corporations in a difficult position. For example, Google faced the choice between censoring the material it made available to Chinese customers and losing its access to the huge Chinese market, and it eventually agreed to create a special Google.cn service tailored to Chinese requirements, though even that was shut down in 2010. Many other high-tech companies have followed suit, one way or another, lest they lose access to the Chinese market. The reach of Chinese political sensitivities extends well beyond its borders. In 2019, the National Basketball Association, for which the Chinese market is an important source of revenue, chose to apologize for a tweet by the general manager of one of its teams that expressed solidarity with pro-democracy protesters in Hong Kong, and researchers and journalists have had to self-censor their remarks in order to ensure that they will be permitted to re-enter China.

Democracies have Internet Issues, Too

Western democracies agree on the basic principles underlying the operation of the Internet, but even they have disagreements over Internet issues like privacy and data localization. First, countries disagree on what kind of information should be permitted to be posted online. Perhaps surprisingly, no country wants complete freedom of online speech. They do differ on what kind of expression should be excluded though, resulting in what lawyers call a conflict of laws. To be sure, fraud is fraud, theft is theft, and slander is slander, online just as offline, and there's international agreement that child pornography should be excluded from the Internet. Beyond these commonalities, however, there are disagreements. Nazi propaganda is illegal in Germany, for good reasons rooted in history, but is protected by the First Amendment in the United States. Similarly, Thailand prohibits criticism of their king (lèse majesté, it's still called), and Saudi Arabia prohibits information on religions other than Islam. No surprises there. But the United States also has restrictions on what can be posted; online gambling is forbidden in many states, a stricture that the Cayman Islands has challenged before the World Trade Organization as an illegal restraint on trade.

A second area of conflict stems from differing approaches to individual privacy, even among democratic countries. The United States does not broadly restrict cross-border data flows. Its legal protections of individual privacy are a patchwork of laws specific to particular areas like financial, educational, and health records. In Europe, by contrast, privacy is a deeply felt value, afforded systematic, high-level attention as a principle enshrined in the General Data Protection Regulation (GDPR) of the European Union (EU). Germany's highest court has even established a constitutional right to the "integrity and confidentiality of IT systems."[34] The GDPR applies to all companies doing operations on data concerning an EU citizen, monitoring EU citizens, or offering services within the EU.

Given the lack of a U.S. law of comparable generality, the GDPR is likely to become a worldwide model. The GDPR provides stronger protection to individual privacy than that in the United States: for example, it requires that an individual receive clear, understandable information about who is processing their personal data and why; that they have access to their personal data and can correct any errors; that they be notified of a potentially harmful data breach; and that they be able to demand that search engines delete links to negative news stories—the "right to be forgotten"—that in the United States would be protected by the First Amendment.[35] It also requires "purpose limitation," meaning that a company cannot collect data for one purpose and then use the data for another. After somewhat difficult negotiations, the EU and the United States negotiated a "Privacy Shield," which they hoped would enable U.S.-based companies to satisfy EU requirements.[36] This was, however, struck down by the European Union Court of Justice, the EU equivalent of the U.S. Supreme Court, in a decision that will have wide-ranging ramifications for firms doing business in the EU.

One of the most challenging conflicts between globalization and national sovereignty concerns the issue of data localization: the requirement in some countries that information concerning citizens of a country be collected, processed, and stored in that country. This requirement is attractive to many European countries because they believe that it gives them more control over data privacy, a concern reinforced by the Snowden revelations of widespread National Security Agency (NSA) spying (we will return to this subject in chapter 9). Skeptics might add that it also could facilitate domestic surveillance, control of information, and protection of the domestic data processing industry. Data localization is attractive to authoritarian countries because it makes it easier for their intelligence agencies to gain access to information on their own citizens and to suppress political opposition.

Data localization would require multinational companies to maintain separate and often redundant databases in each country in which they do business. Critics argue that this would raise the cost of Web- and cloud-based services and make big-data analytics and other kinds of market research much more difficult. Data localization would make it harder for companies to recruit internationally and to establish uniform procedures across their different branches by which to protect the very privacy that data localization is supposed to safeguard. Besides, they argue, data security within a company depends in practice more on its own security procedures than on the location of its servers.

Whatever the merits of these arguments, several dozen countries have enacted or are actively considering laws requiring data localization, at least for health or other personal or sensitive personal data. (The EU has banned its members from adopting such laws except for personal data.) In particular, China has enacted data localization laws as part of its larger effort to establish government control over information and information flow. While the strictness of their enforcement is still unclear, many multinational companies may well be willing to comply in order to protect their access to the Chinese market. Given the importance of data as a key resource for international business, data localization is likely to be a hot issue in negotiations over trade agreements for some time to come.

Governance: Who, If Anyone, Will Control the Internet?

Just as it became clear that the Internet was to develop into the major vehicle for international communication and commerce, its structure was shaken to the core by the unexpected death in 1998 of Jon Postel, who had by then become in essence a one-man system of Internet governance. After some false starts, the U.S. government entrusted authority over top-level domain names and the "address book" that assigns unique URLs to the International Corporation for Assigned Names and Numbers (ICANN), a nonprofit organization incorporated under the laws of California.

As a pragmatic solution, the choice of ICANN made sense—certainly from the point of view of the United States, which had developed the Internet and now controlled it, and of other Western countries that depended on it. The Internet continued to operate and expand smoothly and to reflect Western values of freedom of expression and innovation. To turn the new technology over to the United Nations or some other intergovernmental body, it was argued, would give it formal legitimacy but would risk inefficiency, technical stagnation, and worse, government control. This conflict in values became a conflict over institutional form.

From the legal point of view, the status of ICANN was anomalous. ICANN, after all, was a U.S.-based organization, selected by and on contract with the U.S. government and incorporated under the laws of a U.S. state, with the responsibility for managing the world's most critical piece of communications infrastructure, and with no accountability to any intergovernmental organization. Nor could it be maintained that ICANN's function of maintaining the Internet directory was purely technical. On the contrary, many of the decisions it would have to make were inherently political. Who, for example, should control a country directory after a political coup? Should a country be permitted to create a subdirectory for gambling or for pornography? The answers to these questions are normally the responsibility of governments and intergovernmental organizations.

Governments were not the only organizations interested in the future of the Internet. A vigorous, cacophonous, and outspoken multi-stakeholder community of nongovernmental organizations had grown up around the Internet, dedicated to preserving its unregulated, free-spirited, free-speech- and innovation-friendly character. At the same time, the idea of an anarchic Internet was a threat to the governments of the many other countries for which control of the information reaching their citizenry was essential to political survival. This combination of powerful high-tech companies, rapidly changing technology, democratic governments seeking to protect an open Internet, authoritarian governments accustomed to control of information, developing countries anxious not to be left behind by the new technology, and feisty, not-to-be-denied civil society, posed a unique challenge of governance to the stodgy, turf-conscious international bureaucracies of the United Nations.

A two-part World Summit on the Information Society (WSIS), held in 2003 and 2005 in Geneva and Tunis, respectively, reached a Solomonic solution. It established two units: a Government Advisory Committee, whose members (as the name suggests) are appointed by governments, and an Internet Governance Forum (IGF), a large, spirited, somewhat chaotic gathering of nongovernmental organizations. Both of these had the status of "consultative" (i.e., advisory) bodies without direct operating responsibility. The WSIS reluctantly left the management of the directory in the hands of ICANN.[37] In 2016, more than a decade later, ICANN became an independent, nongovernmental organization, free from U.S. government control, with a complex multi-stakeholder structure that nevertheless allows a considerable degree of local control over Internet content.[38]

The pragmatic evolution of Internet governance had interesting parallels with that of another custodian of a critical global resource that for a long time also lacked a formal structure. As we discuss in chapter 10, the International Board on Plant Genetic Resources (IBPGR) managed and controlled the main international collections of germ plasm (seeds and other forms of genetic material) for the major food crops of the developing countries with financing from the Consultative Group on International Agricultural Research, which is now known simply as the CGIAR, despite the fact that the IBPGR had no formal structure for almost two decades. (In technical legal terms, the IBPGR had no "juridical personality.") The IBPGR was running these gene banks well, and it seemed inadvisable to turn them over to the Food and Agricultural Organization of the United Nations, which is the specialized agency with responsibility for agriculture but which was felt to lack the necessary managerial capacity for so important a responsibility. (In the 1990s, the IBPGR was given a formal structure and is now known as Bioversity International.)

China, Russia, and many other authoritarian governments still take a dim view of the informal state of Internet governance, which they view as a thin disguise for U.S. dominance—the more so given the Snowden revelations of U.S. intelligence activities. In a series of follow-up meetings to the WSIS extending over more than a decade, they have sought to gain international legitimacy for the doctrine of data sovereignty, which justifies government control over their domestic Internet and protection of domestic Internet-related industries.[39]

Behind the disagreement between China and the West is a basic divergence of norms, with democratic countries calling for open access, multi-stakeholder governance, and freedom of expression, association, and innovation, and with authoritarian governments arguing for international control over the Internet and government control over information. The Chinese would no doubt prefer their norms to be widely accepted. On the other hand, they are closely integrated into world markets and have so far been willing to compromise, so the structure of Internet governance has remained fundamentally unchanged. We may expect a further challenge to U.S.-led, multi-stakeholder governance of the Internet as Chinese technology becomes an increasing part of its technical structure.

A Less Open Internet in the Future?

The Internet began as a juggernaut that swept all before it, leaping across borders, creating like-minded communities, disrupting industries, and raising hopes in the West for a new era of universal freedom of expression, democracy, free markets, and innovation. Now, thirty years after the privatization of the Internet and fifteen years after the founding of Facebook, the revolutions in ICT have been thoroughly integrated into all aspects of international relations: diplomacy, business, trade, finance, communications, intelligence, and, as we shall see in the next chapter, crime and warfare. They have added new dimensions to old issues: privacy, civil and human rights, and the rules governing both warfare and conflict short of war.

The hopes of the founders of the Internet have been realized, but not always in the way that they had imagined. The new technologies have given rise to new international communities and new avenues of bottom-up political influence, not only for young people, pro-democracy activists, and environmental advocates, but also for political extremists, conspiracy mongers, and terrorists. They have created new tools, not only for organizing popular movements and demonstrations, but also for surveillance, censorship, and repression. They have provided an international forum for free discussion and creativity and a platform for innovation, but also a vehicle for election manipulation, disinformation, conspiracy theories, and criminal activity. They have opened up new paths by which low-income countries can gain access to world markets but also threaten to widen the gap between low- and high income countries, not to mention a digital divide between high- and low-income people living in the same country, a divide exacerbated by the COVID-19 pandemic.

Finally, rather than allow the Internet to dissolve "information borders," national governments of an authoritarian bent have developed ways to control the information that reaches their people, exploiting this new technology as part of a fundamental, worldwide challenge to democratic values. The Chinese strategy has been to invest in technological development as an underpinning of larger geopolitical objectives. Chinese technology and its strategy of comprehensive, techno-authoritarian surveillance is already being exported to like minded governments all over the world.[40] Much of the next generation of the Internet, the Internet of Things (IoT), is likely to run on Chinese 5G technology, which is cheaper than its competitors. This will enable the Chinese, if they so choose, to build government controls into much of the hardware and software on which the Internet runs, giving them some of the economic and intelligence advantages currently enjoyed by the United States, and leading to a much less open Internet than the one to which we are now accustomed.

The early hopes of the Internet pioneers led them to underestimate the ability and determination of governments to adapt the new technologies to their own ends. To authoritarian governments, the freedoms that were incorporated into the original design of the Internet constitute existential threats. After a stumbling start, these antidemocratic powers, led by China and a resurgent Russia, have turned social media and the Internet into vehicles for mounting a sustained attack on the liberal world order that had been painstakingly erected after World War II. The contest for control of the next generation of the Internet and the IoT is one element of the larger struggle between two alternative visions of the world of the future: one a top-down world dominated by authoritarian governments; the other, a freewheeling, kaleidoscopic, multi-stakeholder, cacophonic, bottom-up world.[41] The contest between these two visions is likely to be with us for a long time. In the end, however, all countries share an interest in the continuing function of the global Internet, and all want to avoid a "digital apocalypse," a strategic attack that could cripple the global economy.

Of the many issues raised by the Information Revolution, some of the most important are cybersecurity and the prospect of cyberwarfare. Cyberweapons are relatively inexpensive and are therefore attractive to smaller, weaker, and lower-income countries, which can use them to attack stronger adversaries and to deter them from responding. The United States and other high-income countries are vulnerable to such cyberattacks because the Internet has penetrated into every aspect of our lives, because much of our Internet-connected infrastructure lacks basic cybersecurity protection, and because much of this insecure infrastructure is in private hands, raising awkward questions about the relation between government and private business. It is to these attacks, these vulnerabilities, and these complications that we turn in the next chapter.

9

Cyberwarfare and Cybersecurity

In 2007, Israeli bombers destroyed a partly constructed Syrian nuclear reactor without being detected by the Syrian antiaircraft radar newly acquired from Russia. Not only did the Syrians fail to detect the incoming Israeli bombers, but when Syrian intelligence reviewed its records after the attack, it found no sign of any incoming aircraft! The Israelis had hacked into the Syrian electronics and fed it a previously recorded picture of the ordinary night sky, long enough for the Israeli planes to carry out their mission and return to base. This attack was the first *tactical* application of cyberwarfare and opened a new dimension of warfare.[1]

The Russian invasion of the Caucasian republic of Georgia the following year integrated cyberwarfare into a broader *strategic* plan. Building on their experience in the cyberattack on Estonian websites the previous year, the Russians used nominally nongovernmental "patriotic hackers" to execute distributed denial of service (DDoS) attacks on military communications and Georgian government websites, hindering them or putting them out of operation by glutting them with useless messages. When the actual "kinetic" invasion began, a Russian criminal organization linked to the government blockaded Georgia from the global Internet and prevented its spokesmen from communicating with the outside world, giving the Russian government time to establish the narrative that the war was the fault of the Georgians.

Since 2015, Russia has used its full armory of digital weapons to demoralize Ukraine as part of its overall strategy in the long-simmering Russo-Ukraine War. In 2015 and 2016, cyberattacks shut down the Ukrainian power grid for a couple of hours at a time. A year later, a cyberattack froze 30% of the business computers in Ukraine, all of which had been using the same venerable and hence vulnerable accounting software. At the same time, an online disinformation campaign did its best to widen social and linguistic divisions within the country.[2]

Stuxnet and its Aftermath

The first high-profile use of a pure cyberweapon to effect *physical* damage on an enemy was the Stuxnet worm in 2008–2010, widely assumed to be a joint U.S.–Israeli effort to damage centrifuges in the Natanz installation that were essential

The Survival Nexus. Charles Weiss, Oxford University Press. © Charles Weiss 2022.
DOI: 10.1093/oso/9780190946265.003.0011

to Iranian production of fissile material for an atomic bomb.[3] (The United States had been using cyberweapons offensively since earlier that year without fanfare, shutting down the servers of Iraqi insurgents and using phony emails to lure them into death traps.) The Stuxnet worm made the sensitive centrifuges either speed up or slow way down, in either case causing vibrations that made them break down prematurely.[4] This attack put about a thousand centrifuges—around a quarter of Iran's total supply at the time—out of commission in ways that baffled and discomfited their operators until the worm was discovered. The Stuxnet victory was short-lived, however, since Iran had about an equal number of centrifuges up and running only a year after the attack.[5]

The Stuxnet worm was an extremely sophisticated and elaborate piece of malware, as shown by the fact that it exploited no fewer than four "zero-day" vulnerabilities, any one of which would have been the closely guarded secret of an intelligence agency or a lucrative product for a criminal hacker. Stuxnet was designed to attack the specialized computers, called programmable logic computers (PLCs), of a specific Iranian plant and no others. Moreover, it was designed to produce premature breakdown rather than immediate collapse, which would have been immediately detected and traced to its origin. For this reason, it might never have been detected, had it not drawn the attention of private cybersecurity services by infecting thousands of PLCs all over the world.

The Stuxnet attack was a watershed in cyberwarfare from at least three points of view. First of all, it was the first use of cybertools to produce physical damage, which if effected by a raid by human commandos would have been an act of war. This, in effect, legitimized the offensive use of cyberweapons and created a precedent for considering them to be an act short of war and a new normal. Second, the quick recovery of the Iranian nuclear program (and of Ukraine from the "Dark Energy" attack on its power grid that accompanied the Russian invasion) confirmed that cybertools were insufficient by themselves to achieve what was in essence a military objective but, rather, needed to be integrated into an overall strategy.

Armed with this insight, the United States developed a comprehensive plan, called Nitro-Zeus, to use cyberweapons to shut down the Iranian military and the complete Iranian economy in case of a conflict, in order to provide time to bomb its nuclear installations. In preparation for the execution of this plan, zero-day vulnerabilities were inserted throughout Iranian infrastructure. (The United States, of course, was outraged when other countries did the same to *it*.) The Nitro-Zeus plan was never put into effect, due to the successful diplomacy that led eventually to the nuclear deal, and was kept secret until the Snowden revelations.[6]

Third, Stuxnet was a wake-up call to the American military, which was forced to deal with the fact that what the United States could do to others, others could do to it, especially since the source code for Stuxnet software was now known and available to hackers of all kinds and motivations. This was not the first time that the vulnerability of the U.S. military to cyberattack had been conclusively demonstrated. In 1997, Operation Eligible Receiver, an internal "red team" from the super-secret National Security Agency (NSA, known to wags as "No Such Agency"), had used software freely available from the Internet to penetrate the entire network of the defense establishment, including the National Military Command Center that is responsible for transmitting orders in wartime. The red team succeeded in disabling military command-and-control systems, reformatting hard drives, and effecting a virtual hijack of a Coast Guard vessel at sea.[7]

A year later, two Californian high school students, coached by an 18-year-old Israeli hacker, were found to have gathered data from a variety of critical unclassified Department of Defense networks in a hack dubbed Operation Solar Sunrise. That same year, the Defense Department discovered that it had been the target of a massive, sophisticated, and sustained hacking operation, dubbed Moonlight Maze. Almost ten years earlier, in 1986, a scientist in Berkeley had identified what was probably the world's first computer hack, which was later traced to a resident of Germany who had stolen the equivalent of three million pages of unclassified but sensitive information from unprotected computer networks over a period of at least two years in the hope of selling it to Russia.[8]

These jaw-dropping events had produced a series of worried reports full of detailed warnings from teams of distinguished experts, as well as some minor organizational changes, but little effective action from a military still wedded to older weapon systems—despite the fact that there were already at least half a dozen daily cyberattacks against military networks.[9] In retrospect, the institutional reasons for this inaction are understandable. Effective defense against a cyberattack of the U.S. military would have required a sophisticated understanding of cybersecurity at every level of a defense establishment whose attitudes had been formed in an earlier era, as well as close collaboration between the army, navy, and air force, which frequently compete with each other for missions and resources. Defense against cyberattacks on civilian infrastructure can be even more demanding, as it requires collaboration by many levels of government with each other and with private firms of all sizes, as well as conflicts with the instincts of the intelligence community to minimize sharing of information.

Many countries, including the United States and China, have had to resolve domestic institutional tensions between government and industry, and between the military and the intelligence agencies, to prepare for the possibility of a cyberconflict. These tensions take different forms and require different solutions in different countries, based on historical interactions and the strength of the government. In the United States, tension between the military and the intelligence agencies stems from the difference between their missions. When the military discovers a vulnerability in another country's software, its instinct is to exploit it immediately, even though this will alert the enemy to its existence, since the vulnerability will become useless if the adversary discovers it by itself. In contrast, the instinct of the intelligence community is to keep the discovery of the vulnerability secret, so that it can be used to gain information and to provide the basis for a zero-day defect to be exploited at a later date.

The U.S. government faces two further problems that complicate its relations with high-tech industry. First of all, many of the private companies that own most of the country's critical infrastructure—financial institutions, power plants, and communication networks, for example—are reluctant to collaborate with government efforts to help them to defend against cyberattacks and, indeed, may be reluctant even to admit that they have been the targets for such attacks, lest they scare their customers or give them reason to think that the information they have entrusted to them is not safe, forcing them to make expensive investments in improved security. Infrastructure for elections is owned and managed by states, raising tricky issues of federal–state relations. Second, the military and the intelligence agencies in the United States have had to overcome reluctance on the part of the technical staff of high-tech companies like Google to collaborate because of revelations of extensive government spying. This is important because the U.S. military now depends on Silicon Valley for the latest and the best technology and expertise, in contrast to the situation during the Cold War, when military technology was generally well ahead of its civilian counterparts.

In China, by contrast, the president has simply ordered industry to collaborate with the military and intelligence and has placed party cadre in a structure parallel to civilian management in strategically important industries, including the local branches of some U.S.-based multinational corporations.[10] While this has been effective in resolving tensions, it has complicated the efforts of Huawei, the world's biggest supplier of telecommunications equipment and a centerpiece of Chinese technological development, to reassure its customers that it is a normal company managed along purely commercial lines and that there is nothing untoward hidden in its software. We shall return to the issues surrounding Huawei in chapter 11.

The Snowden Revelations: The Crown Jewels of U.S. Intelligence Spill onto the Table

In 2013, the casual attitude of the Internet pioneers toward cybersecurity came home to roost in the person of Edward Snowden. Like the pioneers of the ARPANET, the ancestor of the Internet developed by the Advanced Research Project Agency (see Chapter 8), the NSA and the military had assumed that security measures were needed against the outside interloper, but not against the inside job.[11] Snowden was a systems administrator working for a contractor at an NSA installation in Hawaii, from which the Agency's most advanced surveillance and other tools were deployed against North Korea and China. Using a Web crawler program[12] that was available at the time from Amazon for $100, Snowden downloaded at least hundreds of thousands of NSA documents, many of them classified higher than top secret, and passed them on to three major newspapers: *The Washington Post*, *The Guardian* (London), and *Der Spiegel* (Berlin). A few years earlier, an Army private—Chelsea Manning—had downloaded hundreds of thousands of military videos and Department of State cables and handed them over to WikiLeaks. Snowden and Manning were regarded as villainous traitors by the intelligence community and the military and at the same time as heroes to their critics.

The three newspapers filtered out and kept secret any information from the Snowden disclosures that could cause immediate deaths, although they did expose a system that alerted the military to impending attacks with improvised explosive devices (IEDs). Wikileaks was less scrupulous, especially after Manning's arrest and imprisonment by the Army, and revealed the names of Afghan and Iraqi citizens who had been helping the war effort, forcing the military and intelligence forces to rescue them as quickly as possible to keep them from being killed by the Taliban.

The Snowden trove of documents exposed to public scrutiny the stunning scope of overseas NSA spying, as well as the entire "Advanced Network Technologies" (ANT) catalog of tools and methods used by the Tailored Access Operations (TAO) staff, the topmost elite of NSA code crackers and intelligence operatives. This included such delicious details as the use of tiny chips that were surreptitiously incorporated into the thumb drives or hard drives of intelligence targets and could broadcast their contents to receivers located miles away. Despite U.S. protestations that these and the other devices in the catalog were purely defensive, it would have been difficult for any outside observer to tell whether these tools were for offensive or defensive purposes, as the technologies involved for either could be much the same.

Of most direct interest to U.S. citizens was Snowden's revelation that the NSA, in collaboration with major U.S. telecommunication companies, had collected metadata—the telephone numbers of caller and recipient, along with the date, time, and duration of the call, but not the identities of the parties or the substance of the communication—of some five billion domestic and international telephone calls a day, often involving people who were not suspected of any involvement in terrorism or any other criminal activity.[13] This enormous database was said to be a resource for reference, consulted only occasionally to track the contacts of a "probable" terrorist or foreign agent.[14] The sheer scale of this operation made it arguably a violation of the prohibition of "unreasonable search and seizure" in the Fourth Amendment of the U.S. Constitution.[15] The Snowden disclosures also revealed that NSA was systematically hacking into smartphones, emails, chats, videos, photos, voice-over IP, file transfers, and social networks, and most notably into foreign embassies, through a bulk collection program code-named PRISM. From the NSA point of view, if their job was to find a needle in a haystack, they needed to have the whole haystack.

The exposure of the PRISM program was a wake-up call to U.S.-based high-tech companies, as well as to their customers, who were shocked to learn that the personal data they had posted on supposedly secure platforms was, in fact, vulnerable to intrusion. Although these companies were based in the United States, they regarded themselves as global corporations whose customers demanded "end-to-end security" and would get it from their non-U.S.-based competitors if they could not get it from them. Unlike older U.S. telecommunications giants, which had grown up with the Cold War, they felt no patriotic obligation to cooperate with U.S. intelligence unless they were legally required to do so, despite protests from U.S. law enforcement and intelligence agencies that it was dangerous to leave them "in the dark," unable to read messages sent by potential enemies or terrorists. High-tech companies therefore fought the demand of these agencies for back doors through which coded messages could be decrypted.[16] Besides, these companies argued, criminals would inevitably detect and gain access to any such back doors, and authoritarian regimes would insist that they, too, should be given access to them.

To foreign countries, the Snowden revelations were not only an awe-inspiring overview of the vast extent of NSA spying operations, but also a guide on how to evade them, at least until the NSA could come up with new tools. The revelations also provided foreign governments with yet another reason to resist U.S. technological domination insofar as possible, at least within their own markets. At least as important, they gave the intelligence agencies of U.S. allies reason to doubt that their U.S. intelligence counterparts could keep a secret and therefore lessened their confidence in continued cooperation with the United States. Rubbing salt into U.S. wounds, North Korea and Russia used the stolen toolkit for cyberattacks that cost government agencies, hospitals, and major corporations more than $10 billion.

Challenges to Democratic Values and U.S. Dominance

Cyberwar is only one aspect of the technology-based challenge that China and Russia have mounted to the liberal order erected by the United States and its allies after World War II.[17] Neither country shares democratic values. Both seek to restore their honor and status as great powers: in the case of Russia, after the loss of the Cold War, and in the case of China, after more than a century of humiliation by colonial powers. Beyond these commonalities, they employ different strategies regarding the Internet and social media and the philosophies that underlie them, although these strategies are beginning to converge. In brief, China hopes to dominate the technology, management, and policy of the Internet of the future, as part of its general objective of surpassing the United States politically and technologically, whereas Russia is using a broad array of communications tools, including not only Internet and social media but also plain old television, to undermine Western democracies and alliances.

The international strategy of Chinese President Xi Jinping is based on a mixture of geopolitical ambition, technological development, legal and illegal acquisition of intellectual property, promotion of exports and economic growth, job creation, and a commitment to "cyber-sovereignty," which in his words means "the right of individual countries to independently choose their own path of cyber development and participate in international cyberspace governance on an equal footing."[18] In practice, this means increased government control over information, resistance to foreign (i.e., United States) surveillance, and international control of the Internet under the aegis of the United Nations.

Chinese government and industry have made major strides in 5G technology, the latest generation of mobile connectivity that greatly multiplies the speed and capacity of the Internet and makes possible the Internet of Things (IoT), in which connectivity is being extended to more than twenty billion household appliances, vehicles, agricultural, infrastructural and environmental sensors, machine components, surveillance cameras, and even pieces of clothing with built-in sensors—basically, anything with an on–off switch or a wireless reporting device. These technological advances are in no small part due to the skill of Chinese engineers and scientists in mastering technology, adapting it to the Chinese situation, and improving it through innovations of their own—although they are still dependent on the United States and Taiwan for semiconductors. (U.S. efforts to restrict Chinese access to imported semiconductors have increased the pressure to overcome this dependency.) These advances also owe a good deal to major government support in the form of direct subsidies and low-interest loans from government-owned banks, both to the company and its customers.[19] (The United States and other Western governments do much the same thing; U.S. telecommunications and information industries have long benefited from generous government financing at all stages of the innovation process, from basic research to military procurement.)

Huawei technology is relatively inexpensive and has been adopted by more than fifty countries, including a number of NATO allies, although not by the United Kingdom. This equipment has a long lifetime and is unlikely to be replaced for decades. If Huawei designs are adopted by the International Telecommunication Union (ITU) as the basis for international 5G standards, non-Huawei hardware manufacturers would have strong incentives to ensure that their equipment is compatible with them. The widespread adoption of Huawei technology raises serious issues of security for its customers, as Huawei could install back doors that would enable it to tap into communication lines even more extensively than the NSA does with the metadata of Internet messages passing through the United States. Worse, such back doors could enable Huawei to cut off communications, including military communications, during times of crisis. Given China's increasingly expansionist foreign and commercial policy, the security threat posed by this possibility is a matter of serious concern to any U.S. government. While Huawei insists that it is an ordinary private company with no links to the government and that it has done and will do no such thing, the Xi government has required all large companies to establish close links to the ruling Communist Party and to cooperate with Chinese intelligence.[20] Even if the Huawei equipment being sold today has no hidden back doors or other "defects," there is no assurance that the government might not order Huawei to insert such vulnerabilities at some future date, and it is likely that Huawei would have no choice but to comply. This would be technically possible, as the next generation of 5G software is likely to be reprogrammable remotely.

A second potential Chinese threat to the postwar liberal order comes from the export of the model of techno-authoritarianism it has developed to control the Uyghurs and other minorities, as an alternative to Western models that put higher priority on privacy and personal freedom. Behind this sociotechnological package is the concept of cyberspace as a place where people live and in which they can be controlled, using technologies of mass surveillance and artificial intelligence. These are deployed in Chinese cities, where they also reduce crime and improve the management of municipal services. The package is sold to authoritarian countries in Africa, Central Asia, and Latin America, especially those involved in the Digital Silk Road, the part of the Chinese Belt and Road Initiative that strengthens digital infrastructure in Asia and Africa.[21]

The Russian attack on the postwar order is based on a different philosophy. The Russian concept of cyberwarfare starts from the holistic objective of "domination of information space" in both war and peacetime, as part of a more general insight that "the role of nonmilitary means of achieving political and strategic goals has grown and, in many cases, they have exceeded the power of force of weapons in their effectiveness," an idea now embodied in what has become known as the Gerasimov Doctrine.[22] This is much broader than the stated U.S. concept of cybersecurity, which is defined as the defensive protection of both military and civilian communications and connectivity, with limited offensive use of cyberweapons (also known as "defending forward") in order to preempt possible attack, infiltrate an enemy's connectivity so as to gain information, and plant zero-day defects in the critical infrastructure of a potential enemy as a warning at a time of crisis or as a force enabler or multiplier in a military conflict.[23]

Russia uses both social and traditional media in a long-range, systematic, worldwide campaign of *dis*information (deliberate use of misinformation), intended to exacerbate social and political tensions and thereby to undermine Western democracies, the European Union, and the alliances that bind democratic countries and underpin the liberal world order, not to mention the complementary goal of eroding the very idea of objective truth. Russia has employed online interference tactics in at least eighteen countries since 2013, with any of a broad range of political goals: discrediting individual politicians, polarizing the electorate, directly supporting right-wing parties, promoting issue-specific platforms, disrupting elections, and generally undermining faith in the democratic process. In a relatively early example, thousands of Russian automated Twitter accounts promoted anti-immigrant rhetoric in the lead-up to the 2016 Brexit referendum in the United Kingdom.[24]

In the United States, Russian strategy began by "preparing the battlefield." The government-owned RT (originally Russia Today) television network—which attracted a quarter of a million regular listeners—broadcast straight news, spiced with conspiracy theories and other disinformation, framing issues in a way that would heighten social and political tensions, pushing candidates favored by Russia, and eagerly spreading and elaborating on negative material (factual or invented) on candidates they opposed—especially on Hillary Clinton, against whom Russian President Vladimir Putin has special animus. This coverage was intended to be picked up by extremist websites and podcasts and eventually to make its way into mainstream media.[25]

The propaganda broadcasts of the RT network prepared the way for full-scale, multifaceted interference in the 2016 U.S. election, directed by the most senior levels of the Russian government. This effort was in some ways analogous to previous Russian efforts to influence U.S. elections dating as far back as 1952,[26] and to the similar efforts by the United States to influence democratic elections in Italy and other Western European countries in the years immediately following World War II.[27] However, the Russian effort in 2016 was much more aggressive and far-reaching. In addition to the broadcasts of RT and its sister network, Sputnik, it involved releasing embarrassing communications gathered by Russian intelligence agencies at critical moments in the 2016 presidential campaign on WikiLeaks, mobilizing an army of trolls and bots[28] on social media to amplify the distribution of disinformation and conspiracy theories (some of them bizarre),[29] financing campaign advertisements through intermediaries, and probing state and local voter rolls, all in order to promote the prospects of the Republican candidate.[30] The objective of the weirder stories put out by Russian propaganda was to make people unsure that such a thing as a fact even exists. Ironically, this effort continued until it appeared that Trump was certain to lose, at which point the Russian effort tried to undermine the overall legitimacy of the election, of the U.S. government, and of the anticipated administration of Hillary Clinton.[31]

As a result of this history, Facebook and other social media platforms face criticism that they have not done nearly enough to police the spread of malinformation, especially when it emanates from foreign sources. Facebook has undertaken efforts in this direction, for example, by flagging obvious dis- and misinformation and taking down flagrant offenders wherever they are based, but they also point out that they face difficult issues of censorship and freedom of speech, in addition to the sheer technical difficulty of filtering hundreds of millions of daily posts. A more fundamental problem is the fact that social media platforms use algorithms to feed surfers a steady stream of information that is in line with their existing tastes. These algorithms encourage the formation of like-minded online communities that can spread mis- and disinformation faster than social media platforms can suppress it.

History does not disclose its alternatives, so we will never know whether the Russian effort was decisive in determining the results of the close 2016 U.S. presidential election. These and similar threats are likely to get worse. Advances in artificial intelligence, for example, have made possible deep fakes: videos that mimic the appearance, voice, and gestures of public figures so well that it is difficult to tell them from the real thing.

Meanwhile, the cyber-battles have continued. While the United States was focused on preventing foreign interference with the 2020 elections, world-class hackers, presumably based in Russian intelligence services, inserted malicious code into updates of software that had been developed in Eastern Europe by SolarWinds, a Texas-based company reputed to be lackadaisical about its cybersecurity. This stunningly successful "supply chain hack" enabled access to the computer systems of numerous federal agencies, including the Treasury, State, and Energy Departments and parts of the Pentagon, plus no fewer than eighteen thousand private companies.[32] Hackers, presumably Russian, also made off with the "red-team toolkit," the closely guarded investigative software used by Fire-Eye, the U.S. private company most widely relied upon by both government and private industry to detect malevolent cyber-intrusions.[33]

Underlying these events is the fact that the world digital economy depends on trillions of lines of computer code, from which it is humanly impossible to eliminate all errors and omissions that can be exploited by a determined and skilled adversary. What is more, the computer system of any given organization, no matter how sophisticated and well-defended, is itself a system of systems, and is in turn a subsystem of a still larger system that extends outside of itself, encompassing telecommuters, clients, vendors, and resellers. This structure inevitably gives rise to vulnerabilities at its interfaces. The bottom line is that the cat-and-mouse game between hackers and "hackees" is likely to be a permanent fixture of international relations.

Arms Control for Cyberweapons?

The Stuxnet attack introduced the world to a new, cheap, powerful weapon and established a precedent for using it against a hostile power. Since the technology was well within the capabilities of much of the world, other such attacks are sure to follow. Possibilities range in size from mischievous "hacktivism"—defacing or altering a website to further a political or social goal, as by replacing its normal message with propaganda—to a full-scale "cyber Pearl Harbor" intended to destroy the economy and infrastructure of an enemy. There were—and still are—no agreed "rules of the road" for this new kind of warfare.

Like nuclear war, a strategic cyberattack could result in widespread destruction of civilian communications, as well as municipal, health, financial, commercial, and hydrological infrastructure. Like a nuclear attack, a large-scale cyberattack could easily spiral out of control in the absence of agreed guidelines, because a cyberattacker has the incentive to be the first to strike if a strategic conflict seems likely. The attacker needs only to find a single vulnerability in what are likely to be multiple targets, each with many millions of lines of software code, to gain the upper hand over an enemy, whereas a defender must defend everything at once to avoid defeat. For this reason, countries have more incentive to invest in offensive than in defensive programs, and this is, in fact, what the United States and other cyber-powers have done. Both the United States and Russia are thought to have placed back doors and zero-day defects into the software of each other's electric grids, oil and gas pipelines, dams, and other critical infrastructure, sometimes exploiting vulnerabilities in decades-old industrial control systems.

One might therefore imagine a concept of mutual assured cyber-destruction, analogous to the mutually assured destruction (MAD) doctrine for nuclear weapons that we discussed in chapter 5, as the basis for a formal or tacit agreement that no cyber-power will launch a strategic cyberattack against the critical infrastructure of another for fear of "second-strike" retaliation and a resulting cyber-apocalypse.

However, there are differences between nuclear and cyberweapons that complicate the task of preventing unintended all-out conflict (arms control), over and above the obvious fact that nuclear weapons would kill and maim far more people than would a cyberattack. First, like nuclear weapons, cyberweapons are weapons of the weak (asymmetric weapons), but for a different reason. A country that depends heavily on online functions is much more vulnerable to cyberattack than one that is not so dependent, so that cyber-based retaliation by the stronger power in response to an attack by the weaker one might well be ineffective. There might be too little cyber-activity in the attacking country for a retaliatory counterattack to do unacceptable damage. Retaliation could be carried out by other means, but these might bring their own complications.

Second, a nuclear attack would be easy to distinguish from one with conventional weapons, and retaliation would probably be nuclear. In contrast, a cyberattack could come in many sizes, and retaliation could take many forms. Third, nuclear weapons would most likely be delivered by ballistic missiles, which have clear return addresses. (Hypersonic missiles could change this characteristic, at least until a suitable network of detection satellites is deployed.) In contrast, it can take weeks to identify the source of a cyberattack, and even then, uncertainties are likely to remain—the so-called attribution problem. (This problem would also apply to smuggled nuclear weapons, as we discuss in chapter 5.) Fourth, it is not always possible to distinguish whether a given cyber-penetration constitutes an act of espionage, or the test of a cyberweapon that could be used for damaging information systems or for physical destruction of critical infrastructure. Fifth, nuclear weapons are the subject of a long-standing treaty regime designed to minimize proliferation and discourage their use, a regime that is still largely respected even though the treaties that define it have come under attack and their membership has eroded. In contrast, it is far too late to control the proliferation of cyberweapons, since they are potentially within the capabilities of virtually any country in the world. After all, North Korea, one of the poorest and most isolated countries on Earth, has developed world-class expertise in cyber.

The unique characteristics of cyberwar raise new and difficult political questions related to but different from those raised by nuclear weapons. These would have to be addressed in any arms control discussions. Given the wide range in the intensity of possible cyberattacks, what level of attack would constitute an act of war? Given the difficulties of attribution, once a cyberattack has taken place, how sure must a victim be about who or what the attacker might be before retaliation is justified? What would be the appropriate kind and level of retaliation for different kinds of attack? Are there civilian targets against which any attack should be forbidden—emergency response teams, hospitals, dams, civilian airplanes, automobiles, and nuclear power plants, for example—or is virtually all civilian infrastructure to be considered of dual use? What should be the consequences for violating any such prohibition? Should attacks on military command-and-control centers be forbidden, especially those that control the launch of nuclear weapons, lest some lesser officials take matters into their own hands and launch a strategic attack at a time of crisis in the absence or inaccessibility of their country's central authority?

Another important issue arises from the fact that a cyberattack need not have come from a government but, rather, might originate with activists, terrorists, or criminals who may or may not be in formal or informal cooperation with the government of the country in which they operate. This raises yet more attribution problems. It also means that formal treaties among governments must be complemented and often preceded by informal, non-binding norms that define and codify widely accepted ways of behaving to guide and constrain potential attackers and defenders.[34] All these issues require collaborative, multidisciplinary research analogous to the research that developed the concepts and the shared understanding that made possible the multilateral regime for nuclear arms control.

First Steps Toward Rules of the Road for Cyberconflict

Developing norms for cyberwarfare is likely to take time. As in the case of nuclear arms control, the evolution of norms concerning cyberattack is hindered by the deep ideological divide between democratic and authoritarian powers. Democratic powers prefer to concentrate on norms related to attacks on critical physical and digital infrastructure, and to argue that existing laws of war can be extended and adapted to such attacks. To most authoritarian governments, in contrast, the free flow of information is a threat to political and social stability and hence is viewed as a form of cyberattack. Even so, both sides share an interest in avoiding a cyber-apocalypse that causes extensive damage to their cyber-controlled infrastructure. Similarly, the shared interest in avoiding nuclear war drove deeply hostile nuclear superpowers to arrive at nuclear arms control regimes during the Cold War. Nevertheless, it took more than a decade of careful analysis and negotiation, involving academics, military officers, thinktanks, nonprofit organizations, and government diplomats from many countries to agree to these regimes. There is a similar need for norms governing the application of geoengineering and gene driver technologies, although for different reasons that we shall discuss in chapter 12.

The most important contribution of the democratic powers to these issues date is the Tallinn Manual, now in its second, 2.0 version. This is an authoritative, 562-page restatement of existing international law as it applies to the use of cyberweapons, both in war and in peacetime.[35] The Tallinn Manual was developed from a consensus of the personal views of experts at a series of meetings at a NATO cybersecurity think tank in Tallinn, Estonia. It does not have the status of an official NATO document but it has been endorsed by the high-level

Disarmament Committee of the United Nations. On the substantive side, the Tallinn Manual suffers from the important limitation that it considers a cyberattack to constitute an act of war only if it results in physical damage to the target. This does not correspond to historical experience, as, for example, the Russian attacks on Georgia and Estonia described earlier. To date, the only time that war has been threatened in response to a cyberattack occurred in 2016, when President Barack Obama threatened war if Russia attacked the mechanics of the 2016 election.[36]

On the multilateral front, the UN Group of Governmental Experts (UNGGE) on cybersecurity agreed in 2015 on norms calling for international cooperation, governments taking responsibility for acts that take place within the borders of their country, refraining from attacks on emergency response teams or on civilian or other critical infrastructure, reporting vulnerabilities when they are discovered, and providing technical assistance to countries that need and request it. The norms called for respect for the right of privacy but were somewhat ambiguous on the broader question of whether human rights law should apply in cyberspace.[37] When follow-up meetings failed to reach further agreement, the UN established a working group of government delegations to continue to seek common ground.[38]

On their side of the ideological divide, Russia, China, Kazakhstan, Kyrgyzstan, Tajikistan, and Uzbekistan have proposed a "International Code of Conduct for Information Security" that asserts that "policy authority for Internet-related public issues is the sovereign right of States," and that States must be prevented from using their "dominance in . . . communications technology" to undermine the "independent control of information technology goods and services." It further "recognizes" that the "right and freedom to seek, receive and impart information . . . may . . . be subject to certain restrictions . . . necessary . . . for the protection of . . . national security and public order"—all of which is UN-speak for government control of information flows, protection of the domestic information technology industry, and the doctrine of "data sovereignty" we discussed earlier.[39]

In the private sector, the President of the U.S.-based high-tech company Microsoft has proposed a Geneva Convention (i.e., a treaty) on Cybersecurity, with three principles: no attacks on privately owned targets or on critical infrastructure, help to private companies in addressing their vulnerabilities, and an obligation to announce security vulnerabilities detected in software rather than stockpiling them for possible use in a future cyberattack. In return, companies would agree to protect consumers and not to assist governments in developing software for cyberattacks or in carrying out such attacks.[40] A bilateral agreement between the governments of the United States and China in 2015 also addresses damage to the private sector, providing that neither country would "conduct or knowingly support cyber-enabled theft of intellectual property."[41]

The Tallinn Manual, the UNGGE recommendations, and Microsoft's proposed Geneva Convention on Cybersecurity, taken together, could constitute first steps toward an international consensus concerning "rules of the road" for the different levels of cyberconflict. Further progress toward an effective arms control agreement on cyberconflict will require a broad range of actions on both the technical and political fronts, many of which will involve cooperation and collaboration among a broad range of people and institutions, some of whom will be in countries whose governments are hostile to each other. These include research in academic institutions, thinktanks, research institutes and NGOs to work out conceptual and practical problems; informal discussions and formal negotiations among governments to clarify mutual understanding of national interests and how issues are perceived in different countries; and wideranging communication and discussion to ensure public understanding of the issues and support for whatever agreement is finally achieved.

If all goes well, this process could lead to formal agreements that bind those governments that are parties to them, or to rules that become accepted as customary international law.[42] Ultimately, however, coming to a consensus on the "rules of the road" needed to avoid a cyber-apocalypse will require governments to identify areas of common ground and agree to disagree on issues like data sovereignty and control of information and information technology. In these areas, the ongoing geopolitical and technological conflict shows every sign of continuing and intensifying.

Any arms control-type agreement on cyberconflict will require compromise on all sides, and compromise requires giving up freedom of action that some countries might prefer to keep. Many countries seem to be reluctant even to accept the idea that cyberconflicts should be governed by any rulebook, official or unofficial, and in many instances have proved unwilling to acknowledge that the cyberweapons that they possess are anything but purely defensive. Perhaps the international community will find better luck in establishing norms for geoengineering and gene driver technology, which we discuss in chapter 12.

10
Frugal Innovations for the "Bottom of the Pyramid"

Indian amputees are often young and poor. Some of them have lost a limb due to disease, others from accidents like falling off of the roof of the train on which they were riding in order to avoid paying for a ticket. These young people need an artificial leg in order to avoid becoming a burden to their impoverished families—but not just any artificial leg. An artificial leg should look like the real thing. It should be waterproof so that its wearers can walk through muddy fields behind bullock-drawn plows. It should have a working ankle and a natural-looking foot so that its wearer can sit without embarrassment in sandals, cross-legged on the mud floor of a traditional home.

Enter the Jaipur Foot, a textbook example of a frugal innovation—an innovation that applies innovative technology to a problem specific to poor or otherwise disadvantaged people without loss of quality or performance essential to the user. In the early 1980s, I had the privilege of visiting the clinic and workshop of the late Dr. P. K. Sethi in Jaipur, India. Dr. Sethi explained to me how he and his collaborator, R. C. Sharma, had combined several different rubber polymers to create an artificial leg with a working ankle and a realistic foot, sealed to an aluminum tube. The resulting prosthetic is waterproof and looks pretty much like a natural leg and foot. A guide took me to a workshop where prosthetics were being fabricated at a cost of only $50 each.[1] The workshop can be duplicated in a typical Indian a town of a few hundred thousand people. At the end of the tour, my guide climbed a small tree and jumped down from a branch six feet or so off the ground. He then lifted his pants leg to the knee to expose his Jaipur foot.

Sethi and Sharma worked at a government hospital, and their funding came from the Indian government and from an Indian private foundation. But they did not exemplify the stereotype you may have of government-funded researchers. On the contrary, they thought like businessmen. Jaipur is not accessible for the vast majority of Indians, so they franchised the operation, setting up similar clinics and workshops all across the country. They have since transferred their technology to Angola and Cambodia, where wars have left a legacy of land mines that explode beneath the unwary—especially children—and exact a constant toll of lost limbs.[2] Some 1.8 million Jaipur feet and legs have been distributed free in

The Survival Nexus. Charles Weiss, Oxford University Press. © Charles Weiss 2022.
DOI: 10.1093/oso/9780190946265.003.0012

thirty-three countries—an extraordinary record of scale up for a simple but sophisticated technology.

Mobile Finance: The Cell Phone Fills a Banking Vacuum

Here's another example of a frugal innovation. Walk down the few blocks that constitute the main street of Maseru, the capital of Lesotho, one of the poorest countries in Africa, and you will be buttonholed several times and pressed to buy prepaid minutes for your cell phone. Drive down a Lesotho country road, and you will pass huts that double as beauty shops or hardware stores *and* as sellers of prepaid minutes. Cell phones have penetrated to the remotest part of Africa. Once a luxury product for businesses in advanced countries, they have trickled down to people of all income levels and have spread farther and faster than any other innovation in history, far outpacing the spread of telephones, televisions, and electric lights.[3]

It didn't have to happen this way; early cell phones were not technically superior to landlines when they first appeared on the market. Their selling point was that you could get one almost immediately if you were willing to pay for it, in contrast to the five-to-ten-year wait for a landline that you would have faced in many countries. The problem with landlines at the time was rooted in the institutions and regulations that controlled them. Former European colonies in Asia and Africa followed the European practice of combining the post office, the telegraph, and the telephone company into a single government PTT monopoly, funded from tax revenues and usually forbidden to raise capital by borrowing from private sources. Since tax evasion in these countries was widespread, there was never enough money in the public coffers for a PTT to expand coverage or even to carry out routine maintenance.[1] As a result, once you did have a landline, monthly charges were low but service was terrible. Even a dial tone could be a fairly rare event. In the 1970s, I visited the office of the president of a large Indian company, on whose desk sat ten old-style dial telephones. When the president wanted to make a call, a peon (unskilled worker) lifted the receiver of each telephone instrument in turn until one of them yielded a dial tone, at which point he would hand it to his boss. I believe this was the peon's full-time job.

Development theorists, who set the tone for public investments and foreign aid projects at the time, pretty much ignored the importance of communications infrastructure to the economic development of low-income countries. Telecommunications, they thought, were a luxury, not a necessity, so it was not a serious problem if they were starved for capital. Rather than being a critical tool of business and agriculture, they thought that telephones "enabled the prime minister's wife to call her hairdresser." (This myth reflects the sexism of the time.)

The intellectual and financial neglect of telecommunications in developing countries was so complete that the International Telecommunications Union, the United Nations technical agency concerned with telecommunications, was actually making a controversial statement when it stated, in a plaintive major report issued in 1980, that "[t]elecommunications should be regarded as . . . an essential component in the development process which can raise productivity and efficiency in other sectors and enhance the quality of life in the developing world."[5] You'd think this hardly needed to be said, even in 1980!

Enter the cell phone. Few PTT authorities had had the foresight to draft their enabling legislation in such a way as to define their areas of monopolistic control to include telecommunications services based on unanticipated new technology. This fortunate omission allowed the newfangled cell phone to establish itself in a niche market consisting of businesses willing to pay premium prices for service, however limited, that was at least immediately available. This allowed cell phones to attract enough subscribers that there were always enough *other* subscribers to make the price of a phone worthwhile, and in this way to achieve what economists call network economies. (My phone is useful only if there is someone else to talk to; other people's telephones are more valuable because I have a telephone; the more other subscribers there are, the more valuable is my telephone, and so on.) Cash-starved PTTs couldn't expand fast enough to compete with the new service.

The cell phone quickly improved its performance and became the dominant means of telecommunication. Its ubiquity made possible any number of further useful innovations: reminders to AIDS patients to take their many pills each day, telephoned advice from doctors or agricultural extension services, easy reporting of the findings of field surveys of household income or nutrition back to a central office, and telephone calls to different market towns that let coffee farmers or fishermen find out where to get the best price for their products.

The most important of these "secondary" innovations was mobile finance, or M-PESA in Swahili, the lingua franca of Kenya (where M-PESA originated) and most of East Africa. M-PESA addressed a long-standing problem in Kenya: securely and cheaply remitting money from urban centers back to rural villages. By the time of M-PESA's launch in 2007, many Kenyans had left their native villages for Nairobi or other cities and needed to send money home to their families at regular intervals. But most Kenyans did not have bank accounts, few had credit cards, and most business was transacted in cash. The post office was unreliable, so the best way to get money home had been to entrust cash to a relative or friend and have them carry it to their village by bus and deliver it in person—a slow, expensive, and dangerous process in a country where theft is common.

The idea that became M-PESA was first developed with public money by a team of researchers working from a grant from the development assistance agency of the British government. The researchers' original idea was to create a system for expediting the granting and repayment of microfinance loans to small-scale entrepreneurs. They soon discovered that their customers were using M-PESA as a general-purpose system for sending remittances home and paying bills. They did this by using cell phone airtime as a kind of currency, buying minutes in one place and transferring them to a creditor or a relative in another place, who could then sell them for cash or credit.[6]

Thinking like businesspeople, the researchers recognized that this was a bigger opportunity than microfinance. Working with Vodaphone, a British-based telecommunications company with investments in several African countries, they created a network of trusted agents throughout Africa and West Asia through which people can reliably transfer money, pay bills, make deposits in and withdrawals from formal banking institutions (if they do have an account), and carry out other banking transactions that we take for granted but that were unavailable to most Kenyans. Here they benefited from a ruling by the Kenyan government that M-PESA was not subject to the cumbersome rules then governing banks, even though it was carrying out characteristic banking functions.

M-PESA is an example of a technological innovation that benefits the poor directly, although it was built on a platform—the cell phone—that trickled down from an application that had first benefited the better-off. The story illustrates many of the secrets of success for such innovations. First, the idea for M-PESA began, not with research motivated by the hope of private profit, but with a development aid grant to a group of researchers who were hoping to improve the lives of "unbanked" East Africans. Second, the researchers did not try to commercialize their idea by themselves but, instead, collaborated with private business to scale it up. Third, the innovators and their business partners responded to market demand and changed their product into something quite different from their original conception. Fourth, M-PESA was built on a platform, the cell phone, whose initial success depended on loopholes in government regulations that might otherwise have squelched the entire innovation. It later benefited from being exempted from a government regulation that applied to other organizations that carried out similar functions. We shall see these secrets of success play out in different contexts as we recount the histories of other innovations that have benefited the most economically disadvantaged around the world.

Some Technologies Won't Trickle Down; They Need Special Attention

Most innovative new products are first marketed to high-end consumers who are willing to pay extra for the latest and the best and gradually make their way down to cheaper products. By then, the costs of research and product development have been amortized, the product has been standardized, its bugs have been worked out, its manufacture has been scaled up, competition has increased, prices have come down, and more and more people can afford the product.

This trickle down of innovation from the well-off to the poor works fine for cars and computers and, indeed, for most products. But some requirements of the poor—a malaria vaccine or an emergency food ration for starving children, to cite two examples—require specific innovations for which there is no "Cadillac." We call these orphan technologies because the people who need them do not constitute a market sufficient to attract private investment. Developing them is an ethical imperative but is not usually a profitable undertaking. These products must sustain themselves in the marketplace from the beginning, despite the fact that their intended users do not have the resources to amortize the costs of innovation.[7] This means that public-spirited intervention by a government or a private organization is typically necessary at one or another stage of product development if such products are ever to come into wide use. In some cases, a product already on the market for one purpose can be repurposed for another, as in the case of cell phones and mobile finance. In another example, Ivermectin, a deworming medicine for livestock, was repurposed as a cure for river blindness and elephantiasis, a mosquito-borne parasitic disease whose victims' limbs swell up enormously and leave them disabled.

In Western countries, the idea that special products may be needed for the specific needs of low-income communities began with the "appropriate technology" movement in the 1970s, which was inspired and led by E. F. Schumacher. In his classic book, Schumacher criticized the large-scale, capital-intensive, "advanced" imported technologies he found in newly independent developing countries and advocated technologies "intermediate" between these and the inefficient, traditional, small-scale technologies that were then common in places into which more modern technologies had not yet penetrated.[8] Such "appropriate" technologies, he argued, would be small in scale, locally produced, easily repaired, environmentally sustainable, and culturally compatible.

The first issue faced by the appropriate technology movement was whether such small-scale technology was even technically possible. From the strictly engineering point of view, most manufacturing technologies benefit from economies of scale: the higher the capacity of the machine, the cheaper it is to produce each unit of product. As proof-of-concept examples, George McRobie, Schumacher's engineering colleague, showed that it was, in fact, possible to design and produce two proof-of-concept machines for the efficient, small-scale production of egg crates and palm oil.[9] These, he argued, were well suited for small-scale manufacturers in remote areas where poor transportation and a limited supply of raw materials made large-scale production impractical.

But it's a long way, filled with hurdles of all kinds, from such a prototype to the widespread use of a new machine, which is what distinguishes an innovation from an invention. Some of these hurdles are technical, like developing a practical, attractive, durable product that is easy to use and maintain; others have to do with marketing, distribution, business management, commercial context, and economic policy. On the technical side, like any other commercial product, "intermediate" technologies have to undergo a period of experimentation, failure, redesign, and market development before they become robust, manufacturable, easily usable and repaired, and widely accepted. All this has to be paid for. On the business side, communications and marketing techniques in pre-Internet days could not reach enough isolated and widely scattered potential customers to allow the manufacture of a small machine to achieve economies of scale. (Small machines need to be sold in large numbers in order to bring their price down, just as big ones do.) In any case, many "appropriate" technologies were developed by nongovernmental or community development organizations that were not accustomed to doing the market research needed to understand the practical requirements of potential customers.

The policy and commercial context for small machines posed especially daunting obstacles. Small machines faced competition from well-entrenched, imported technologies that were often made artificially profitable in developing countries by subsidized loans to well-connected factory owners. These stacked the odds *against* small-scale technology that employed large numbers of people even in countries with very cheap labor. Few small machines were able to overcome these obstacles.[10]

A compelling lecturer and advocate, Schumacher inspired the formation of the U.K.-based Intermediate Technology Development Group (ITDG, now called Practical Action). As ITDG branched out beyond small-scale machines, it found itself entering more and more into issues of economic policy, in effect encountering the Survival Nexus that is the theme of this book. In the energy sector, for example, the choice between renewable energy and fossil fuels or between centralized and decentralized generation of electricity has as much to do with economic policy as with engineering, as we saw in chapter 4. Similarly, safe and energy-efficient transportation is as much a function of city planning and hence of municipal politics as it is of vehicle design. To deal with this mix of technology, economics, and politics, ITDG spun off client-oriented consulting firms, IT Power and IT Transport, respectively.

"Appropriate Technology": Philanthropy Plus Engineering

The engineering-oriented nongovernmental organizations (NGOs) that formed the core of the appropriate technology movement have left a legacy of useful devices. Of these, the one in the most widespread use is the improved wood stove, which has been reproduced in the by the millions in countries as widely separated as China, India, Kenya, Senegal, and Guatemala.[11] One version of these stoves, the so-called smokeless chula, was originally developed by NGOs in El Salvador and the Netherlands with the initial objective of reducing deforestation by making the use of firewood more efficient. The original version of the chula was made of cement and installed in a kitchen by local artisans. In addition to burning less wood, it boasted a chimney, meaning that women and girls in cold, mountainous regions no longer had to spend hours each day wheezing in smoky kitchens, rendering them susceptible to asthma and to lung and heart problems. (Doctors were major champions of smokeless chulas in both India and Nepal.) In Southeast Asia and East Africa, by contrast, housewives preferred the more modern-looking stove developed in an applied research laboratory in Thailand. This consisted of a metal shell and a ceramic insert suitable for burning charcoal. Known in East Africa as a jiko stove, this model caught on in Kenya after it was marketed as a gift that a man could bring home to his wife on payday.

The experience with improved wood stoves revealed problems in both technology and marketing. The first models of the smokeless chula actually used more wood than the traditional method of placing a cooking pot over an open wood fire, supported by three stones. Dutch engineers solved this problem by careful redesign of the firebox, but only after two or three generations of product development. On the marketing side, the first instinct of developers was to market improved stoves in rural areas where deforestation was taking place. They soon discovered that their real market was in the cities and towns, where firewood had to be purchased with cash and people were willing to invest money in order to conserve it. (In rural areas, unpaid women and girls gathered firewood—a serious problem, since it discouraged families from sending their daughters to school, but not a problem of cash, and hence not one that improved wood stoves could solve.)

From this kind of experience, NGOs learned that despite their nonprofit motivation, they, like the inventors of the Jaipur foot and the developers of M-PESA, needed to learn to think like business people. If the first generation of a given product did not sell, either because of poor technical design or because of inadequate attention to customer preferences, additional product generations would be needed before these problems could be overcome. Nevertheless, subsequent experience with the adoption of improved stoves has varied from country to country.[12] In China, hundreds of millions of improved stoves are known to be still in use. In India, by contrast, hardly any of the more than thirty million improved stoves distributed free by the government are still in use, in large part because they did not meet the culinary, social, and cultural needs of housewives as they themselves saw them.[13]

Some "appropriate" products run up against deep-seated cultural obstacles, as Arunachalam Muruganantham discovered when he set out to design a sanitary napkin that could be sold for 2.5 rupees (five U.S. cents), an eighth of the price of the pads that were then commercially available in India. Such pads were to be more than just a convenience to Indian women, who faced multiple restrictions on their lives during their monthly periods and risked infection from the unsanitary and unsightly rags that they had no choice but to use. Teenaged girls would miss or even drop out of school because of the embarrassment—a major impediment to girls' secondary and higher education. During six years of product development, Muruganantham produced a cheap, comfortable, biodegradable sanitary pad by reverse engineering (i.e., figuring out how to reproduce) the manufacture of the compressed fluffy cellulose that enables sanitary napkins in advanced countries to absorb large quantities of menstrual blood.[14] The son of a hand-loom weaver, Muruganantham followed the Gandhian tradition and developed small-scale machines to enable the pads he had designed to be made and marketed locally, often by local women's cooperatives.[15]

Can Frugal Technology be a Business Opportunity?

If NGOs that develop frugal innovations for the poor must think like businesses, can businesses think like NGOs and develop profitable, frugal innovations? Perhaps low-income communities constitute an underserved market, and businesses can profit from making and selling them the things that they need. This was the theme of the influential book by the late C. K. Prahalad. Writing for a business school audience, Prahalad documented the aggregate market power of poor people around the world and argued that there lay a "fortune at the bottom of the pyramid" for those businesses who understood how to find it. It is possible, he said, to "eradicate poverty through profits."[16]

Despite the fact that they often lived from hand to mouth, Prahalad stressed, lower-income customers want the same products as wealthier folks and insist on high quality—but they need the product, marketing, and distribution to be custom designed to fit their needs. For example, cash-short customers need to buy soap and catsup in small packages costing a few cents each so that they can purchase each day's supply separately. The sales efforts of private companies, he found, can extend to the promotion of public health. Lifebuoy India, for example, promoted the sales of its hand soap by an extensive educational campaign extolling the sanitary virtues of handwashing.

A prime example of the Prahalad approach is PlumpyNut, a tasty, nutritious, peanut-based emergency ration for severely malnourished children, many of whom are fleeing from wars or natural disasters. PlumpyNut was developed by Andre Briend and produced by Nutriset, a French company, as a more stable and better-tasting alternative to the milk-based product that had been developed for the same purpose by the World Health Organization. In 2017 alone, companies produced 87,000 megatons of PlumpyNut, treating up to 6,600,000 children.[17]

The major changes in the world economy since Schumacher and McRobie wrote in the 1970s have altered the economic context for appropriate or frugal technology. India and China have emerged as "innovative developing countries"; they have enough lower-income people to constitute a sufficient market to attract investment in technological innovation and new product development to meet their specific needs, as well as firms capable of developing new products to meet those needs. The globalization of the world economy, combined with the advent of the Internet and of containerized shipping, has made it easier to export low-cost manufactures—cheap motorcycles and electrocardiograms, to cite prominent examples—to developing countries all over the world.

Two successful models of such frugal technology are the coolers produced in India by the Mitticool and Chotukool companies for customers with unreliable sources of electricity or none at all. Mitticool is not a refrigerator but is, rather, an evaporative cooler that does not rely on electricity. Chotukool is a simple, small, lightweight, portable refrigerator with a battery option. It uses a cooling fan in place of an expensive compressor. Chotukool plans to upgrade its coolers for upscale markets in India and Africa in hope that their lower-income customers will look to these upscale purchasers as examples. If successful, this could cut into the market for full-scale refrigerators, putting it onto the classic trajectory of a disruptive technology (see note 7).

On the other hand, frugal technology has also seen much-ballyhooed failures. Of these, perhaps the best known is the Nano car produced by Tata, a well-established Indian automobile and truck manufacturer. Originally conceived as a $3,000 "people's car" and the recipient of numerous early awards for frugal innovation, the Nano was plagued by production and safety issues and never came close to its expected sales. Customers often preferred to save up for more expensive and prestigious brands rather than settle for a "poor man's car." The end of the Indian version of the Nano came in 2017, when the company decided not to make the investments necessary for the Nano to pass anticipated new safety and emissions requirements in India and the European Union. In the meantime, other cars with superior quality were steadily coming closer to the Nano's cheap ticket price.[18] The unfortunate experience of the Nano shows how good ideas for frugal technology can be defeated by lack of attention to mundane aspects of marketing and engineering.

Another instructive experience with frugal technology is that of the One Laptop per Child (OLPC) project, a pioneering effort by the Media Laboratory at the Massachusetts Institute of Technology to develop a stripped-down, $100 computer specially designed for use by schoolchildren in low-income countries. Intended to last five years under demanding field conditions, the OLPC computer had no hard drive and featured thick protective walls, a sealed, rubber-membrane keyboard, a minimum of failure-prone internal connectors, and a rechargeable battery adaptable to any of a variety of power sources. OLPC received a great deal of favorable publicity and won several large orders from the governments of low- and medium-income countries for distribution to elementary schools. In practice, however, it has received mixed reviews. In medium-income countries like Peru and Uruguay, OLPC gave promising results when carried out by well-trained teachers. Despite its best efforts to reduce costs to a minimum, however, OLPC came out at double its projected price. It turned out to be unnecessary in middle-income countries and too expensive for the lowest-income countries, which could have spent their money on more and better teachers, classrooms, and school equipment.[19]

In sum, then, two decades of experience have shown that profits do exist at the bottom of the pyramid, but that it's harder to make them than early optimists had predicted. It takes time and costs money to redesign products and packaging, to develop new marketing and distribution channels, to recruit and train staff, to educate consumers, and to scale up production. Poor infrastructure still makes it difficult to sell products to widely dispersed customers in remote areas. By the time these obstacles are overcome, established commercial alternatives (smartphones, in the OLPC case) may have trickled down to lower-income customers. Because of these heavy initial costs, relatively few private business efforts have been able to achieve their customary profit margins from selling to the bottom of the pyramid without government or foundation support at some stage. Companies have generally accepted the inevitability of low rates of return on capital, either transferring their frugal innovations to a designated internal department of social responsibility or seeking external finance from government or nonprofit sources.

Businesses Without a Profit Motive

We have just seen that nonprofits intending to develop frugal technology for lower-income people need to think like businesses, and that businesses intending to develop such technology can't make much profit. Hmmm. What about a nonprofit business: one that thinks and acts like a business but judges itself by its social impact rather than its bottom line?

Such a business is Equalize Health (formerly D-REV), which describes itself as "a non-profit product development company focused on the needs of the poor [that] design[s] and deliver[s] world-class, affordable health technologies."[20] D-REV thus combines the motivations of an NGO with the mentality and methods of a profit-oriented business. These characteristics are especially important when an innovative product is scaled up from prototype and first market development to widespread distribution and impact.

D-REV's flagship products are an artificial knee and a small, portable lamp that applies phototherapy to infants with neonatal jaundice, an easily treatable condition that affects about 60% of newborns but nevertheless causes tens of thousands of deaths each year in India alone.[21] The Brilliance phototherapy lamp is capped at only $400, as opposed to the $3,000 price of a comparable appliance in the United States. D-REV estimates that it has saved the lives of some two hundred thousand infants.[22] The knee, marketed as ReMotion, complements the Jaipur foot and was initially fitted by the same foundation as the Jaipur foot.[23] Its price is $80, about one-fifth of the price of the corresponding device sold in advanced countries. Each of D-REV's products is evaluated by an independent consulting organization, which estimates the impact of the product, as measured, for example, by the number of disability-adjusted life years (DALYs) avoided, and which divides that number by the total cost of identifying, designing, developing, manufacturing, and distributing the product. (The idea of a DALY is explained in Annex B). Evaluators found that the cost-effectiveness of D-REV products was an impressive $600 per DALY avoided. [24]

Another nonprofit business is the Aravind Eye Care Center. Across India, people with impaired vision can be treated at one of Aravind's forty local vision centers for an extremely low price, or even for free. Thanks to the Aravind technique of convenient and quick diagnostics and assembly-line cataract surgery, people emerge from these centers dazzled by the light and awed by their newfound ability to enjoy things they had not been able to see for many years. The Aravind vision centers are an outstanding example of how frugal technologies can be scaled up to meet the needs of large numbers of poor people. Aravind centers and hospitals accounted for 350,000 operations for cataracts and other eye diseases in 2017—a remarkable scale up. Still, even this extraordinary number pales in comparison with the two hundred million Indians who suffer from serious eye conditions.[25]

Aravind has reduced the cost of cataract surgery to $50, about 1/60 of the U.S. cost, by rigorous standardization of procedures, multiple sets of equipment, multiple teams of carefully trained nursing assistants drawn from local communities, and high productivity—more than two thousand operations per surgeon per year, or six to eight per hour (one every eight to ten minutes!). Aravind also reduces the cost of the artificial lenses used in the surgery by manufacturing them itself. It uses telemedicine to identify patients with more serious problems and refers them to one of its operating centers or to affiliated hospitals located all over the country. Aravind makes no profit but is financially self-sufficient; it receives a small government reimbursement of $10 per procedure. Patients who can afford to pay cross-subsidize those who can't.

"Orphan Technology": A Need But No Market

The health problems of poor people in developing countries are not generally attractive business opportunities for profit-oriented companies, as explained earlier in this chapter. Malaria, for example, afflicts a billion or so people in developing countries, but a billion people with no money do not constitute a market sufficient to attract risk-oriented private investment in research and development. Vaccines or cures for malaria and other tropical diseases are therefore some of the most dramatic examples of orphan technologies—technologies for which there is an urgent and acute need but that lack a commercial market. For this reason, research on malaria, as well as on schistosomiasis ("snail fever"), filariasis (including elephantiasis), trachoma, leishmaniasis, and other scourges afflicting over two billion mostly low-income people, languished in underfunded obscurity for decades.

Fortunately, governments and NGOs have paid fresh attention to these and other health problems of low-income people, sometimes in collaboration with the nonprofit arms of private companies, and major research programs are now at last underway to address them. Chapter 7 describes how the Bill and Melinda Gates Foundation sponsors research and trials of malaria vaccines. Two examples of this kind of public–private international cooperation are effectively combating little-known scourges in Africa: river blindness and Guinea worm. In each case, a technology developed by a private firm in the United States for an entirely different purpose turned out to be effective against a neglected tropical disease, and large quantities were donated to an NGO for use in Africa.

In the case of river blindness, Ivermectin, a profitable veterinary drug developed by Merck, turned out to be a cure for a disease that had caused almost everyone who tried to farm the most fertile (or more accurately, the least unfertile) parts of the African Sahel to go blind by the age of thirty-five or so. Merck's donation of billions of doses of Ivermectin complemented the use of pesticides to spray the habitat of the black flies that spread the disease, and it greatly increased the effectiveness of the river blindness campaign of the World Health

Organization (WHO) and the World Bank, so that the disease is now substantially under control.

The case of Guinea worm is somewhat similar. This parasitic disease is caused by a roundworm whose intermediate host is a tiny crustacean that infests unprotected wells, chiefly in West Africa. After the crustacean is swallowed in drinking water, it develops into a worm that can grow to a length of several feet and disable its unfortunate host. The Atlanta-based Carter Center, the retirement project of former U.S. President Jimmy Carter, discovered that a fine-mesh plastic filter developed by the Dupont Company for an entirely different purpose could protect African wells from this infestation. Funded by a grant from the Sasakawa Foundation in Japan, the Carter Center succeeded in its thirty-five-year campaign to protect thousands of African wells and reduce the incidence of this disease from 3.5 million cases to a few dozen.[26]

Sometimes, a preexisting technology cannot be used to solve a particular health problem, so a specific innovative effort, sometimes based on sophisticated basic research, is needed. For example, cholera was a major cause of death in most low-income countries, especially for young children. For the first half of the twentieth century, it seemed like there was little that could be done to help. Then, in the 1960s, researchers in Dhaka, Bangladesh, discovered that cholera patients died not from the disease itself, but from the dehydration and malnutrition it caused by upsetting the biochemistry of the intestinal lining. Further study of this basic biochemistry led to the realization that a simple solution of salts and sugar—technically named oral rehydration therapy (ORT)—could sustain cholera patients until their immune systems could deal with the infection on their own. The WHO, UNICEF, and many other organizations now routinely distribute sachets of sugar and salt in correct proportions, together with instructions in many languages, to mothers all over the world, thereby saving more than a million lives each year.

The Best You Can Afford: The Ethics of Toilets

In many cities in low-income countries, the prevailing "sanitation" technology is the flying toilet: put the excreta in a plastic bag and throw it out the window. Children playing in the street below contract diarrheal diseases and die from them at the rate of approximately 750,000 every year.[27] Flush toilets would solve the problem, but they are expensive, especially when you count the cost of the water needed to flush them and the sewage pipes needed to carry away the resulting "black water." Indeed, while the history of water-borne sewage extends back some five thousand years to the Harappan kingdom of ancient South Asia, only the wealthiest empires in history could afford them—the Romans, the Victorian British, and the high-income countries and people of today.

In a major research project, the World Bank of the 1970s looked for an alternative to water-borne sanitation that would be affordable to low-income people in developing countries in the sense that it was the best and most convenient service that could be provided at a manageable fraction of their per capita income. A team of engineers and economists surveyed the world for low-cost alternatives to conventional water-borne sanitation and sewage treatment. They discovered, first, that research on low-cost sanitation suitable for tropical countries was sadly neglected. The major research on latrines at the time, for example, dealt with the problems of campsites in national parks under the very different physical and economic conditions of high-income temperate countries. Second, they found that a laboratory in what was then Rhodesia (it's now Zimbabwe) had developed an innovative design for a low-cost, non-waterborne, non-smelly, non-fly-ridden, ventilated improved pit (VIP) latrine.[28]

Somewhat later, projects in Brazil and Pakistan applied a low-cost approach to water-borne sewerage, the so-called condominial sewerage system, which permits the use of flush toilets but saves costs by using shallow, relatively small-bore pipe, whose contents are eventually discharged into the regular municipal sewage system (assuming that one exists).[29] Condominial systems require strong community organization, since a blockage anywhere in the system cuts service to all customers downstream and must be unblocked by the consumers themselves, an unpleasant task. This is a prime example of an innovative, inclusive technology that involves the community in its planning and deployment and fails in its objective if this aspect is neglected.

The success of these low-cost solutions leaves important philosophical and ethical questions unanswered. Numerous United Nations resolutions have held that every person has the right to health, clean water, and a clean environment simply by virtue of being human. If sanitation is, indeed, a human right, why should poor people have to settle for a VIP latrine or a condominial sewer when their middle-class neighbors enjoy flush toilets, water-borne sewerage, and subsidized drinking water to boot? At a practical level, the answer is obvious: the best is the enemy of the good enough. Public money for more sophisticated services for poor people is not going to be available in many countries anytime soon, and people have no practical choice but to settle for what they can get. Better to dig the latrine or organize the community for condominial sewerage than to wait forever for politicians to make good on their promises for a "proper, high-class" service.

Analogous questions could be posed regarding many other applications of "intermediate" or "appropriate" technology. Why must poor people settle for basic technologies that may be minimally adequate, but that are manifestly inferior to those available to the better off? The only possible answer is that life is unfair. But at least some of the technical problems affecting poor people in developing countries are now attracting serious attention from scientists and inventors, and knowledge is increasing about the practical problems connected with the scaling up of the needed innovations. We now turn to the CGIAR, originally known as the Consultative Group on International Agricultural Research, the largest and longest lived of these efforts.

The CGIAR: Almost $1 Billion a Year for Research for the Small-scale Farmer

Cassava is the main ingredient of tapioca pudding, one of my favorite desserts when I was a kid and still a beloved comfort food. But to poor farmers in Africa's drought-prone Sahel region, cassava is a main source of food energy and the last defense against famine—a root crop that can stay in the ground for a couple of years until times are tough and you need to dig it out, soak it to remove a chemical that causes cyanide poisoning, cook it, and eat it. That's why a pest called the cassava pink mealybug threatened tens of millions of Africans with starvation when it started to eat its way across the continent by way of the cassava that was their underground reserve food supply. The story of how researchers avoided this catastrophe is one of the great examples of technology applied for the benefit of some of the poorest people in the world.

The pink cassava mealybug is an immigrant, an interloper. It has plenty of natural enemies to keep its population under control back in South America where it originated, but none in Africa, where it found itself able to wreak havoc without opposition. So, the best countermeasure was classical biological control: go to South America, find a natural predator, and bring it to Africa for a good meal. But there's a catch. You have to be sure that the predator eats only the pest you're trying to control and nothing else, lest it start dining on other species after it has finished with the mealybug, get out of control, and make matters worse.

Back in Paraguay, a number of tiny wasps make their living at the expense of pink cassava mealybugs. The wasps lay their eggs inside the mealybug. When the wasp larvae hatch, they eat their way through the body of the mealybug, killing it and leaving behind an empty husk. Then they go looking for another mealybug on which to repeat the performance. Wasps like this are called parasitoids. Hans Rudolf Herren, an entomologist working at the International Institute for Tropical Agriculture (IITA) in Ibadan, Nigeria, went to Paraguay to find such a parasitoid. He sent samples of several different species of wasps to Great Britain and had them tested there under controlled conditions to find one that wouldn't eat beneficial African insects. Then and only then could he test it out in Africa to make sure it could survive and prosper there.

Bingo! Herren found a particular wasp, *Anagyrus lopezi*, whose only food was the cassava mealybug. He introduced it into test sites in Africa and found that it spread rapidly on its own without any further action on his part. The Sahel was saved from starvation, and Herren won the World Food Prize, the agricultural equivalent of the Nobel Prize. Indonesia, Thailand, and other countries in Southeast Asia used the same method to save their cassava crops.

The IITA where Herren did his research is one of fifteen international institutes funded by the CGIAR, the worldwide network of research institutes responsible for most of the improvement in technology for smallholder agriculture in low-income countries since World War II. The history of the CGIAR is the story of practical and organizational challenges of funding, mounting, and sustaining one of the few successful programs of international technology- and development-oriented research whose size and scope are comparable to those of the major global problem that it addresses. It also shows unexpected parallels with those of international efforts to promote global health, as recounted in chapters 6 and 7.

The CGIAR had its origins in the 1940s, when the Ford, Rockefeller, and Kellogg Foundations—three of the great private foundations of the time—fearing that famine would not only be a humanitarian catastrophe but might also lead to social and political disruption that could result in increased support for communism,[30] realized that the technology available for growing staple crops in poor countries was insufficient to support their growing populations and decided together to support research on improved varieties of wheat and rice. The researchers they funded took technology for high-yield, short-strawed, fertilizer- and irrigation-responsive varieties of corn and rice that had been developed in the United States and Japan and adapted them to conditions in tropical countries. The short, stiff straw of the new varieties enabled them to benefit from fertilizer because it could support a larger edible seed without keeling over to the ground (lodging) and rotting there or becoming food for rats and other pests. The new varieties were also bred to be resistant to common pests and diseases, to have a shape more efficient for capturing sunlight, to mature more quickly so as to make possible multiple crops per year, and to mature independently of day length so that they could be grown at any geographic latitude.

The dramatic increases in yield per acre that resulted from the adoption of these varieties became known as the Green Revolution. It averted the worldwide food shortage and famine that had been authoritatively predicted,[31] and it won a Nobel Peace Prize for plant breeder Norman Borlaug. The new varieties of wheat and rice spread rapidly, dramatically increasing yields per acre but also offering harbingers of complications to come.[32]

The first of these complications stemmed from the transition of small farmers from multiple to single crops—from polyculture to monoculture. Traditionally, farmers in developing countries had hedged their bets by planting landraces of staple crops: fields that looked uniform but were, in fact, a mixture of genetic types. In this way, any given pest or disease could destroy some but not all of a farmer's crop, so their family would always have something to eat. The higher yield of the new varieties led them to plant a genetically uniform monoculture that could be wiped out when a new pest or disease emerged. The international centers had thus assumed the responsibility of developing crop varieties resistant to new pests and diseases as fast as they appeared.

A second complication stemmed from the fact that the new varieties owed their increased yield to their responsiveness to irrigation water and chemical fertilizers, as well as to the use of pesticides. Farmers who wished to take advantage of the new varieties had to take out loans at planting time to cover the cost of these inputs and repay them at harvest time. In the absence of crop insurance, this entailed the risk of bankruptcy if there was a crop failure.

To stimulate the adoption of the new varieties, governments in many developing countries subsidized these inputs to encourage farmers to use them. Some also put in place incentives that encouraged the use of tractors and the pumping of groundwater. These policies made sense at the time—these inputs were little used then because they did not increase the yields of traditional varieties—but led to environmental problems later on from overuse of inputs. The new staples also tended to displace a number of nutritious traditional crops.

The biggest economic benefits of the Green Revolution went to consumers, who paid lower food prices for basic foods, demonstrably improving their nutrition. Among producers, the first economic benefits accrued to larger farmers in Asia and Latin America, who adopted the new varieties while grain prices were still high, took early advantage of the input subsidies, used their increased incomes to enlarge their holdings and took advantage of cheap loans to mechanize production. Smaller farmers benefited later and to a lesser degree. African farmers did not benefit from these first varieties, however, in part because of weak national agricultural research systems; in part because land was cheap in most parts of Africa, so increased yields were less important than in Asia; and in part because it had been incorrectly assumed that varieties of staple crops developed in Asia and Latin America could be adapted to African climate, pests, and diseases.

Building on the success of the new varieties, the three foundations established four international agricultural research centers in Mexico, the Philippines, Colombia, and Nigeria. They then approached the World Bank and the United Nations for funding to maintain these institutions and to make them the nucleus of a network that would continue to support their work and would expand it to support research on the food crops and agroecological regions that had not benefited from the Green Revolution. The major seed companies made no objections to this proposal, since their markets in developing countries were for cash crops like coffee, cotton, and rubber; research on food staples did not threaten their businesses.

Responding to this request in 1971, the World Bank joined with the United Nations Development Program (UNDP), a development finance arm of the United Nations, and the Food and Agricultural Organization (FAO), the technical arm of the United Nations with responsibility for agriculture, to form the CGIAR.[33] The CGIAR was an informal meeting of donor countries, analogous to the informal groups that "coordinate" development assistance to poor countries from multiple donors—that is, they avoid duplications and fill gaps. This enabled donors to support the CGIAR and still avoid a permanent commitment to do so.

To help it to define its program, the newly formed CGIAR organized an independent Technical Advisory Committee (TAC) of distinguished natural and social scientists and relied on its recommendations to decide what research should be included in the CGIAR program and what new research centers should be created. The centers themselves retained their independence and separate boards of directors. The result was a remarkable informal structure that provided central oversight and strategic direction without a cumbersome central bureaucracy. In the ensuing decades, the CGIAR was able to face such new challenges as the globalization of agricultural markets, the need for policies to deal with genetically modified crops, the privatization of much agricultural biotechnology, collaboration with private agricultural research laboratories and with newly strengthened national agricultural research services (NARSs), especially those in China and Brazil, and the consequent need for new policies concerning the ownership of intellectual property.

As was expected, the TAC's first recommendation was that a new research center be established to deal with non-irrigated (rain-fed) crops, such as cassava, millet, sorghum, pigeon peas, and chickpeas. These mainstays of the poorest farmers in low-income countries had not benefited much from the Green Revolution. For several of these crops, little was known even of basic genetics and production technology, since most colonial agricultural research laboratories had concentrated on export crops and had paid insufficient attention to the needs of subsistence farmers. By the following year, a fifth CGIAR-supported center, the International Crop Research Institute for the Semi-Arid Tropics (ICRISAT) was up and running in Hyderabad, India—an extraordinary accomplishment for the usually slow-moving United Nations system. By the mid-1980s, research results from ICRISAT and other CGIAR-supported institutes were benefiting small-scale farmers in Africa who had missed out on the first Green Revolution.

Fast-forwarding to 2018, the CGIAR network has expanded from the original four to its present fifteen centers. These cover the gamut of food crops and agroecosystems and have a total budget of nearly a billion dollars—a remarkable achievement, considering the many demands on the limited amounts of development assistance money from governments of wealthy countries. The CGIAR donor base has expanded from the original six organizations to some sixty-five countries and foundations, including $150 million in contributions from the Bill and Melinda Gates Foundation, a private philanthropy. Newer centers study fish, livestock, food policy, water management, forestry, and agroforestry. Networks spanning the centers coordinate work on crosscutting topics like nutrition, farming systems (how agriculture fits in to the life of farm families), gender in agriculture, big data, and climate-smart agriculture.

Since its establishment, the basic CGIAR strategy has been to increase the productivity of food crops, i.e., their yield per acre. In the typical pattern, researchers in one of the international centers develop a new variety with internationally desirable characteristics, such as resistance to a newly emergent pest or adaptability to drought or flood, and then work with NARSs in different countries to adapt it so that it can thrive in the local climate, resist local pests and diseases, fit in with local farming systems, and satisfy local tastes. This strategy has had its brilliant successes and has saved millions from starvation. On the other hand, it depends critically on the strength of NARSs and national extension systems, which is unfortunately highly variable from country to country and which suffered from the de-emphasis of agriculture and the budget cuts enforced by the Washington Consensus in the 1980s.[34] The CGIAR did try to help the NARSs by establishing the International Service for National Agricultural Research (ISNAR), but this experiment did not go well and was eventually shut down.[35]

The ISNAR experience has its parallel in the history of international efforts to aid global health. Indeed, ISNAR's mandate to help the NARSs was analogous to the task that donors to global health sought to *avoid* when they refused to become involved in "horizontal" efforts to build health delivery services accessible to the entire population and capable of treating a broad range of ailments—an attitude that was especially strong during the 1980s, when all government or service organizations (or parastatal organizations that worked closely with government) were viewed with distrust. From this point of view, agricultural research on specific crops constitutes "vertical" efforts analogous to the campaigns against specific diseases.

These two experiences indicate that building local infrastructure, whether for the delivery of health services or for agricultural research, may be difficult to sustain by external initiatives and, instead, requires initiative and commitment from the country itself, sometimes even in opposition to donor desires. If a project initiated by external donors reflects the particular concerns of the donor rather than local needs and priorities, it is likely not to be continued by the host country once the donors have turned their attention to other problems and their money has run out. Conversely, sound projects, whether initiated by donors or the country itself, tend to attract outside money, since everyone likes to be associated with success.

The informal structure of the CGIAR has not stopped it from taking on enormous responsibilities. One of its most important functions is the funding and management of the gene banks of the developing world's major food crops: rice, wheat, barley, cassava, corn, and ten others. A gene bank of a given crop is a curated repository of seeds gathered from all over the world that embody its biodiversity—the variation in genetic heritage of a crop that enables it to thrive under a broad range of environmental conditions and to defeat a broad range of pests and diseases. Whenever a threatening new pest or disease arises, gene banks provide researchers with seeds whose genes for resistance can be bred into international and then into local varieties. The resulting free exchange of genetic material has benefited the entire world.

Gene banks thus are responsible for preserving and managing critical, irreplaceable world resources. Surprisingly, gene banks supported by the CGIAR successfully fulfilled this responsibility without a formal intergovernmental structure for the first forty-odd years of their existence. They derived their legitimacy from an international declaration that the CGIAR was the authorized steward for humanity of the biodiversity of food crops. This pragmatic arrangement was in many ways parallel to the informal arrangements for the governance of the Internet we discussed in chapter 8. Its basic justification was that it worked. In the end, however, gene banks came under the aegis of a formal organization, Bioversity International, by the terms of the 2006 International Treaty on Plant Genetic Resources.[36]

The overall structure of the CGIAR was also formalized in 2008, as it was felt that the informal structure described earlier had not scaled well.[37] The term "consultative group" was dropped, and only the CGIAR initials were retained as the name of the system. As the CGIAR had expanded, three-quarters of the overall research budget had come to be financed as individual projects negotiated with donors, rather than from central funds allocated according to priorities and recommendations set by the TAC, whose status was reduced to that of an advisory body. This greatly reduced the ability of the CGIAR to guide and manage the overall work of the system. This experience parallels that of the WHO, three-quarters of whose budget now comes from projects individually negotiated with individual donors rather than from the organization's own funds, as recounted in chapter 7. Both cases reflect the desire of donor governments to maintain control over the activities of an international undertaking, despite its nominal independence. For CGIAR, a further complication results from the centers' increasing practice of hiring scientists on short-term contracts, hindering recruitment and making it difficult to undertake long-term research that could lead to transformational results.

The stated objective of the restructuring of the CGIAR was to restore a measure of central control in order to facilitate the development of overall strategy in response to changing conditions—an issue that frequently crops up in the management of decentralized research networks of all kinds. The centers, however, resisted the proposed reduction of their independence and emerged even stronger and more independent than before the restructuring. This reinforced the power of scientists to choose and seek donor funding for the specific projects they believed to be the most promising. On the downside, it did not address the long-standing problem of scientists being forced to spend an inordinate amount of time preparing proposals rather than doing research.

Ironically, the very success of the Green Revolution has made calorie-rich foods so cheap that low- and middle-income countries face an epidemic of obesity similar to that in high-income countries. Hunger in the sense of a lack of food energy (calories) is still with us—but as the result of poverty, war, natural disaster, and in some cases, deliberate public policy, not because of any worldwide lack of physical food. Meanwhile, vegetables that could improve nutrition and prevent deficiencies in vitamins and minerals remain relatively expensive. This would logically call for the acceptance or establishment of at least one new center devoted to fruits and vegetables, and a suitable candidate for CGIAR membership does exist: The World Vegetable Center. (Fruits other than bananas would require yet another center.) However, this excellent center is located in Taiwan and is barred from membership in the CGIAR by a Chinese veto—an unfortunate example of how geopolitical considerations can have a decisive effect on the development and impact of science and technology, and in particular on science and technology for low-income people.

A variety of economic analyses have shown that the successes of the CGIAR have more than paid for their investments. Economists have calculated a sustained rate of return of 40–60% on the investment in CGIAR-financed research, a remarkable achievement.[38] In a more recent example, researchers at the International Potato Center in Lima, Peru (known as CIP) and the International Food Policy Research Institute (IFPRI) have developed a vitamin-rich, orange sweet potato suited to local tastes and convinced East African families to eat it in preference to the less nutritious white varieties to which they had been accustomed.[39] Orange sweet potatoes are now grown all over East Africa and are reducing the scourge of childhood blindness caused by diets deficient in vitamin A.

This innovation won the World Food Prize for the multinational, multidisciplinary team of researchers that developed and popularized it: two African plant breeders, Maria Andrade of Cape Verde and Robert Mwanga of Uganda, and two American researcher/evangelists for orange sweet potatoes: Jan Low, an economist-nutritionist who proved that orange sweet potatoes could effectively reduce the incidence of vitamin A deficiency in African children and promoted it tirelessly to African families, and Howarth Bouis, an economist who pioneered and championed the idea that enrichment of staple crops with micronutrients (biofortification) would be an effective way to address the hidden hunger caused by vitamin and mineral deficiencies. Bouis waged a successful campaign within the CGIAR, whose culture has been strongly oriented toward the biological rather than the social sciences and on increasing crop yield per acre rather than on nutritional quality.

The CGIAR has also wrestled with the classic problem that affects all development-oriented programs that are effectively controlled by their sources of funding, namely, that of ensuring that the voice of its presumed beneficiaries is heard and influences its research and its priorities. We have already encountered this problem in our discussion of global health. In agriculture, such consultations have led to increased attention to issues of nutrition and gender. These are especially critical issues in South Asia, where agricultural extension has tended to focus on male farmers, and in Africa, where women do much of the farming and crop marketing and where vitamin deficiencies are increasingly important public health problems.

The world is very different from the 1960s when the CGIAR was formed, and even more different from the world of the 1940s and 1950s in which the Green Revolution varieties of wheat, corn, and rice were developed. The CGIAR has survived six decades of global change—including flagging donor interest in agriculture in the 1980s and 1990s—and is still making important contributions to poor farmers. This in itself is an enormous accomplishment. The CGIAR is now a legacy system facing problems of maturity.[40] Notwithstanding its spectacular success with sweet potatoes, it finds it hard to shift from its near-exclusive focus on crop yield rather than nutrition and has not attempted to resume efforts to help NARSs, two of the major strategic problems facing world agriculture. Its history reflects the organizational problems associated with any large research program requiring the management and coordination of multiple institutions, with extra complications stemming from its worldwide scope, its focus on orphan technologies, and the multiplicity of its sources of financial support.

In one sense, the CGIAR is a victim of its own success. Due in substantial part to the work of the CGIAR-supported laboratories, the world has enough food to eat. For the CGIAR, this implies a substantial change in strategic focus toward issues of micronutrients (vitamins and minerals), fresh fruits and vegetables, environmental sustainability, farming systems, and climate change, even as it continues its historic missions of adapting staple crops to increasingly difficult agroecological and climatic challenges and of defending them against emerging pests and diseases. These new mandates, without new resources with which to tackle them, raise the problem of "bricks without straw" at a time of plateauing or decreasing resources for development assistance and continued underinvestment in agricultural research on the part of low-income countries, which are likely to depend on agriculture for much of their future economic growth and development.

"Digital Green": Localized Videos on Sustainable Agriculture

A common criticism of the Green Revolution has been its emphasis on agricultural technology requiring irrigation and substantial inputs of fertilizers and pesticides, which can lead to environmental damage. The advantage of this approach is that it lends itself to the development of uniform technology packages that can be developed and delivered over a wide area. In contrast, productive low-input agricultural technology may be more sustainable, but it tends to vary with local circumstances and hence places heavy demands on the typical agricultural extension worker, who would face the difficult task of developing and delivering a large number of different messages to different farmers in different places at more or less the same time. This would pose a large burden on the extension services of low-income countries, which are at capacity delivering a standard message to all farmers in a given region.

Digital Green, a San Francisco–based NGO that spun off from Microsoft in 2008, has developed a promising solution to this vexing problem. In consultation with farmers and organizations already working in the community, it enables local smallholder farmers to share locally specific, sustainable solutions to practical issues. It trains local people to make videos on familiar terrain and in their own local language, sharing local experience and best practices. The work of Digital Green responds to the environmental problems created by the fact that Green Revolution crop varieties largely rely on regionally uniform solutions based on irrigation water and on chemical fertilizers and pesticides.

Digital Green has made more than six thousand such videos in more than fifty languages in India and Ethiopia, in collaboration with local extension services or other partner organizations. They have also used cell phones, which, as we have seen, are now widespread in even the poorest and most remote regions, to disseminate information in local languages on agriculture, weather, market conditions, health, nutrition, family planning, and pre- and postnatal care. Their efforts have reached more than two million people, of whom 90% are women, in more than fifteen thousand villages.[41]

Technology for the Poor: From Invention to Innovation

Technological innovations are not a substitute for broad-gauge economic growth—for "a rising tide that raises all ships." Poor people still need jobs and money. But as we have seen, innovations can help solve serious, well-defined problems: curing or preventing diseases that cause millions of deaths a year, reversing blindness that afflicts hundreds of millions of people, providing clean water and sanitation to hundreds of millions of people, and increasing the productivity of small farms. Some of these solutions, like cheap medical instruments and sanitation systems, have come from downscaling and re-engineering well-established technologies. Others, like vaccines for tropical diseases, can emerge only from basic scientific research.

But research is not enough to turn an idea into an innovation. An idea doesn't become an innovation until it is in widespread use, and this requires scale up, management, and marketing. One of the most dramatic advances in public health, oral rehydration therapy for diarrheal disease, began with discoveries in basic science, but its success required acumen in product design, packaging, marketing, distribution, and culturally sensitive consumer education—yet another example of the many links among technology, economics, and culture. Assembly line cataract surgery, to cite another example, used imaginative organization to adapt a well-established technology to local conditions and deliver it to large numbers of blind people who had never before had access to it. Low-cost sanitary napkins needed still more culturally sensitive marketing. Technology is thus only part of the story.

Innovative technologies for the poor and disadvantaged go by many names, depending on the context. We call them frugal technologies—or (in Hindi) *jugaad* technologies—when we wish to emphasize the engineering inventiveness needed to keep costs down and attention focused on the special needs of the intended user. We call them orphan technologies if we wish to emphasize the lack of commercial interest and the need for support from governments or private philanthropy. We call them inclusive innovation if we wish to emphasize the fact that users must be involved in the innovation process, sometimes as implementers, sometimes as marketers, and sometimes (as in agricultural research) as both research collaborators and participants in research strategy and policy. We use the older terms "appropriate" or "intermediate" technology if we wish to emphasize simple, labor-intensive, easily repairable and maintainable equipment made of local materials.[42]

The good news is that technological innovation for poor people is getting increasing attention, can boast many successes, and is increasingly originating in the low-income countries, as ideally it should. India and China count hundreds of millions of people newly liberated from extreme poverty, as well as firms with enough capacity to design and manufacture higher-quality alternative products for this new market: bicycles or motorbikes to replace walking, cinder block houses instead of tin-roofed shacks, meat in addition to beans, and latrines instead of open defecation and flying toilets. Foundations and universities around the world are sponsoring conferences and contests to encourage engineering students and professional inventors to invent new concepts to address long-neglected needs.[43] Blockchain and other information technologies promise to empower poor people to register their land and other property and actually to receive the social benefits for which they qualify.

Not all frugal technological innovations succeed, just as not all commercial innovations succeed. The keys to the success of any frugal technology, like that of any technology, are first, responding to the felt needs of the consumer, and second, scaling up—the process by which an invention that worked on a small scale becomes an innovation in wide use. This is where thinking like a business is essential. In an expanding organization, the problem is management: being sensitive to the needs of the market, creating distribution channels, maintaining quality control, keeping costs down, hiring and training people, and continuing to improve the product. Successful frugal organizations therefore must combine the skills and mentality of a business with the public spiritedness of an NGO or a foundation. In most cases, they must either accept losses or a relatively low return on capital, or else seek subsidies or grants from governments or private foundations.

When all is said and done, technology for the poor is fundamentally an issue of equity. A frequent criticism of technology is that it benefits people who are already well off, who learn about technology when it is still new, who want and can afford to buy the latest, and who can take the risk that it may fail. Defenders of technology respond that the high prices paid by these relatively well-off people enable inventors and innovators to reap the rewards of their labor, and that low-income people eventually benefit when the technologies they have invented trickle down to cheaper houses, cars, or computers. In many cases, however, this is not sufficient. The stories in this chapter illustrate the heroic combination of business acumen and public spiritedness that can ensure that the fruits of scientific research and technological innovation do reach the poor and otherwise disadvantaged.

11

Jobs, Competitiveness, and Inequality

Up the Down Escalators in Manufacturing

The revolution underway in advanced manufacturing technology will make possible extraordinary new products and processes. If its promises are fulfilled, cars will talk to each other and make traffic lights and stop signs unnecessary. New methods for working with light metals will make possible lighter, more fuel-efficient, less polluting airplanes and trucks. Manufactured biological tissues will help heal wounds and repair genetic defects. Nanoparticles circulating in the blood stream will broadcast vital statistics and other health indicators. Wearable electronic circuits will speed healing by measuring the level of oxygen on wounds.

Some of these innovations may be on the market by the time you read these words. Other manufacturing innovations will take place behind the scenes in factories, out of sight of the consumer: artificial intelligence and improved sensors will save money and improve worker safety by monitoring the soundness of bridges and buildings and enabling machines to sound the alert when they need maintenance or repair. Interactive goggles will guide workers through complicated tasks. "Additive" 3D printing will speed the commercialization of new products by facilitating the fabrication of prototypes. Improved scientific understanding of the molecular workings of chemical catalysts will make the synthesis of industrial chemicals cheaper and more energy efficient. Photons (light particles) will replace electrons in many electrical circuits, making them faster and more reliable. Improved high-power switches using wide-bandgap semiconductors will do the same for the long-distance electric power grid.

These innovations pose challenges to competitiveness and employment in all parts of the world that go far beyond the challenges posed by traditional automation and robotization. Will there still be jobs for people when these innovations are in place? If so, what qualifications will these jobs require? Where will these jobs be located? What will happen to the people who cannot qualify for them? How will the new technology affect the recovery from the worldwide shutdown from the coronavirus pandemic? What would be the impact of increased tariffs, should these become permanent? Will China still be the world center of manufacturing? Will manufacturing jobs return to the United States? Is it too late for

The Survival Nexus. Charles Weiss, Oxford University Press. © Charles Weiss 2022.
DOI: 10.1093/oso/9780190946265.003.0013

the lowest-income countries to industrialize? If not, what will their people do for a living? Where will they go to find jobs?

Every country, from the United States to Ethiopia, is trying to improve its competitiveness in today's globalized, technology- and information-intensive world market. The COVID-19 pandemic has suspended much of this process, but it is likely to resume soon. Technology and markets are changing fast, and more and more countries are learning to produce goods and services more and more cheaply. The United States, Germany, the United Kingdom, China, and other technologically advanced countries are vying for leadership in advanced manufacturing by investing in research, development, and education, and by helping businesses master new production technologies. Middle-income countries, including China, face the challenge of advancing from managing well-established or imported technology to accepting and managing risk and innovation. Low-income countries will need to make major investments in infrastructure and connectivity and strengthen the capacity of their educational systems to encourage creativity and teach digital skills.

We can think of all countries, rich and poor, as climbing up the same down escalator, so that, like the Red Queen in *Through the Looking Glass*, everyone has to run uphill as fast as they can just to stay in the same place. Governments, businesses, and educational institutions—the "triple helix"—all have important roles in this uphill scramble.

Global Value Chains and Networks

Not so long ago—say in the 1970s or 1980s—multinational firms typically started to manufacture a new product close to the place where it was designed and developed, to make it easier to iron out the bugs in its design and production technology. Once the product was standardized, manufacturing was "outsourced" to a developing country where labor was cheaper. For higher-income countries, innovation was a reliable source of well-paying "first manufacturing" jobs, in addition to the jobs directly linked to research, development, and product design and launch. To low- and middle-income countries, outsourced manufacturing and the technology transfers that went with it were key to jobs and economic growth.

Global production chains and networks added a new dimension to this pattern.[1] Low tariffs, combined with cheap transportation via containerization, made it possible to break up the production of products of all kinds—from blue jeans to consumer electronics—into separate steps, and to locate each of these steps wherever it was most convenient and profitable. Computer-driven, precision manufacturing equipment made it possible to make components anywhere in the world and snap them together into an assembled product.[2] As a result, a manufactured product—or even a food product like filleted fish—might cross borders many times on its way to the final purchaser. The resulting system is typically optimized for low cost, quality control, and speedy response to changing market conditions—although not for resilience to possible disruption, an oversight that raised acute problems during the COVID-19 pandemic.

This "organizational decomposition" may take one of two forms. In the case of coffee, the growth, transportation, roasting, and grinding of coffee beans and the serving of the resulting beverage in an attractive ambience together constitute a linear system called a *value chain*. In the case of products like automobiles that are assembled from thousands of parts, the resulting system of parts suppliers and subassemblers has multiple branches and constitutes a *production network*. To low- and middle-income countries, value chains and production networks have the major advantage that they offer to producers that can meet global standards of quality and price the chance to participate in global markets that might otherwise be inaccessible. Once established in a value chain, locally owned firms can acquire a continuous stream of knowledge about technology and markets from their clients. If the country is open to foreign investment, subsidiaries of international firms train local managers and workers and create competitive pressures that may stimulate locally owned firms to improve and expand their participation up or down the value chain.

In both chains and networks, a "lead company," usually a multinational headquartered in a high-income country, sets the design and marketing of the final product, its quality specifications, and the technologies needed on the way from raw material to final product. The lead company can take the lion's share of the value added from the chain or network by squeezing the profit margins of its vendors.[3] The extreme case is that of Apple, which has become one of the most valuable companies in the world by dint of its ability to design dramatically new products and market them at high margins. As an illustration, almost two-thirds of the selling price of the iPhone X is gross profit margin representing value added in the United States, despite the fact that all of its manufacturing takes place overseas.[4]

In a global production chain, the lead company controls the location of the steps in the chain and the nodes in the networks and will naturally insist that the suppliers it chooses be able to master the necessary technology. Where the suppliers are located depends on many other factors: business climate, nearness to expanding markets, the cost and productivity of labor, the state of physical infrastructure and connectivity, whether the product is light or heavy, whether the production process is easy or difficult, and whether that process can be codified in software or is "tacit," i.e., can be understood only by experienced humans.[5]

Environmental issues can also enter into these location decisions. There is a strong temptation to locate polluting elements of value chains in countries with weak or weakly enforced environmental regulations. Many low- or middle-income countries have long been willing to accept these activities in exchange for the jobs and economic benefits they provide. However, a backlash has begun. China, for example, has refused to accept many kinds of trash imports, which were a main input to a recycling industry that was a major source of both profits and domestic pollution.[6]

These decision criteria of lead companies are key considerations for suppliers in low- and middle-income countries that want to participate in a value chain or production network; benefit from the technology, marketing, and management skills they acquire from this participation; and later increase the sophistication and the value added of their contribution and the income that they derive from them. Value chains and production networks create space for suppliers to add value, cut costs, and improve product quality. In the case of agricultural commodities, for example, a firm in a low- or middle-income country might advance from primary producer to processor and then to purveyor of branded consumer products and developer of new ones. It may also try to develop some unique capacity within the value chain through improved agronomy, food technology, or transportation capability. It may try to establish a unique role in in the organization or the structure of the value chain, or even to define and market the ultimate consumer product, as Colombian coffee growers once did by promoting the special virtues of their product.

If a supplier is a producer of industrial products, it may seek to advance up the chain from component or subcomponent manufacturer to partial assembler to assembler or manufacturer of consumer products to be sold under someone else's brand name. It may develop a unique capability that is unavailable elsewhere, as, for example, the wafer fabrication capabilities of Taiwanese companies or the engineering skills of Chinese firms for scaling up medium- and high-tech manufacturing. The final step in this progression takes place when the supplier makes the fateful decision to strike out on its own and to produce products for the global market under its own brand name, taking full responsibility for their performance, perhaps in competition with the original lead company.

Global value chains thus offer opportunities to firms in low-income countries to expand their markets and increase profits if they can meet the challenge of upgrading quality and cutting costs. The challenge lies in the fact that production technology is becoming steadily less labor intensive, so that low-cost labor is no longer a decisive competitive advantage. In addition, high-income countries are realizing that in many industries, technological innovation is closely linked to manufacturing, so that by offshoring manufacturing ("innovate here, produce there"), they are offshoring their ability to keep products and processes up to date and to develop new ones.[7]

The backlash against globalization, the recent increase in tariffs and other trade barriers, and the near-global shutdown caused by the COVID-19 pandemic are substantial but still undefined threats to the integrated global manufacturing system. So far, the main response of multinational corporations to tariffs and other obstacles to the free movement of intermediate goods has been to shift the locations of the links in the value chains and the nodes in the networks that they control from one country to another. This does not necessarily work to the advantage of the country that imposes the tariffs. We may anticipate that the pandemic will raise awareness of the need to incorporate into value chains and networks resilience and multiple suppliers in the face of possible disruption, even if this adds to costs.

If the backlash against free trade continues to the point where the globalized economy fragments into regional trading blocs, the global result is likely to be less competition, less pressure on companies to cut costs, lower productivity, and therefore more manufacturing jobs for the same amount of output. It would also mean fewer new products and more expensive consumer goods, which would reduce overall demand for manufactured items, decrease manufacturing production, and hence lessen manufacturing employment—the opposite of the purpose of the barriers to free trade. The net effect of these conflicting trends will become apparent as the world emerges from the virus-induced slump and will depend on the details of the trade regime that emerges.

Some Jobs Will be Lost as Productivity Increases, But Others Will be Created

Changes in technology have historically increased worker productivity. Futurists have predicted that artificial intelligence and other new technologies will replace nearly half of all American jobs over the coming decades.[8] There have even been predictions that human work will become virtually obsolete, and that our descendants' biggest problem will be to figure out what to do with their new leisure time.[9] These predictions are probably overblown, but the new technology nonetheless has serious implications for the worldwide future of work.

Historically, new technologies have created more jobs than they have destroyed. Agricultural mechanization has enabled less than 2% of the U.S. workforce to produce more, cheaper, and better food than 40% of the workforce produced a century ago. New jobs making, selling, maintaining, and repairing agricultural machinery, transporting and selling the increased production, and doing research and development (R & D) on new and improved foods offered opportunities for some of the displaced agricultural production workers. The others found their places in the expanding industrial and service economy. To be sure, these new jobs required new and different skills, a phenomenon known as "skill-based technological change."

Will this time be different? On the one hand, the new manufacturing technologies will probably require many fewer workers than the technologies they replace to produce the same number of products. On the other hand, new technology will make possible new products with hitherto undreamed-of capabilities and will improve old products so that they will be better and cheaper. Unlike agricultural products, for which there can be only so much demand, there is almost no limit to the demand for the industrial products that future factories could turn out. Consumers will be happy to buy more and more of them, and manufacturers will respond by expanding production and hiring more workers.

In the United States, well paid, unionized manufacturing jobs once provided the classic road to middle-class status for American high school graduates. Manufacturing supported not only jobs on the factory floor, but also "upstream" employment in domestic suppliers. (Service jobs also support upstream jobs, but not nearly as many.[10]) These jobs are now much scarcer. U.S. manufacturing employment has dropped from over 25% of the domestic workforce in 1970 to under 10% in 2012.[11] Some jobs were lost to automation and others were "offshored," leaving behind depressed Rust Belt cities throughout the Midwest. Outside of manufacturing, advances in artificial intelligence have already replaced many routinized jobs, including white-collar jobs in retailing, law, insurance, accounting, and finance. On the other hand, highly paid jobs that require expertise, creativity, and personal judgment, like data scientists or division managers, or low-paid jobs that are tied to localities, like bellhops, barbers, or beauticians, probably will not be automated anytime soon.

Paradoxically, the total value added from U.S. manufacturing has remained roughly constant over the past few decades because of the slowly increasing productivity of the factory workers who did manage to hang onto their jobs.[12] This increased productivity has led to higher pay for high-skilled workers but stagnated pay for low-skilled workers. In other words, the workforce has become increasingly split into a highly paid tier and a low-paid tier with few workers in between, a phenomenon known as "job polarization." The mean (average) wage has increased, but the median (middle) wage has remained about the same.[13] The recovery from the Great Recession of 2009 dramatically illustrates this trend. Between 2010 and 2014, while most high-wage occupations were adding jobs, blue collar and education-related positions *decreased* by 250,000. Of the nearly three million "good" jobs created during the 2010–2014 recovery (defined as paying more than $53,000 a year annually), 2.8 million went to applicants with at least a college degree.[14]

The social and political costs of these changes have been severe around the world. In the United States, big cities with large pools of well-educated professionals have attracted high-tech industry and boomed. In smaller cities and rural areas, workers who lost their jobs became unemployed or found new jobs that generally paid less than the jobs they had lost. Some of these cities and towns did succeed in reinventing themselves. In many "Rust Belt" cities, a coalition of businesses, higher education institutions, and state and local governments underwent a period of self-reflection, identifying the key skills left over from their halcyon days and building new industries that exploited these skills.[15]

Not every town has been able to achieve such a revival, yet many people do not want to leave their hometowns and move to where the jobs are. In less successful former factory towns, the middle class shrank ("hollowed out") as its population split into those with high incomes and those with low incomes, with few people in between. The vulnerability of disaffected workers displaced by technological change to angry messages of xenophobic nationalism has been a major factor in the backlash against globalization and the rise of far-right, xenophobic parties in the United States and Western Europe and is one of the major influences of technological change on international relations.

The New Jobs Will Demand Different Skills

None of this means that human workers will all be replaced by robots; apocalyptic predictions of wholesale replacement of humans by robots are unlikely to be borne out.[16] First of all, advanced manufacturing technologies, once they are developed, are likely to take a good many years to spread to companies all across the world. Second, advances in technology will in themselves create new jobs. Someone has to make, program, deliver, sell, maintain, and repair the new machines, and to design, make, transport and sell the products that they make possible.

Third, in many areas, new technology will complement and augment rather than replace human skills. In manufacturing, collaborating robots (cobots) are already assisting human workers and increasing their productivity. An automated machine, for example, now lifts the heavy roof of an automobile and holds it in position so that a human worker on the assembly line can attach it to the chassis. To be sure, humans working with robots may need to master new skills. Fourth, advances in technology are at least as likely to create new jobs and to upgrade existing jobs as to destroy them. People are better than computers at creativity, at entrepreneurship, at pattern recognition, at solving complicated problems, and at teamwork, complex communication, and other interpersonal skills, and are likely to remain so for some time. As a result, the more interesting and creative tasks will be left to humans, even though routine tasks will be automated.

Here the classic examples are bank tellers, who were widely predicted to be replaced by automatic teller machines (ATMs). Defying the predictions, there are more tellers today than there were when ATMs were first introduced, but they no longer spend their time cashing checks and taking deposits. Instead, they are part of the bank's marketing and advisory team—a role that requires interpersonal, communication, and analytic skills, and a certain amount of creativity, not just facility with numbers. Computers have not replaced tellers (who may now be called "ambassadors") but, rather, are tools that help them analyze the needs of clients and match them to the bank's product lines. The jobs of tellers and the skills they require have in this way been upgraded and their salaries have been correspondingly increased. There are fewer of them in each branch, but this allows banks to open more branches. The net effect is more jobs for tellers, not fewer.[17] We may expect a similar pattern to take shape in other professions. CAD/CAM tools facilitate the work of architects, for example.[18] In jobs that involve pattern recognition, like radiologists and interpreters of satellite imagery, automation will make it possible for humans to spend less time scanning images for anomalies, which computers can do more effectively than people, and more time assessing the significance of these anomalies.

Finally, and most important, advances in biotech, nanotech, and artificial intelligence will create whole new categories of jobs, many of which will require whole new sets of skills. These will change more and more rapidly as technology and markets undergo more and more rapid evolution, forcing human workers to get more and better education that continues throughout their lifetimes. This, in turn, poses a challenge to educational institutions in all countries and at all levels, from kindergarten to community college and university.

We conclude that there will still be jobs, some despite and some because of the revolutions in manufacturing. But where will these jobs be located? We will return to this question after a brief historical excursion.

South Korea: A Classic Model of Industrial Development

Historically, export-oriented manufacturing has been key to the growth of the relatively few low-income countries that have attained advanced industrial status. As a job creator, manufacturing has major advantages. Manufacturing jobs give low-income people leaving rural areas valuable skills and prepare them to participate in a modern economy. Manufacturing jobs tend to have higher productivity and hence to pay more than low-skilled jobs in agriculture or services.[19] Urban manufacturing jobs have historically offered income and personal freedom to young people from rural areas. Moreover, it is easier to provide city people with education, health services, infrastructure, and access to world technology and market information. Unlike agriculture, manufacturing is easily scalable; there is almost no limit to how much can be produced and sold.

I first visited Korea in 1971, just eighteen years after the end of the Korean War, which had ended in a partition of the country that had left most prewar industry in the communist North. As late as 1960, when a corrupt dictator was forced to resign, South Korea was widely seen as a hopeless case. When I visited, the country was still recovering from the damage wrought by the war, although it had benefited from generous U.S. aid and from the roads, ports, and other infrastructure the U.S. military had left behind.

Today, South Korea is a developed country, a thriving democracy, and a full-fledged member of the Organization of Economic Cooperation and Development (OECD, the "club of rich countries"), with a per capita GDP of over $43,000 in 2020.[20] This dramatic achievement makes South Korea an attractive model for today's poor countries. But as we shall soon see, changes in manufacturing technology and in the structure of global markets have made the path followed by South Korea much more difficult today.

Korea's early industrial strategy was simple, at least in concept: launch Korean businesses in industries from which post–World War II Japan was soon to exit, and in which Korean firms could compete at the low end of the market. Japan, after all, had followed the United States in much the same way. Korea began by assembling electronics, producing textiles and clothing, and selling them cheaply on world markets. The major cost in producing these products was labor. Unions were repressed so as to ensure that the workers were poorly paid. Working conditions were substandard and often dangerous.

Everyone in Korea when I visited understood the importance of highly trained workers. Across the main street of Seoul, the capital city, hung a huge banner in Korean and in English, "Welcome to the Champions of the World Technician Olympics," which had just concluded in Germany. Everyone understood that education was the key to a better future. The demand for education was so great that I was told that a good way to make a fortune was to start a private school—despite the fact that class size in elementary schools could be in the '60s.

Central to Korean industrial strategy was government support to private conglomerates, called *chae-bols*, which were granted privileged access to finance and to the domestic market on the condition that they meet demanding export targets. As in Japan, the Korean government strongly discouraged direct foreign investment and encouraged exports by keeping the value of the Korean currency, the won, artificially low, so that Korean products would be extra cheap on world markets. These exports made it possible for Korean firms to expand their markets and to achieve economies of scale, and at the same time it forced them to meet international standards of quality and price. Korean companies gained a reputation for being fast and agile learners. Their customers were therefore willing to keep them up to date on changes in markets, product design, and manufacturing technology. This kind of technical assistance was supplemented by extensive licensing from Japanese firms, and by technology transfer from moonlighting Japanese engineers who commuted to Korea on weekends.

Korean science and technology policymakers, led by Minister Choi Hyung-Sup, established a publicly funded applied industrial research laboratory, the Korea Institute of Science and Technology (KIST), modeled on the Battelle Memorial Institute, and a science- and engineering-oriented university, the Korea Advanced Institute of Science and Technology (KAIST), modeled on the Massachusetts Institute of Technology. Minister Choi encouraged the engineering consulting profession, established an extension service for small and medium-sized industry, and launched the developing world's first venture capital firm, the Korea Technical Assistance Corporation (KTAC), originally intended to commercialize KIST inventions in which local firms did not express interest.[21]

Korean firms invested money in mastering imported technology, i.e., in learning first to operate it and then to adapt it to local conditions and to improve it. After gaining experience supplying finished articles (original equipment) for sale under the brand name of United States or Japanese companies, a few of the more ambitious Korean companies (especially Samsung and Goldstar) began to market internationally under their own brand names. This made them responsible for satisfying their consumers, a major watershed in the history of the technological development of Korean industry. In a second watershed, Korean firms increased their investments in research and development to the point where they exceeded those of the government.[22]

Korea was one of the Gang of Four (also known as the AsianTigers; the others were Hong Kong, Singapore, and Taiwan) that followed similar technological trajectories.[23] These economies took advantage of the relatively easy credit, the expansion of world trade, and the relatively slow pace of technological change that characterized the 1970s, to ride export booms to achieve high-income status by the turn of the twenty-first century. With a few exceptions, the other low- and middle-income countries missed this golden period. They protected their industry with high tariffs and other obstacles to trade throughout the 1970s, overvalued their currencies, and did not encourage exports. Many of them restricted imports of foreign equipment and technology. The idea at the time was to enable local infant industry to learn to thrive on its own and to give it the chance to develop its own technology, after which the tariffs were supposed to be lowered to force it to face the rigors of foreign competition.

The firms that benefited from this protection did prosper. They supported relatively well-paying jobs, which were, in turn, protected by unions that were important supports for democracy and civil liberties, which in many countries were under intermittent pressure. However, the companies that benefited from this protection grew rich and politically powerful without having been forced by competitors to produce things of international price and quality. They used their political influence to press the government to continue to protect them from foreign competition long after the tariffs were supposed to have been phased out. When these countries belatedly opened up their economies to international competition in the 1980s, their manufacturing firms faced a much tougher world environment. Now, decades later, globalization and the revolutions in manufacturing technology make the path blazed by the gang of four even tougher for low- and medium-income countries to follow than it was in the 1980s.

A Worldwide Scramble to Create Good Jobs

Countries of all levels of income and technology face the task of creating enough well-paying jobs for their people in the face of changes in technology that require fewer and fewer workers with higher and higher levels of technical and digital skills. To this end, all countries face the same basic challenges, although from different starting points. All of them must educate their people to meet the demands of new technology and must restructure their educational systems toward greater creativity and lifelong learning. All must invest in physical infrastructure and connectivity. All must create a stable macroeconomic environment and business climate that attracts both foreign and domestic investment, limits the power of monopolies, facilitates the launch of new businesses, welcomes new knowledge and technology from abroad, and encourages local innovation. All need to provide basic security, reasonable personal freedom, a basic legal framework, honest courts, financial institutions that can evaluate risk, respect for intellectual property, and basic technical services. And they need to do all this within a governmental framework that also controls environmental pollution, limits the emission of carbon dioxide and other greenhouse gases, protects against the spread of infectious disease, and ensures worker health and safety.

These are tough enough challenges for high- and medium-income countries; they are vastly more difficult for low-income countries. We earlier spoke of countries and firms doing their best to scramble up a down escalator. We now need to refine this image by imagining a sequence of three such escalators, to be climbed one after the other as countries advance from low to medium to high per capita incomes, and from the beginnings of industrialization to the mastery of imported technology to cutting-edge, information-intensive innovation. These two sequences typically coincide but do not always do so.

We might imagine firms in the United States, Europe, and Japan, together with China and the Gang of Four, running hard up the top escalator, investing in research, development, and technical assistance to industry in order to be among the first to develop new technologies, master advances that take place elsewhere, and turn new technologies into profitable business opportunities. Behind them on this top escalator are India and the three "new little dragons" (Indonesia, Malaysia, and Thailand), scrambling to stay within hailing distance of the leaders.[24]

On the second escalator are middle-income countries, whose firms are seeking to strengthen their industry and learn to innovate, in many cases in order to free themselves from their dependence on the export of oil, minerals, agricultural products, or other raw materials. The challenge for the firms in these countries is to master foreign technology as it becomes available to them by import through commercial channels or through foreign investment: to become rapid learners, choose the machines, design the factories, adapt production systems to local conditions, and start to improve them and to develop new products and production techniques. Getting started on the bottom escalator are firms in most of the countries of sub-Saharan Africa, plus a few countries in other regions like Haiti and Myanmar, which are learning to manage and to operate technology.

Manufacturing Employment at the Cutting Edge

Let's start with the topmost of the three down escalators. For the United States, the major challenge is to restore lost manufacturing jobs and to create new ones by making new products at home rather than offshore, and by being among the first to develop and adopt new production technologies and to develop and commercialize the new products that they make possible. American manufacturing jobs have not disappeared, but they now move to foreign countries at a time much closer to the date of the initial introduction of a product than they once did. In many industries, new products are manufactured overseas from the very beginning, Apple computers and other products being the best-known and most dramatic example.

The innovative strength of the United States is the result of both its entrepreneurial culture and its government policies. Contrary to the common belief that the U.S. government supports only basic research and leaves application to private industry, the United States, in fact, provides generous public support to applied research in defense, aerospace, health, biotechnology, and agriculture, which, not coincidentally, are among its strongest exporters.[25] The United States maintains a comfortable lead in pathbreaking research and innovation based on this research due to the high level of its support to basic research—it has been the only country large enough to aspire to being number one in every scientific field—and equally as much to a cultural climate that celebrates novelty and individualism and that accepts the risk of failure as an inevitable part of experimentation and innovation. Its venture capital markets are well equipped to finance the risk of innovation in information technology, biotechnology, and other advanced technologies, although less so in fields with longer time horizons, like manufacturing and energy.[26]

These are areas of high technology in which many other advanced economies are still struggling. In most of Europe, innovation is less of a cultural value than it is in the United States, and family background and being a graduate of the right schools are still considered to be important qualifications.[27] Failure is more severely punished, so that people tend to take fewer risks. Labor tends to be less mobile, and job protections more rigorous—an advantage to those with jobs, but a disadvantage to the young, to the unemployed (since businesses may be reluctant to hire workers who would be hard to fire or lay off), and to businesses that are trying to keep up with a fast-changing market.[28] The result has been that European countries have lagged in information technology, a deficiency they are hastening to make up because they realize that it threatens their ability to master the new advanced manufacturing.

Manufacturing as a Source of Innovation

Until about 1980, U.S. policymakers viewed the loss of manufacturing jobs as the natural and inevitable result of the workings of the free market, analogous to the shift from agricultural to industrial employment in the twentieth century. They assumed that the service sector would create new jobs to take their place and displayed little concern for the possible loss of capacity for innovation or for possible security issues. The best engineering students regarded careers in manufacturing as less prestigious than those in design, suitable for the sons of foremen who were the first members of a family to go to college (note the implied condescension and sexism), but not for themselves. Top engineering universities tended to neglect their curricula in manufacturing in favor of seemingly more exciting fields that were more likely to attract students.

Implicit in these events was the idea that the essence of innovation is the development and design of new products based on research and development, and that their manufacture can be left to others. A good deal is lost, however, when innovation is defined in this way. Designing and prototyping a new product for production and developing the technology for its initial production are all part of the innovation process and must be repeated whenever the product is improved. The wholesale offshoring of U.S. manufacturing, especially to China, has led to a decline in U.S. innovation in manufacturing processes that are critical to maintaining future competitiveness.[29]

The Japanese challenge to American manufacturing in the 1970s and 1980s abruptly changed the complacency that had built up in the previous decades.[30] Japanese automakers like Toyota and Honda become powerful competitors to General Motors, Ford, and Chrysler by pioneering new approaches to business management, quality control, time management, and the integration of manufacturing and design.[31] Sony and other Japanese firms used similar methods to drive long-standing industry leaders like RCA from the consumer electronics market. The Japanese government helped with financial and tax incentives, and by encouraging exports by maintaining the yen at an artificially low exchange rate against the dollar.

The result was to briefly awaken U.S. business and academia to the importance of manufacturing. Engineering and business schools strengthened their manufacturing curricula. U.S. automakers slowly caught up with their Japanese competitors in productivity and quality, although not in market share. It took the 1990s revolution in information technology, a technology uniquely suited to the U.S. start-up culture, to rescue American competitiveness. Attention to manufacturing waned, and manufacturing jobs continued their inexorable movement to China and its rapidly expanding market.

As production shifted away from the United States, so, too, did the "ecosystem" of specialized services that once supported manufacturing—applied research laboratories, technical services, consulting firms, trade associations, newsletters, headhunters, and more—making it difficult for those U.S. small- and medium-sized manufacturing firms that did remain to expand and hire additional workers. At the same time, the countries to which production has been offshored—especially Taiwan and China—have themselves become strong in manufacturing technology and are developing their own capability to innovate, especially in scale-up engineering and in technologies like chips, batteries, and displays, in which innovations frequently emerge from improved production techniques.[32]

In contrast to the history of manufacturing in the United States, the experience of Germany shows that it is possible to have a strong and robustly competitive manufacturing sector, despite high wages paid to workers.[33] Germany benefits from its long tradition of strong government and public support to manufacturing. German small and medium manufacturing firms specialize in high-end capital equipment whose buyers are willing to pay top prices for top quality. Factory jobs are highly esteemed. Strong unions ensure high wages and effective worker training. Universities and community colleges offer strong programs in manufacturing engineering. German banks are typically locally owned and understand the importance of manufacturing to the local economy. Inheritance laws facilitate the continuation of family ownership from generation to generation.

On the technical side, professional networks, trade publications, consulting firms, and other service organizations cluster around German factories. The German government supports a double system of laboratories: The Max Planck Institutes for basic research, and the sixty-five applied research laboratories of the Fraunhofer Gesellschaft (Institute) for applied research and technical assistance. On the other hand, Germany largely missed the revolutions in information technology, artificial intelligence, and "big data" and is now working hard to catch up in these areas.

All industrially advanced countries, Germany and the United States included, face the social and political consequences of the continuing loss of well-paying manufacturing jobs in once-prosperous regions, a loss that will only be intensified by the ongoing revolutions in manufacturing technology. It is no longer enough to rely on the hope that advanced manufacturing will restore the competitiveness of the manufacturing economy, that dramatically new products and increased production of older ones will lead to the creation of new jobs, and that displaced workers will move to where the new jobs are. This hope may fit the economic theory, but it ignores the sociology, the politics and the new circumstances.

Many of the workers displaced by digitalization and globalization do not want to move, both in the United States and in Europe, even after the enforced immobility of the coronavirus pandemic. Worker mobility in the United States is hampered by high housing prices in the expanding high-tech cities, and by the phenomenon of "bowling alone"—the decline of unions and of social organizations like the Elks, the Moose, and the American Legion, which once held working-class communities together and supported newcomers while they were getting settled in a new area.[34] European workers, for their part, have traditionally been more tied to their locations than their American counterparts, although the new generation has become much more comfortable with their identity as Europeans and are more willing to take advantage of the mobility afforded them by the boundary-less (although now U.K.-less) European Union. Besides, if less-educated workers do move, there may not be high-paying jobs for them even in thriving technology-oriented cities. On the contrary, many well-paying, high-skilled jobs for workers with only high school education in production and administration in traditional industries have been automated or offshored and hence have largely disappeared from cities in both the United States and in Western Europe. The less-educated workers who do remain in U.S. cities have been forced into lower-paying, lower-skilled jobs that do not pay enough to enable them to afford living in the city.[35]

The aging of the U.S. work force will open up jobs for younger workers, and if advanced manufacturing technology is adopted, it will bring some manufacturing jobs back to the United States. Still, many of these jobs are likely to be "upstream" in research, development, design, and marketing. Jobs on the factory floor in the new manufacturing industries are likely to require skills different from those of the jobs that were lost when the old manufacturing was automated or moved offshore, continuing the trends we discussed earlier. The new familiarity with online communication and collaboration could facilitate the development of a worldwide market for workers whose skills lend themselves to telecommuting, further exacerbating what is already a skill-based job market. All this suggests that the "reshoring" of manufacturing jobs back to the United States will not do much to restore the role of factory jobs as a road to middle-class status for American high school graduates.

The United States faces a special problem because of its long devotion to the concept of shareholder value as the main criterion of successful management, and the preference given by its tax system to financial rather than to human capital, i.e., to investment in people. Western European countries are experiencing many of the same impacts of technology that we have seen in the United States but have devoted more resources to investments in worker skills and social safety nets.[36]

The technologically advanced countries therefore face triple challenges: investing heavily in advanced manufacturing to maintain their competitiveness[37] and value added, directing science and technology policy toward regions devastated by the loss of these jobs so as to help them to recover, and facilitating lifelong education for those who need to build and maintain skills for high-wage jobs. This will require not only a general defense of overall national manufacturing competitiveness in advanced manufacturing technology, but also "place-sensitive" policies that pay specific attention to the special needs of these depressed areas on the part of governments, educational institutions, and the business community.

A Tougher Path for Low- and Middle-income Countries

On the middle escalator are countries (or in the case of China, parts of countries) that have reached middle-income status, say around $12,000–$15,000 per capita income. These face problems known in the economic development literature as the "middle-income trap." At this level, industries have achieved competence to manage well-established, usually imported, technology. Their factories know how to operate machinery and to produce and sell products requiring low or medium levels of standard technology on domestic and sometimes on global markets. What they lack is the capacity to innovate—to develop new products and processes and launch them on local or world markets. This requires a rather different business strategy based on research and development, an educational system that teaches analysis and creativity rather than rote learning, and sophisticated technical services (consulting firms, geological and natural resources surveys, standards institutes, and technical publications). It also requires a culture, a business climate, and a financial system that encourages innovation and that accepts and can manage risk, since one never knows for sure whether an innovation will succeed.

Many middle-income countries have achieved their current status through the export of oil or other natural resources and now need to diversify and to overcome the challenges posed by the up-and-down markets that characterize natural resource exports (the natural resource curse), and by the Dutch disease that makes it difficult for any other economic activity to compete with the profitability of low-cost extractive industries like mining and oil production.[38] For natural resource exporters, the additional challenge is to reinvest much of the income from the exploitation of these natural resources into productive and sustainable industry, as well as into education and research, and to encourage the start of new businesses in new industries. This requires the difficult political task of freeing up the economy so that new businesses can compete on equal terms with established and politically powerful legacy firms. This is unlikely in kleptocracies in which the chief motivation of the government is to steal from the public and in which vested interests are particularly well entrenched. The revenues from natural resource exports make it easy for governments to relieve pressures for democratization by buying off the population with social benefits and patronage jobs, and to repress dissent via well-paid police and military.[39]

For the lowest-income countries, whose industrialization is just getting started on the bottom escalator, the revolutions in advanced manufacturing pose the biggest hurdles. First, many industries that depended on cheap labor in the 1970s have been automated to such an extent that labor is now only a small percentage of their overall cost structure. Profitable labor-intensive industries like textile or clothing manufacture still do exist in low-wage, low-skill, low-productivity countries, but these opportunities tend to disappear after a relatively small increase in average wages, a process known in the technical literature as premature deindustrialization.[40] The result is that much production is being reshored back to the high-income countries where the products are to be sold so that producers can respond more quickly to changes in market demand, although not so much as to restore manufacturing to its previous prominent position in the high-income economy.

Second, many low-income countries lack the roads, containerized ports, and other transportation infrastructure needed to get manufactured products to export markets reliably and in sufficient quantity. Third, many African countries have discovered oil or other mineral resources and are suffering from the Dutch disease that we discussed earlier in connection with the middle-income trap. This would make it difficult for them to sustain even a competitive export crop like rubber or cocoa, let alone internationally competitive, export-oriented manufacturing.

Fourth and most important, most low-income countries lack the educated workforce needed to support advanced manufacturing. Even though education accounts for the largest portion of public spending in most low-income countries, ministries of education are typically weak. Corruption and teacher absenteeism are widespread. Elementary education is typically particularly neglected and emphasizes rote learning rather than creative problem solving. In many countries, gender-based discrimination is a serious problem. As a result, illiteracy is widespread, productivity is low, and workers often do not have even basic skills, let alone the digital skills they will need for the new machines.[41] All these issues are likely to become even more important as advanced manufacturing technology becomes more widely practiced.

The revolutions in manufacturing technology thus promise to exacerbate the youth unemployment crisis in sub-Saharan Africa, which is expected to add more population to that region in the next couple of decades than today's entire population of Western Europe.[42] African agriculture missed the Green Revolution (see chapter 10) and is already suffering from climate change. Many parts of Africa do not have access to electricity. If Africans cannot find jobs at home, they are likely to do their best to migrate to wherever the jobs are. The flow of these migrants may well dwarf the migrations that are already roiling politics in Western Europe and in the more prosperous parts of Africa.

The Rise of China

China is doing its best to climb up all three of our metaphorical down escalators at once. China is making major investments in advanced manufacturing in hope of dominating the higher levels of manufacturing technology. At the same time, Chinese competition has complicated the prospects of all other low- and middle-income countries that seek to follow the traditional path of using export-oriented, labor-intensive manufactures as the jumping-off point for industrial development.[43] The good news for other low- and middle-income countries is that China has also been a huge market for raw materials—at least until the recent relative slowdown in its growth and the subsequent virus-induced slump—and a source of low-cost consumer goods. The bad news is that China's export industry is a source of almost insuperable competition.

Like Japan and Korea, China began with low-tech, labor-intensive industries. Until about 2015, it also depressed the value of its currency to make its exports cheaper and protected its domestic market as a first step in building up exported manufactures and information-intensive services. Unlike Japan and Korea, China not only encouraged foreign investment but also used the lure of its huge and expanding market to force foreign investors to accept domestic partners, to pressure them to share advanced technology with Chinese firms, to establish cutting-edge research laboratories, and to acquiesce in the theft of intellectual property—all in violation of World Trade Organization (WTO) rules—as the price of market access. The result was four decades of extraordinary economic growth, albeit at a significant cost to public health and the environment.[44]

Today, Chinese firms are competitive in artificial intelligence and the 5G communications technology that is key to the greater speed, quicker reaction time, and vastly expanded capacity needed for the Internet of Things (IoT) that will soon accommodate billions of connections in homes, transport, industry, energy, medicine, agriculture, and the military. China excels at adapting product design to the needs of the market and at engineering manufacturing scale up, for both low-end, medium-tech products like motorcycles and sophisticated products like the latest model of smartphone.[45] This has enabled them to dominate technologies like wind turbines, liquid crystal displays, and lithium ion batteries, in which leadership in innovation flows directly from skill at manufacturing.[46] China's difficult geography has forced Chinese engineers to develop skill at designing and constructing roads, railroads, tunnels, airports, and ports under challenging conditions, skills that have served as the technical underpinning of the Belt and Road Initiative (BRI) that finances transportation infrastructure linking Western China to the Mediterranean Sea.[47] Despite these high points, however, the bulk of Chinese industry is at a much lower technological level and will require major upgrades in management and logistics if Chinese industry is to fulfill the ambitions of its government.[48]

The wages of Chinese workers are rising, but its huge rural population provides a ready supply of workers to continue staffing the textile and other low-tech, labor-intensive industries that are the traditional starting point for poor countries seeking to industrialize. As worker productivity and wages rise in the coastal provinces where Chinese industrialization got its start, production is automated or is moved to the less-developed inland provinces or offshored to nearby Southeast Asian countries. This last pattern is likely to intensify in response to the increases in tariffs on Chinese exports, the aging of the one-child generation, and the increasingly acute shortage of skills. Much depends on the speed of the world's recovery from the virus-induced economic shutdown.

To encourage cutting-edge innovation, the Chinese government is investing many billions of dollars in a major effort to dominate ten key technologies of the future, including advanced manufacturing and artificial intelligence. This sum is comparable to those being invested in these fields by the United States, Germany, and the United Kingdom. The report *China—2025* sets forth a comprehensive, ambitious, $250-billion plan to dominate ten major new technologies, including several areas of manufacturing.[49] China is already strong in a few of these areas, such as artificial intelligence and 5G communications, but has a ways to go to catch up to the leaders in most of the others. To achieve this leadership position, the government intends to invest in infrastructure and connectivity, promote venture capital, use government procurement as an incentive to Chinese firms to improve their technology, and require technology transfer from investors and foreign exporters as a condition of market access.[50] It has also engaged in a broad-gauge program of semiofficial, high-tech industrial espionage.[51]

As part of its drive to raise the technological level of its industry, the Chinese government has mounted a systematic effort to promote worldwide acceptance of technical standards that favor the widespread use of Chinese technology and equipment.[52] The stakes for companies and governments in the standards-setting process can be considerable. A technical standard that depends on patented equipment or technology that is available only from a single company can be a lucrative source of income for that company. If its fulfillment requires information or communication technology from a particular country, it can be a valuable intelligence resource for that country, as we have seen in chapter 9. For these reasons, technical standards are likely to be an important arena for future competition between China and Western economies.[53]

For the Chinese government, worldwide acceptance of Chinese standards is a step toward the goal of technological dominance in artificial intelligence, 5G, and the IoT. Some Chinese standards are likely to be widely accepted as a direct consequence of widespread use of Huawei and other Chinese 5G technology and equipment, in part because of its extensive commercial sales overseas and in part because of the Digital Silk Road, the program of the Belt and Road Initiative that installs Chinese digital infrastructure in countries from Uzbekistan to Guinea. China is also devoting time and resources to increase the role of Chinese experts in the International Organization for Standardization (ISO) and the many other international organizations involved in developing technical standards.

China's plans to reach the cutting edge of high-technology innovation face a number of obstacles. The Chinese educational system still stresses rote learning rather than analytic and creative skills. Chinese funding for higher education skews away from technological institutions, so that skilled professionals are likely to be in short supply in the future.[54] Public funding for research in science and engineering is relatively generous—China will soon pass the United

States in research and development spending, although much of this is for applied research—but the scientific community is still taking shape and has faced problems of fraud and plagiarism.[55]

Almost three-quarters of Chinese investment in R & D occurs in industrial firms, with less than 10% going to universities and research institutes.[56] This could, in principle, indicate a relatively smooth connection between successful research results and eventual commercialization, as is the case in the United States and Korea, whose statistics are similar on this point. However, much Chinese industrial research takes place in government-owned businesses that receive preferential access to credit from government owned banks and are well positioned to resist competitive pressure to improve their products and their production technology. This makes them less likely to put innovations into practice—even those that emerge from their own research. The end result is that China may not be getting its money's worth for its investments in R & D.

Political Control Versus Technological Innovation in Chinese Industry

The Chinese government has yet to confront the tension between the need of high-tech industry to innovate freely, on the one hand, and the traditional emphasis of Chinese rulers of all political stripes—but especially the communists—on political control, on the other. President Xi Jinping is consolidating his own power and that of the Communist Party, insisting on loyalty and discouraging discussion.[57] Xi is likely to tolerate dissent on technical matters, but not on big policy issues like environment, social security, limits on the independence of private business, and the allocation of credit to favored businesses, industries and projects.

President Xi may be defeating his own purposes by pressing private Chinese high-tech companies to incorporate a parallel structure of Communist Party officials into their management structure, and by continuing the long-standing practice of channeling capital from state-owned banks to less efficient and less innovative state-owned enterprises.[58] The power of party cadre is increasing, so that a report of disloyalty from the party cadre can result in a private company being cut off from its preferential access to capital and government procurement. (A big "private" Chinese company is closely tied to government, somewhat like Lockheed in the United States, only more so.) The new Chinese strategy may well be effective in encouraging the development of advanced technology for the military, espionage, and surveillance of the civilian population, for which government is the main market, but it is likely to be less effective in creating a climate of innovative civilian technologies oriented to market needs. It has also begun to exacerbate fears in Western countries that the Chinese government is pressing exporters to introduce hidden vulnerabilities into the equipment they export, despite the vigorous insistence of these companies that they are strictly nonpolitical and free of government interference.

The proud announcements of ambitious Chinese plans for high-technology development, coupled with Chinese muscle flexing in the South China Sea and the Indian border, have strengthened the impression that the competition between the United States and China has become "geo-economic" (or more accurately, "geo-technoeconomic"), which is to say that it has strong political, military, economic, and technological dimensions.[59] This fear has already made it more difficult for Chinese firms to purchase American high-tech companies and to penetrate United States and U.K. markets for sensitive communications equipment.[60] These restrictions had already begun to contribute to the slackening, even before the COVID-19 pandemic, of the extraordinary rates of growth that the Chinese economy was once able to sustain.[61]

These and other restrictions on the import of technology have brought home to Chinese leaders the extent of their dependence on imported technology and have led them to put science, technology, and innovation at the heart of the fourteenth Five-Year Plan, the basic government policy blueprint for the years beginning in 2021, in hope of achieving an increased level of technological self-sufficiency in anticipation of the possibility that relations with the United States will continue to deteriorate. This drive is likely to begin with support to the design and manufacture of relatively unsophisticated microchips and other electronic components, with the intention of catching up later in more difficult areas of advanced technology. In support of this goal, the government is providing generous support to innovation incubators and technologically innovative start-up firms.

The Future of Work

Prediction is difficult, especially when it concerns the future,[62] and even more in the presence of so many crosscurrents: rapid technological change, the trade war, and the shutdown from the coronavirus pandemic. But let's try anyway.

Once the virus-induced global shutdown ends, we should expect that manufacturing productivity will increase worldwide, meaning that the same number of products will be produced more cheaply with fewer workers. This sounds like there will be fewer jobs—i.e., technological unemployment. On the other hand, new, cheaper, and better products normally lead to higher demand, increased sales, and expanded production. This would mean *more* workers to fill entirely new kinds of jobs, in addition to those needed to fabricate, transport, and sell the increased production, and those needed to do the same for the new production machinery. By the time you read this, there may be signs of which of these two conflicting trends will win out. In any case, these changes will probably happen sufficiently slowly that work will not become entirely obsolete anytime soon.

But where will this advanced manufacturing take place? Where will the new jobs be created? Again, it depends. Americans hope that the new technologies will bring back manufacturing jobs from the low-wage countries to which they have migrated. This is likely to happen to some extent, but the reshored jobs will be different, and there will not be enough of them to restore manufacturing employment to its previous levels. The new jobs are likely to require education beyond high school and will not restore the old path to the middle class for the high school graduate. The workers thus displaced are likely to continue to form a political force that threatens democracy at home and the international order abroad.

The impact of the revolutions in manufacturing is likely to be even more serious in countries with low and medium per capita incomes. The new technologies are likely to demand increased literacy and digital skills on the part of the factory worker, as well as expensive investments in connectivity and communications infrastructure. This, in turn, will require politically difficult reforms in the education systems of many countries. Most important, the new technologies, combined with overwhelming Chinese competition, are likely to reduce the need for labor in many of the industries on which the lowest-income countries, especially in Africa, rely to create jobs for the increasing numbers of unskilled workers who are migrating away from farms and into growing cities.

Perhaps the displaced African workers will find jobs in their home countries in agriculture (although climate change puts even existing agriculture in danger in many parts of the continent) or in the service sector. A hopeful sign is the increased private investment by Chinese entrepreneurs in small- and medium-sized firms, which will help to mitigate African unemployment problems. These entrepreneurs often find that there are less competition and more profitable opportunities in Africa than back home, and that the situation there resembles the situation they remember from their own experience decades ago in China. They are comfortable with the less settled business climate in Africa, take up residence there, and make it their permanent home.[63]

But if African workers cannot find jobs in their own country, they will migrate in search of employment—first to neighboring countries, but eventually to countries in Western Europe that are having their own unemployment problems and that are not eager to absorb large numbers of immigrants from different cultures and religions. The resulting migrations are likely to be far larger and more disruptive than those that Europe and the United States have already seen.

The prospect of dramatically increased migration from sub-Saharan Africa due to climate change and technological unemployment is a recipe for political turmoil in Europe that could complicate the world's efforts to deal with the global problems we discuss in this book. If Europe and the United States respond with immigration restrictions, the turmoil will shift to Africa and other low-income countries, with unpredictable results. The employment and demographic effects of the revolutions in manufacturing technology are thus likely to have worldwide ramifications, and to bring into high relief the nexus between technology, economics, demography, and domestic and international politics that is the focus of this book.

The most serious threat to the global trading regime, and to the system of global value chains and value networks that it makes possible, stems from the fact that China and the United States are geopolitical rivals whose economies are nevertheless closely interconnected, so that economic, technological, and geopolitical issues are closely intertwined. This gives rise to a number of issues that are best addressed separately. As we have seen in chapter 8, the governance of the Internet lends itself to a live-and-let-live compromise, with China and the democratic powers maintaining a global Internet while each manages its domestic networks in accordance with its own political philosophy. Other issues, like the exclusion of foreign firms from important Chinese markets, are basically trade disputes best handled by existing (though now threatened) WTO dispute settlement procedures. Still others, like the competition between 5G communications systems, involve substantial security risks for the United States and its allies and require stout defense.

The wild cards in this prognostication are therefore the path of economic recovery in different countries from the virus-induced shutdown, and the question of whether there will be a continuation of the current efforts to break down the world trade regime governed by the WTO—which in the extreme could mean that the United States and China split the global economy into competing trade blocs—or whether frank dialog and tough negotiations will restore a global trading system that, despite its flaws, has brought the world unprecedented prosperity.

12

New Technologies, New Issues

Gene Drivers and Geoengineering

On November 26, 2018, Dr. He Jiankui, a senior, well-respected Chinese scientist, shocked the world by announcing that he had edited the genomes of twin babies in the hopes of defending them against HIV/AIDS. Dr. He had used the latest "gene driver" technology to introduce a genetic change that will be inherited by the babies' descendants and that is likely to spread throughout the world. This could affect everyone's descendants and make a permanent change in the human gene pool.

Dr. He had violated a consensus of the world scientific community that it was premature to edit the genes of human germ cells—sperm and eggs—at least until a lot more was known about the safety and effectiveness of such germline editing, and until its many social, economic, philosophical, and religious implications had been thoroughly examined from all points of view and in all parts of the world. Dr. He thought that his pioneering work would be universally applauded and that he would be a good candidate for the Nobel Prize. Instead, he was roundly condemned by leading scientists as "irresponsible" and his work denounced as "horrific."

Chinese officials, sensitive to their reputation as a "wild east" that pays only casual attention to scientific ethics, were quick to condemn Dr. He's many ethical violations, not the least of which was that he had done a risky experiment on human babies before trying it out on nonhuman animals and had failed to properly inform the babies' parents about what he was going to do.[1] What is more, Dr. He's work was scientifically sloppy. His gene editing very likely didn't protect the babies against HIV/AIDS after all. Besides, there was no way, short of intentionally infecting the babies, of finding out whether they had, in fact, been protected and if they had been, whether the gene modification had been responsible for the protection.[2]

Dr. He was subject to condemnation and discipline for his violation of professional norms, both by his scientific peers and by the Chinese government. He lost his job and his laboratory and was sentenced to prison. But there are likely to be governments, foundations, and private entrepreneurs that are willing or even eager to support experiments along these lines, and scientists willing to carry them out. What is more, technology similar to what Dr. He used to edit the

The Survival Nexus. Charles Weiss, Oxford University Press. © Charles Weiss 2022.
DOI: 10.1093/oso/9780190946265.003.0014

babies' genes is already being sold in kits that are available to amateur hobbyists and high school students, who may or may not be trained or socialized in the ethics of medicine and science. It is thus only a matter of time until private companies offer such services to anyone willing to pay for them, regardless of their effect on the human gene pool. Governments may offer incentives to ensure that companies in their country succeed in what they believe will inevitably become a global competition. Clearly, these issues are not the sole purview of scientists but affect people of every race, religion, and culture all over the world, all of whom have the right to be in on the discussion.

Dr. He's misadventure brought into sharp relief issues that had long simmered among bioethicists but now have to be faced directly. How do we weigh potential benefits against unknown risks when both may be existential? How do we balance individual and collective responsibilities and interests for technologies that have global implications but that can be put into practice by individual states, by professionals, or even by amateurs? And ultimately, who gets to decide if and how these technologies are to be deployed?

These governance questions are shared by many new technologies. In another example, geoengineering technologies could create negative emissions that mitigate the effect of carbon dioxide emissions. Like gene editing, they could and, indeed, would be intended to have global effects but can be put in place on a large scale by private firms or even by wealthy individuals. You do not have to be a government, for example, to have enough money to develop and deploy a fleet of airplanes to scatter reflecting material over high-altitude clouds, to hire a ship to dump iron filings into the ocean in international waters, or to crowdsource reflective balloons, roofs, or ground cover. Your motives may be the best, but you may be creating ecological or climatic effects that neither you nor anyone else can reliably predict.

Gene Drivers: A Technology that Controls Evolution

Dr. He's scandalous behavior cast a shadow on an exciting scientific advance with immense potential for practical applications from medicine to agriculture.[3] Gene editors and gene drivers have become powerful tools with which to address a wide range of scientific challenges. They promise new approaches to the cure of sickle cell anemia, Huntington's disease, Tay-Sachs disease, muscular dystrophy, and other intractable genetic diseases. They can make it possible to contain the insect vectors of deadly diseases by protecting them from infection or by eliminating them entirely. (How this might work will be explained shortly.) Agricultural scientists are using the technology to develop varieties of oranges, rice, and other crops that are resistant to diseases that could otherwise be controlled only by large quantities of chemical pesticides or fungicides.

CRISPR-cas9, the technological system used by Dr. He, is both a gene *editor* and a gene *driver*.[4] Gene editing—the process of changing or deleting a specific piece of genetic material in somatic cells, cells that are neither sperm nor eggs—has been around for many years, but the genetic changes are not inherited, so the long-term effects are relatively small. Gene drivers, on the other hand, increase the probability that a given gene will be inherited, which can impact the genetic makeup of future generations. To understand the implications of these technologies, let's start with the science.

Gene *editing* changes a gene in four steps: finding and marking the DNA that corresponds to the gene to be edited, clipping that gene out of the stretch of DNA of which it is a part, replacing it with the modified gene, and repairing the surgery. In the first step, a piece of "guide RNA" finds the DNA gene to be edited and hybridizes with it, marking it for modification.[5] In the second step, enzymatic "molecular scissors"—more properly known as an endonuclease—clips out the gene or the part of the gene that it is supposed to delete. In the third step, the endonuclease adds whatever it is supposed to add. This may be a modification of the gene that was removed or the addition of one or more new genes. In the fourth step, normal repair mechanisms that are present in every cell stitch the edited DNA back together. The gene editor then disappears from the scene.

Applied to a somatic cell, gene editing is an extraordinarily useful tool for research. To give an oversimplified example, if researchers want to learn the function of a particular gene, they can delete it and see what happens, or they can make deliberate changes to the gene and make it do something other than its original function. The edited genes will affect only the organism whose genes are being modified. They will not be inherited, so future generations will not be affected.[6] If, on the other hand, the genes of germ cells—eggs or sperm—are edited, the results can be inherited like any other genetic change. If the modified gene makes it more likely for those who carry it to survive, it will spread slowly; if not, the gene will sooner or later disappear from the population.

When CRISPR-cas9 is used as a gene *driver* as well as a gene *editor*, the edited genes are much more likely to be inherited and to spread inexorably, permanently changing the gene pool of the species. This happens because of a neat trick that ensures that the descendants of the carrier of the modified gene will carry that gene in both of a pair of chromosomes and will bequeath this "homozygous" condition to *their* descendants, whether or not the modified gene makes it any more likely that these descendants will survive.[7] This means that the modified gene will gradually spread to much or all of the population, perhaps with a major impact on the species and on the ecology of which it is a part. For example, CRISPR-cas9 might spread a gene that causes sterility in malarial mosquitos and lead to their extinction. Alternatively, a new gene might immunize the mosquitos against the parasite that causes malaria, perhaps by erasing the information needed to synthesize the receptor protein that the parasite needs as its entry point for infection. Sounds great, either way.

The problem lies in the fact that the application of gene drivers involves risks that need to be resolved by further research and experience. Could the "molecular scissors" accidentally clip out the wrong gene? Could the gene driver somehow give rise to an unintended mutation? Might an incomplete gene edit result in a mosaic—progeny with a mix of cells, some with edited and some with unedited genes? Most important of all, suppose scientists release a gene-driven organism accidentally and want to get it back. Could the action of the gene driver be reversed?[8]

Happily, it looks like some important ethical issues can be left to the future, if only because gene drivers are likely to be the therapy of choice for only a limited number of human ailments, since there are less risky ways for doctors to prevent the spread of most of the single-gene afflictions that current gene driver technology could potentially cure.[9] It seems unlikely that today's gene drivers will soon enable parents to create prettier, taller, smarter, more athletic, or more congenial children, however much they might desire to do so. All of these characteristics are the result of hundreds of genes instead of one or two and are therefore beyond the reach of present-day technology. This makes it possible for us to postpone thorny ethical questions concerning what constitutes a disease or disability worthy of the risks of a hypothetical genetic therapy (short stature? pattern baldness? color blindness?), as opposed to the "enhancement" of a desirable characteristic (beauty? athletic ability? congenial personality?). Nor do there appear to be immediate prospects for the use of gene drivers to develop more lethal or race-specific bioweapons, to cite two of the more dystopic fears.

Nevertheless, gene drivers might still be the therapy of choice for those diseases for which other approaches are ineffective, for example, if it is the only path for parents to have a child free of some terrible genetic disease, or if they desire a "savior" child that could supply a transplant organ for an afflicted sibling. These difficult situations pit the interests of today's parents against those of future generations, presenting classic problems of intergenerational equity. Once effective gene driver therapies have been developed for such situations, it will be very difficult to deny them to the relatively small number of people who need them, despite the fact that they may create risks for the entire future human gene pool, i.e., for everyone's descendants.

The Pandora's Box Congress

The dawn of the biotechnological era in the 1970s raised issues of similar gravity. This was a time of tremendous scientific promise and excitement. It was also a time of concern that scientists might create genetically engineered bacteria that could escape from the laboratory and create unpredictable havoc. Wishing to forestall such a calamity and at the same time to gain the public's trust that they were aware of and were dealing with the risks inherent in their work, researchers agreed to observe a moratorium on certain kinds of experiments until they were sure that the bacteria they were creating could not survive outside of the laboratory.

This issue came to a head in February 1975, when a who's who of top researchers in molecular biology met at Asilomar, a beach resort near Monterey in Northern California, to "assess the risks of recombinant DNA technology and devise ways of reducing them."[10] For three hectic days, 140 scientists from different subfields of molecular biology worked hard to master the basic concepts of each other's disciplines and to agree on what, if anything, needed to be done to protect the public. Some fifteen journalists observed the proceedings.

A good many of the scientists in attendance at Asilomar continued to insist throughout the conference that no extra precautions were needed over and above usual laboratory practices. This attitude changed when the small group of lawyers present laid out the legal consequences of failure to take adequate measures to ensure the safety of laboratory workers and the general public.[11] On the final morning of the meeting, its organizers pushed through a consensus report calling on researchers to continue the moratorium until experiments had proven that bacteria could be genetically engineered in such a way that they could not survive outside the laboratory. The report also set forth a four-tier classification of future experiments, the safest of which could be carried out in an ordinary laboratory, the next two with increasing levels of containment, and a top tier—work with bacteria known to cause disease—that required the high levels of containment customary for work with dangerous organisms. The recommended experiments having been successfully completed, scientists at another meeting at Wye College in England four years later relaxed the stringent guidelines associated with the moratorium, and research resumed at its previous furious pace.[12]

In retrospect, the Asilomar Conference (more formally, the International Congress on Recombinant DNA Molecules) was a pioneering effort by scientific researchers to assess the risks of a new technology well before the public was aware of its promise and its inherent risks. On the technical level, it was a complete success. There have been no mishaps of the kind that the precautions prescribed at Asilomar were designed to prevent. As a precedent for CRISPR and geoengineering, however, the Asilomar Congress has important limitations. First of all, the technical problem confronting molecular biologists in the 1970s was to make sure that the newly engineered bacteria *do not* escape from the laboratory. In contrast, the whole point of gene driver technology is to *release* modified organisms into the environment in such a way that their altered genes *do* spread throughout the population of the affected organism. Similarly, the ultimate objective of geoengineering is to influence the world's climate on a large scale. What is more, the genetic engineering techniques of the 1970s were accessible only to elite researchers with sophisticated equipment (although this situation did not last long), so participants assumed that the relevant actors were still bound by the same code of professional scientific ethics. For these reasons, the ethical problems associated with gene drivers and geoengineering are much trickier than those addressed at Asilomar.

The procedural issues are no less difficult. Except for the relatively few lawyers and journalists, the participants in the Asilomar discussions were all carefully selected scientific researchers, eager to resume their research and their pursuit of the Nobel Prizes that awaited the most successful of their number. Their discussions centered on safety, not on the possible philosophical or religious objections to mixing DNA from different species—a problem that soon bedeviled the technology of genetically modified foods. The organizers maintained control of the discussion and the final report.[13] For gene drivers and geoengineering, this restriction is neither possible nor desirable.[14]

The Ethics of Do-it-yourself Biotechnology

Gene driver technologies give researchers, or even rogue individuals, the power to change the genetic heritage of humankind all over the world. At a more philosophical level, we can think of the sum of all human DNA—the human gene pool--as a legacy that we leave to future generations and hence as a resource common to all humankind, present and future, like the atmosphere, the climate, or the ocean. We therefore need internationally agreed rules to prevent, or at least to discourage, the misuse of the awesome, useful, but potentially dangerous new power to alter this precious resource.

Well before the invention of CRISPR-cas9, scientists realized that biotechnology raises ethical issues that go beyond those addressed at Asilomar and whose resolution would require discussions in which a much broader range of interested parties should participate. The spread of gene driver technology far beyond the community of credentialed scientific researchers gave rise to the further concern that any of thousands of businesses, hobbyists, or even high school students could use it to do serious and irrevocable damage. How, then, to guide people to use this powerful, widely accessible new technology responsibly? Any ethical guidelines would have to be understood and respected not only by researchers, but also by much of the general public.

As it happened, two of the first draft codes of conduct for the use of gene drivers were developed in 2011, not for professional researchers, but by and for high school students participating in an annual contest that was intended to encourage applications of the new technology, and whose organizers and participants were eager to demonstrate their intention to promote its responsible use.[15] A flurry of policy declarations from organizations of professional researchers, the UN Educational, Scientific, and Cultural Organization (UNESCO) and other intergovernmental organizations, bioethical thinktanks, conservation groups, theologians, and nongovernmental biotech watchdogs followed, climaxing in two international summit workshops, organized in 2015 and 2018, respectively, by three of the most prestigious international scientific bodies: the U.S. National Academy of Sciences, the Chinese Academy of Science, and the Royal Society in the United Kingdom.[16]

The resulting declarations and codes of conduct by and large came to unobjectionable conclusions, some of which are easier to implement than others. All of them agreed that it was too early to use gene drivers on human embryos that were intended to be implanted in a uterus and brought to term as babies, until the safety and efficacy of such methods is adequately demonstrated. There was also agreement on the need for strict oversight, transparency, and careful assessment of benefits, costs, and risks, as well as agreement that gene drivers should be used on humans only in cases of compelling need. The codes further agreed that the benefits of gene driver technology should be widely accessible; that the technology should not be used to create any kind of superior race; that there should be enforceable norms of professional competence and behavior acceptable across nationalities, cultures, and religions, including those of indigenous peoples; that gene editing and gene driving should be used only with the consent of the patients or prospective parents; and that there need to be serious consequences for violating whatever norms were eventually adopted.

To these shared recommendations, the Academy-sponsored summits added the proposal that a continuing forum be created for worldwide, multidisciplinary discussions on the social, economic, ethical, and religious ramifications of the new technology, involving not only scientists, but also a broad variety of stakeholders and the general public.[17] The hope is that such discussions will lead to a worldwide consensus that can be translated into internationally agreed professional codes of conduct and regulatory standards that are widely understood by the interested public and that could ultimately be codified in national legislation and international treaties.[18] Behind many of these declarations lay the fear that the development of a technology with enormous potential might be hampered by excessive or groundless public fears, as happened (in the opinion of many of the scientists involved) to technology for genetic modification of food crops.[19]

Despite the widespread agreement on these general principles, however, there remains a fundamental disagreement between those countries and professional and advocacy organizations that favor the careful deployment of the new technology, on the one hand, and those who argue for a moratorium on the application of the new technology on any organism or urge that it be banned entirely, at least as applied to human beings, on the other hand. Of those who favor deployment, the organizers of the second Academy-sponsored summit did their best to outline an acceptable path to a "clinical transition" that would enable gene editing and gene drivers to be incorporated into the regular armamentarium of modern medicine.[20] In contrast, the biotech watchdog nonprofit ETC Group, along with other critics of gene driver technology, argues that no possible level of prospective benefit can justify the risks of applying gene drivers to the human genome, citing the precautionary principle as a reason why the world needs a moratorium on gene drivers.[21] These intractable disagreements are likely to make it difficult to arrive at universally accepted guidelines regarding the possible implementation of gene driver technology whenever the broad-based, multicultural discussions recommended by the Academies take place.

If gene drivers are ever to be employed on the human genome or, indeed, on any species, philosophical issues with major practical implications will have to be directly confronted. Which is more important, risks or potential benefits? How are one person's (or one species') gains to be balanced against another's costs or risks? How is a benefit to ourselves and our contemporaries to be assessed against costs to future generations? How are we to assess an economic benefit to one group that offends another group's religious or cultural values? What is to be done about rogue operators who apply gene driver technology in an unauthorized manner, perhaps in a country with lax regulations?

At a minimum, the precautionary principle—and indeed, common sense—calls for using gene drivers on any species in a way that minimizes risk. On humans, they should be used only in cases where there is no alternative and where quality of life—or life or death itself—is at stake. But if gene drivers are ever to be used for any purpose, it will at some point be necessary to accept the risk that something could go wrong. This means we must learn as we go, or in more formal language, we must apply the principles of adaptive management discussed in chapter 4: to maintain a close watch on short- and long-term consequences, to continue research on safety and effectiveness, to maintain a robust discussion of social and ethical ramifications, and to apply the lessons learned as practical experience accumulates. But risks will remain. These issues will reappear in our discussion of geoengineering, to which we now turn.

Negative Emissions Technologies

Imagine that a government—or even a province of a large country—offers research grants or financial incentives to stimulate innovation in technology to reflect solar energy or to remove carbon dioxide from the atmosphere in hopes of dominating the commercial market that it is convinced is sure to develop, and then does its best to ensure that the technology it has supported is deployed on a large scale, regardless of the risk. Or imagine that a billionaire, despairing that the world will never reduce greenhouse gas emissions enough to prevent catastrophic global warming, develops and deploys a fleet of airplanes to cool the Earth by spreading reflective powder over high clouds to reflect some of the sun's energy back into space so that less of it will reach the Earth.[22] Should we cheer these events? What could possibly go wrong?

"Negative emissions" technology, also known as "geoengineering," is an umbrella term that encompasses several technologies that have been proposed to deliberately manipulate the planetary environment on a large-scale so as to counteract human-caused climate change.[23] These technologies are either still experimental, or else are now being operated at a scale much smaller than would be needed for a significant impact on the climate. Geoengineering is very different from gene drivers but raises similar issues. Both technologies promise major benefits but pose large and unpredictable risks. Both are within the means of individual countries or even of individuals, yet are capable of substantial global impact. In both cases, even research to establish safety and efficiency would have to be done on a large enough scale to pose unpredictable risks to the planet and to future generations.

Some geoengineering technologies, known collectively as solar radiation management (SRM), would reduce the amount of the sun's energy reaching the Earth's surface, for example, by placing mirrors, reflective nets, or balloons high in the sky. The most commonly proposed technique involves spraying sulfate aerosols into the stratosphere to reflect solar energy back into space as a sort of artificial volcanic eruption. Another proposal involves increasing cloud cover by "spitting" sea salt into the atmosphere.[24] Either of these technologies could be implemented by a fleet of specially designed aircraft using currently available technology at a large but not unbearable financial cost. Other SRM techniques would reflect the sun's energy back into space *after* it has reached the Earth's surface, a process known technically as increasing the Earth's albedo. This could, in principle, be accomplished by covering large areas with reflective vegetation or plastic, or by convincing large numbers of people to paint the roofs of their houses white.

The second class of geoengineering technologies, known collectively as carbon dioxide removal (CDR), would pull out of the atmosphere carbon dioxide that is already there. The least risky way to do this would be to establish large forest plantations for fuel wood, burying ("sequestering") the CO_2 produced when the resulting wood is harvested and burned for fuel, and then replanting. Most climate models predict that this technique of "bioenergy with carbon capture and sequestration" (BECCS) will be an essential element of any plan to limit global warming to 1.5–2°C by the year 2050, as called for in the Paris Agreement described in chapter 3.[25] BECCS technology would call for large-scale changes in land use, possibly displacing a good deal of food crop production, raising food prices worldwide, and causing serious social and political disruption.

It would also be possible to sequester increased quantities of CO_2 in the soil by changing agricultural practices, changing crops, and converting agricultural wastes into biochar, a form of charcoal that is a valuable soil amendment. Other bioengineering technologies for generating negative emissions include establishing algal fuel plantations, also coupled to carbon capture and sequestration (CCS), and fertilizing the ocean so as to increase the growth of algae, which then die and take the carbon dioxide they have absorbed with them to the bottom. There are also chemical techniques that absorb atmospheric carbon dioxide into various reagents so as to allow it to be sequestered or used for commercial purposes.

The environmental impacts to be expected from a geoengineering program of sufficient scale to affect the climate are still uncertain and the subjects of research. The best models to date suggest that with proper implementation, geoengineering might actually improve regional climates.[26] There remain, however, significant environmental risks that must be weighed against the possible benefits: reduced solar energy for agriculture, increased particulate pollution, and reductions in stratospheric ozone that would increase the Earth's exposure to carcinogenic ultraviolet radiation. Given these uncertainties, if a decision to begin applying geoengineering techniques is to be made, it would be best to apply the principles of adaptive management, starting on a relatively small scale, carefully monitoring effects, and carrying out research on cause and effect.

That geoengineering technologies are being seriously discussed at all is a sign that scientists are becoming increasingly discouraged about the possibility of mitigating climate change and are turning to the consideration of riskier methods. Indeed, even the discussion of these technologies was for a long time discouraged for fear that geoengineering would be used as an excuse to allow greenhouse gases to continue to be emitted at existing rates, a problem known to economists as moral hazard. These gases would then accumulate in the atmosphere at ever increasing concentrations, imposing a major and increasing cost on future generations.

This moral hazard is probably the greatest risk associated with negative emissions technology, one even greater than the direct environmental risks of the geoengineering measures themselves. First, if the investments in forests, reflective powders, chemical reagents, or other geoengineering materials were ever to be discontinued, the Earth's temperature would quickly rise to what it would have been if geoengineering had never been applied in the first place, causing a sudden "climate whiplash" that would be even more dangerous than the more gradual rise to the same temperature that would have taken place if geoengineering had never been put into effect. These investments would therefore have to be maintained into the indefinite future, possibly at ever-increasing rates. Second, geoengineering technology would counteract some of the effects of greenhouse gases but not others. For example, since atmospheric carbon dioxide concentrations would continue to increase, the oceans will continue to acidify, endangering coral reefs, shellfish, and ocean ecosystems.

Research Collaboration and Commercial Competition

The technologies both for geoengineering the Earth's climate and for editing the human genome are still in the research and development stage. Neither is ready for deployment. In both cases, scientific knowledge and technological innovation are needed from both public and private sources to make possible science-based decisions to maximize the effectiveness and minimize the inevitable risks. At the same time, even research and development involve substantial risks to the planet.

Research on both gene drivers and geoengineering can be pursued both for private profit and for humanitarian purposes. In technical terms, it is both a public and a private good.[27] It can involve either scientific and "pre-competitive" research (research that does not directly result in proprietary products), competing investments in proprietary innovation by private companies all over the world, or both. Much of the nonproprietary research will have to be funded by governments and private foundations or individuals. There is every reason for the results of this kind of research to be shared openly among researchers in all countries regardless of politics, and for the training of scientists and engineers to be the subject of widespread international cooperation. To private firms, on the other hand, gene drivers and geoengineering represent both a business opportunity and a global responsibility. The world needs to encourage private investment and innovation in both fields, and this will happen only if there is a reasonable prospect that the technologies will actually be financed and deployed. Governments are likely to encourage such private investments, as a contribution both to the management of climate change and to their countries' technological and economic development. Many of the practical results of such research will be proprietary.

We may expect great pressure to develop and deploy both of these technologies: gene drivers in special medical situations in which there is no alternative therapy, and geoengineering if mitigation of global warming falls short of agreed targets. There is also some danger that the go/no-go decision could become the subject of a major private or intergovernmental lobbying campaign, and it is important that private profit and national competitiveness not become major drivers for the decision to develop and deploy these technologies with insufficient attention to the risks that they entail.

The Ethics of Climatic Intervention

The philosophical and ethical issues raised by geoengineering are broadly analogous to those raised by gene drivers. In both cases, the scientific community has taken the lead in setting forth broad principles to guide research and possible eventual implementation. In the case of geoengineering, the most influential of these statements is the Oxford Principles, which were drawn up by a small group of British academics and were substantially echoed by professional groups and even adopted by the British House of Commons. Like the many declarations of principle concerning gene drivers, the Oxford Principles call for transparency and broad public participation in decision-making, independent assessment of environmental and socioeconomic impact, prior informed consent of those affected, and open publication of research plans and results. Similar principles have been adopted by the World Meteorological Organization, a technical agency of the United Nations, by the American Meteorological Society, and by a sequel Asilomar conference.[28]

The Oxford Principles open with the fundamental principle that the climate is a public good and a key resource for all humankind, so that decisions regarding its possible modification by humans should be taken in the public interest. If this were not the case, decisions regarding geoengineering could, in principle, be governed by purely commercial considerations. The Oxford Principles further hold that "regulations" should be made by "appropriate bodies," and that decisions regarding the implementation of geoengineering should not be made until a "robust governance structure is in place."

Geoengineering research requires a tricky balance of competing considerations, over and above those involved in full-scale implementation. The instinct of the scientist is that more knowledge is always desirable. In the case of geoengineering, however, even scientific experiments can cause unpredictable environmental effects. There is also the "slippery slope" concern that geoengineering research can divert attention from climate change mitigation and inevitably leads to pressure for implementation. Carefully navigating between these conflicting considerations, a U.S. National Academy of Sciences panel has recommended a program of research intended to "enhance policymakers' understanding of climate response options." This is to take place "alongside a robust portfolio" of multidisciplinary research on climate mitigation and adaptation. Like the Oxford Principles, its recommendations stress the importance of "governance mechanisms that include roles for government, universities, independent institutions, research teams, and other nations."[29]

It is not clear what such a "robust governance structure" would consist of, whether such a structure would be politically possible today, or whether it would be effective if it were to be established. As we have seen in chapter 3, even binding national commitments under the aegis of the Kyoto Protocol did not result in reduction in carbon dioxide emissions sufficient to meet internationally agreed goals. In the absence of such an international agreement, decisions regarding geoengineering are likely to be made by individual governments or groups of governments, by private companies seeking competitive advantage, or by nongovernmental organizations or wealthy individuals. We would then have no choice but to rely on broadly accepted norms and standards of behavior to influence the decisions of this multiplicity of actors and to induce in them a sense of accountability to the broader public. In elaborating on these norms, we would have to confront the questions concerning benefits, costs, and risks that we posed at the beginning of this chapter in connection with gene drivers, not to mention the broader philosophical question of whether humans should be meddling with the climate even more than we already are. At the more practical, operational level, any norm needs to satisfy the criterion defined by an admonishment I received from a senior colleague early in my tenure at the World Bank when I made an mushy proposal for a policy on some long-forgotten subject: "Suppose you were the official in charge of putting the policy you propose into practice, what would you do on Monday morning?" If a proposed norm doesn't point you toward an answer to that question, it needs more thought.

Responsible Innovation

The technologies discussed in this chapter pose the question: What constitutes a responsible approach to technological innovation? Both geoengineering and gene drivers promise great benefits but also involve large risks whose magnitude is difficult to assess until the detailed capabilities and socioenvironmental impacts of these technologies are better understood. This is a catch-22, since it's impossible to know these capabilities and impacts until the technologies are implemented on at least a small scale, a process that itself involves considerable risk and is likely to give rise to strong pressures to expand the technologies to full scale. Both technologies have global ramifications but depend on local decision makers for their development and implementation. Both could be developed and deployed not only by governments or international organizations, but also by private firms, by individual scientists or doctors (gene drivers), by wealthy individuals or organizations (geoengineering), or even by smart high school students (CRISPR-cas9).

We are far from a governance scheme to guide such decisions. As things now stand, any country, company, or individual with the money and technical ability is free to develop and implement these technologies, regardless of possible danger to the global climate or to the human genome. It is unlikely that countries will agree to binding treaties on the subject any time soon. Our best hope, then, is for the gradual evolution of norms based on cross-culturally acceptable principles rooted in the common interest in survival. Such norms are likely to be first articulated by professional organizations, think tanks, and independent groups of ethicists and theologians, who have only moral suasion and whose declarations are not legally binding. If consensus is achieved, these norms may gradually work their way into national policies and eventually into binding national and local legislation and international treaties, a pattern we have observed in such diverse issues as nuclear arms control and stratospheric ozone depletion. The success of these norms will depend on the support of governments, which, in turn, depends on the support of the public, which, in turn, depends on public understanding of the implications of the new technology and the need to guide it into socially beneficial directions.

Someone, somewhere, is likely to face the decision relatively soon as to whether to develop and deploy these technologies. We need norms to remind decision makers that decisions to intervene in a global public good in ways that promise major benefits but that also pose enormous risks are to be taken in the interest of all humanity, not to satisfy personal, political, or commercial interests.

Even if proper procedures are followed, decisions regarding these new technologies are likely to be determined by attitudes toward risk. As a practical matter, major disagreements among people and among countries are unlikely to be resolved. Should we be willing to accept large risks in the hope of gaining great benefits—or should we follow the precautionary principle? How are the costs, benefits, and risks accruing to one country or to one generation to be weighed against the costs, benefits, and risks to another country or to future generations? Finally, are there actions that are inherently unethical and should not be carried out under any circumstances, even if they are beneficial on balance? Our answers to these seemingly abstract philosophical questions will ultimately determine our approach to the development and deployment of gene drivers and geoengineering, and no doubt other new technologies as well.

13
Playing with Fire

In the years since World War II, diplomats and policymakers from many countries have worked together to encourage economic growth, improve public health, reduce poverty, and control the threats of nuclear war, global pandemic, and environmental degradation. Scientists have played important roles in many of these efforts. The regimes they have created were not perfect and have benefited from luck at critical moments. They have needed to be augmented, patched, and improved as the world gained experience with what worked and what didn't, and as the issues they address have evolved. But they have succeeded in their broad objectives and have protected us against calamity. There have been no nuclear wars since 1945. The world economy has doubled several times over. Global life expectancy has increased by over 50%. Hundreds of millions of people have been lifted from abject poverty.

In recent decades, these regimes have come under steadily increasing pressure. Politicized obstacles to the fight against COVID-19 have cost hundreds of thousands of lives. Arms control and nonproliferation agreements that have reduced the number and limited the spread of nuclear weapons have been slowly undermined or dismantled, and countries are pushing against the relatively few restrictions that remain. Efforts to limit greenhouse gas emissions are falling far short of what is needed to prevent catastrophic global warming within the lifetimes of our children and grandchildren. Jobs lost from technological changes in manufacturing and services have stimulated a backlash against globalization and the rise of a brand of populism that is hostile to expertise and governance. The resulting revival of protectionism and economic nationalism threatens the trade regime that has brought prosperity and technical progress to much of the world.

We are needlessly allowing technology to take the world to the brink of disasters from accidental climate disruption, nuclear war, and pandemics—at the same time that we are allowing the means for controlling these technologies to erode. In effect, we are edging closer and closer to cliffs from which we have removed the guardrails. Fortunately, people are beginning to develop—or at least to think about—protections against some of these dangers.

A technology-based model of authoritarian government is being promoted and exported as a technological, economic, and political challenge to Western liberal democracies. The Internet and social media, conceived as vehicles for free

The Survival Nexus. Charles Weiss, Oxford University Press. © Charles Weiss 2022.
DOI: 10.1093/oso/9780190946265.003.0015

exchange of information and platforms for untrammeled innovation, in some countries have become instruments of repression. At the same time, a systematic, worldwide campaign of misinformation and disinformation, spread via social and mass media, has deliberately sown distrust in the democratic process, in government, in international institutions, in science, in expertise of all kinds, and in the very idea that there is such a thing as truth. In the United States, the scientific consensus concerning climate change has been overwhelmed by *mis*information and *dis*information spread by political and business interests that it threatens, while public messaging regarding the COVID-19 pandemic has been at best mixed. All this has made it difficult for people to understand the complex and dangerous new threats to the environment, to their health, to their security, and to democratic government.

Stimulating and Guiding Technological Change

Science and technology have much to contribute to the resolution of all these issues. We need better scientific understanding of the climate, of the ways in which diseases spread, and of the likely impact of gene drivers and geoengineering. We need technological innovations to prevent or cure noncommunicable diseases, alleviate malnutrition, conserve resources, defend against cyberattack, and restructure our economies along more sustainable and equitable lines.

But we cannot depend on science and technology to address these issues on their own. We also need policies and institutions that not only support research and encourage technological innovation but also guide scientific research and technological change in responsible directions. This will require both "top-down" measures by national governments and "bottom-up" pressures from public opinion, from nongovernmental organizations (NGOs) and from state and local governments that are often more directly responsive to public pressure. It will need ideas and inspiration from businesses, universities, research institutes, individual inventors, and ordinary citizens, and support from public opinion and from the actions and advocacy of individuals and communities. In some cases, it will require sustained effort to resist commercial, political, and military pressures to ignore broader social and environmental problems when investing in research and innovation and to deploy technologies before at least some of their consequences are understood and anticipated. The response to the COVID-19 pandemic and the denial of the reality of climate change in a number of countries show how even the best scientific advice can be undercut by the words and actions of shortsighted, self-interested political leaders.

We need to devise limits on technological innovations like autonomous weapons and hypersonic missiles that can lead to unintended nuclear war. We need to restructure our economy and redesign our cities to end, or at least to limit the use of fossil fuels and to increase the efficiency of our use of energy and materials. We need controls on technologies like artificial intelligence, gene drivers, and geoengineering, all of which both promise major benefits and involve great risks. We need to defend ourselves against technology-based attacks on the values of freedom of thought, communication, innovation, and access to information that were embodied in the original design of the Internet and social media but are now under attack.

At the same time, we need new science and technology to develop novel approaches to environmental sustainability: improved sources of renewable energy and innovative approaches to efficient energy use. We need both to expand and to restructure our electric energy grids to electrify our economy and to improve their efficiency, reliability, and security. We need to restructure our agricultural economy, our urban infrastructure and our transportation systems. We need explicit measures to ensure that everyone can participate in the benefits of technological change. We need research and development on orphan technologies, and vaccines and cures for the diseases that mainly affect people in low-income countries. Not a small menu.

Some of these issues, like climate change, nuclear weapons, and global health, are governed by long-standing regimes, norms, and institutions that now need strengthening and refurbishing to meet new political, economic, and technological challenges. Nonproliferation and anti-missile agreements need to be restored and extended to limit or ban development and deployment of hypersonic missiles, destabilizing weapons that are now under rapid development in many countries. Voluntary national limits on greenhouse gas emissions need to be urgently increased. Time is running out, and costs will be much greater the longer we take for effective action to mitigate and adapt to climate disruption. In the global health sphere, the system for emergency preparedness and pandemic control needs increased and sustained political and financial support to replace the long-standing pattern of crisis-to-crisis, feast-or-famine funding. Most low-income countries still need basic health infrastructure, both to provide health services to their population and to identify and control epidemic disease that could spread beyond their borders.

Cyberweapons, too, can quickly get out of control and wreak enormous damage on critical infrastructure, including the systems of command-and-control on which we would depend to prevent accidental escalation of a localized conflict to full-scale war. Like hypersonic missiles, the response to these weapons depends on artificial intelligence. Unlike hypersonic missiles, cyberweapons do not lend themselves to the type of verifiable arms control regime that has so far been successful for nuclear weapons. What is more, many governments oppose any limits on the use of cyberweapons, despite the risk they pose of accidental escalation. Continued research and international discussion on how these weapons can be controlled are urgent priorities.

Geoengineering and gene drivers also take the world into uncharted territory. Their governance is complicated by the fact that they are accessible not only to governments, but also to private businesses, NGOs, and individuals. It is far from clear how decisions should be made as to whether and under what circumstances these technologies should be developed, and what criteria should govern any such decisions. Advocates for these technologies have developed roadmaps for deploying them in a way that minimizes risks. Still, there are fundamental disagreements over whether geoengineering and gene drivers should be developed and implemented at all, and there is substantial support for the idea that one or both should be banned outright. Critics have called for moratoriums until the broader questions can be explored by a broad range of worldwide stakeholders, disciplines, and cultures. However, the disagreements we have outlined will probably never be totally resolved to everyone's satisfaction, and deployment decisions will eventually have to be made one way or the other.

The governance of the Internet and social media involves measures to preserve the global Internet and to address those issues of cybersecurity that are of common concern to countries with very different political systems and very different ideas concerning civil liberties and human rights. The competition between authoritarian and democratic governments over freedom of information and innovation is likely to go on for a long time, but both sides have a strong interest in maintaining a functioning global Internet and in avoiding catastrophic damage to information and telecommunication systems.

Several academies of science, research institutes, NGOs, and religious organizations have proposed codes of conduct and declarations of principle to deal with the difficult philosophical, ethical, and practical issues involved in all these issues. These represent a useful beginning, and one may hope that they will reach the level of consensus that would allow them to be codified into national legislation or international agreements.

Dealing with these issues will require respect for expertise in the fields to which we have often referred: science, technology, politics, economics, business, law, and culture. We will need to incorporate scientific advice into decision-making processes, and to acknowledge and manage the risks and uncertainties in our understanding of the science and the technology we are trying to manage, as well as in their ramifications for the larger society. There is also a need to educate governments and the public, both on the underlying science and technology and on their links to the broader context. Finally, I would urge that there is a need for an international obligation to identify areas for scientific research and technological innovation that can help to resolve these issues and to support this research with adequate financial, human and institutional resources. This last requirement should become a general principle that should become part of the accepted framework for thinking about global issues and incorporated into formal agreements on these matters as a matter of usual practice.

Dealing with complex global, technology-intensive issues like these is a tall order, especially when they require democratic countries to find common interest with countries with whose governments they are otherwise deeply at odds. Nevertheless, the world has faced such issues successfully before under the arguably more difficult conditions of the Cold War, which pitted two ideologically opposed superpowers against each other that nevertheless managed to agree on elaborate and technically demanding measures to avoid nuclear holocaust.

There have been major innovations since the days of the Cold War that will help with these new challenges. Civil society is a source for new ideas and new institutional arrangements. The Intergovernmental Panel on Climate Change (IPCC) and epistemic communities provide scientific advice and education on climate to international negotiators and to the public. Multi-stakeholder meetings develop "bottom-up" codes of conduct, norms, and standards that sometimes make their way into more formal government and international declarations and eventually into national legislation and international agreements. State and city governments combine forces across borders to deal with issues when their national governments shrink from action. Flexible, informal, or voluntary arrangements facilitate international cooperation in situations where binding obligations and formal agreements would be cumbersome or slow moving. Pragmatic, informal institutional arrangements arise to manage international resources when more formal institutions or agreements would be cumbersome or impractical. Private foundations support research in areas neglected by governments and intergovernmental organizations. "Innovative developing countries" like India and China develop and market profitable products for low-income people. Multinational corporations and NGOs support innovations to meet the needs of people without market power. And a billion-dollar network of agricultural laboratories develops improved technology for small-scale farmers in low-income countries.

All these innovations are valuable additions to the international scene. They add ideas, flexibility, dynamism, and vigor to a system that had become dominated by legacy institutions that had become stodgy and unresponsive. But it remains to be seen to what extent they can be substitutes for the multilateral regimes that served the world well in the past.

The Rise of Russia and China

These issues must be addressed against the background of the rise of China as an economic, political, and technological power, and the resurgence of Russia from the disintegration of the former Soviet Union. Both nuclear powers have emerged from historic periods of humiliation at the hands of foreign powers. Both see themselves as geopolitical rivals to the United States and its allies. Both do their best to perfect asymmetric weapons with which to challenge the dominance of the Western powers.

Like the United States, Russia maintains a large stockpile of nuclear weapons, almost a thousand of which are on launch-on-warning, hair-trigger alert more than thirty years after the end of the Cold War.[1] Furthermore, Russia has stated that it would threaten the use of nuclear weapons in order to deter outside intervention in a local conflict, a doctrine that greatly increases the danger of nuclear war. (U.S. doctrine is deliberately ambiguous on the question of first use.) More immediately, Russia has undertaken a systematic, disinformation-based attack on democratic institutions and alliances, coupled with a worldwide campaign of interference with democratic elections. This threat to democracy needs to be met with stiff resistance.

China is an ambitious, technologically advanced, resource-poor, demographically challenged, rapidly growing, middle-income country of 1.2 billion people, almost all of whom have seen the quality of their lives vastly improved in the last seven decades. Unlike Russia, China is both a geopolitical rival to Western powers and closely integrated into the world economy. It is an essential partner in fighting pandemics and climate change, in minimizing the danger of nuclear war, in controlling the ambitions of the North Korean pariah state, and in international efforts to work out norms of behavior for new technologies like gene drivers, hypersonic missiles, cyberweapons, autonomous weapons, and artificial intelligence.

To countries devoted to the liberal world order, China is a repressor of its own people and a promoter and exporter of a techno-authoritarian alternative to the free Internet and to the model of a society based on individual liberties. To its neighbors, it is a source of worrisome political and military pressure. On the climate front, it is a major source of both greenhouse gas emissions and the technologies needed to limit them and to mitigate climate change. It is also an active exporter of coal-burning energy technology.

To industrially advanced countries, China is both an able participant in high-tech value chains, an energetic acquirer of intellectual property by both fair and dubious means, and an ambitious investor in research and development, especially in the advanced technologies that it needs to become independent of external pressure. It is expert at engineering scale-up, both high- and low-tech. Its exports of information and communications technologies are commercially and technically attractive but carry with them eyebrow-raising security risks.

To low-income countries, China is a major market for raw materials, a producer of low-cost manufactures for low-income markets, and a welcome investor in infrastructure and small industry, but it is also a dominant competitor in labor-intensive and medium-tech industry. This novel combination requires from the Western powers a careful mix of cooperation, collaboration, competition, deterrence, and conflict, based on a clear understanding of the constructive Chinese contributions to global security and sustainability, its challenge to democratic institutions, and its strategy for the development of military, industrial, and information technology.

Science and Technology as Pervasive Influences on World Affairs

The issues addressed in this book illustrate the many ways in which science, technology, and world affairs affect each other.[2] We have seen, for example, how economics strongly influences whether an innovation will take place, how fast it will spread, and what form it will take. For example, the failure to price greenhouse gas emissions and the worldwide subsidies to fossil fuels have slowed innovation in alternative fuels and efficiency in energy use. Similarly, the specific needs of low-income and disadvantaged people do not necessarily constitute a profitable commercial market, so valuable innovations serving their needs often result only from targeted, public-spirited intervention.

The influence of politics and economics on science is more indirect but no less real. The emphasis on nuclear and environmental issues, for example, led to the growth in importance of scientific fields ranging from seismology to human geography. The end of the Cold War shifted research funding in the United States from the physical and engineering sciences to those concerned with health.

On the flip side of the coin, technology is a particularly ubiquitous influence on world affairs. We have seen how information and transportation technologies make globalization possible, and how they revolutionize geopolitics, warfare, commerce, and intelligence. We have seen how the possession of nuclear weapons or the mastery of information or manufacturing technology has improved the relative economic and political power of some countries and diminished that of others. We have seen how the Internet and information technology has changed the relative power of governments and civil society and created a whole new arena of global information warfare, pitting democracies against authoritarian states.

We have seen advances in science and technology raise new issues, risks, and uncertainties that transcend national boundaries, transform hitherto domestic policies into international concerns, and leave governments and businesses struggling to keep up with rapid change. We have seen how collaboration between scientists and foreign affairs specialists has made it possible to achieve international agreements on stratospheric ozone, climate change, and nuclear weapons that bind countries of very different and often hostile political orientations. We have seen how scientists and ethicists in many countries have begun efforts to devise norms and standards to govern the balance of costs, risks, and benefits posed by technologies for gene drivers and geoengineering. We have seen the unexpected effects of new technology, as the Internet has been used not only for the free global exchange of ideas, but also for techno-authoritarian repression and a worldwide campaign against democratic values and institutions. We have seen how the destruction of jobs by capital-intensive manufacturing technology has led to worldwide populist backlash and the prospect of large-scale international migrations.

We have seen how technology has changed the meaning of some of the fundamental concepts of international relations. Climate change, health security, and nuclear nonproliferation have made hitherto purely domestic subjects like national energy policy, epidemic surveillance, and nuclear fuel cycles into "intermestic" issues of global concern. Information, communications, and transportation technology have limited (but not ended) the power of national governments to control their economies and the flow of information across their borders, powers that were once hallmarks of national sovereignty. The thousands of daily cyberattacks have blurred what was once a fundamental distinction between war and peace. The prospect of cyberattack blurs the distinction between military and civilian infrastructure and personnel because in a market economy, critical infrastructure, such as banks, insurance companies, credit agencies, and public utilities, is privately owned, so that a breach in a private database may endanger national security.

Most ominously, we have seen how the heedless pursuit of technology can risk catastrophe. The long-standing failure to rein in the use of technology based on fossil fuels makes catastrophic climate disruption much more likely, and the race to develop hypersonic missiles and autonomous weapons dramatically increases the risk of nuclear war. In a last nod to the blurring of previously valid distinctions, we may all hope that the world will never fulfill the prediction that nuclear weapons will erase the practical distinction between winners and losers in war, and that we may never see the day when the "survivors envy the dead."[3]

Crossing and Uniting Disciplines

The ubiquity of science and technology in world affairs forces scientists and diplomats into unfamiliar territory. International affairs experts must grapple with concepts of molecular genetics, artificial intelligence, nuclear fuel cycles, climate geophysics, electric power grids, Internet security, and stratospheric photochemistry, to cite but a few examples. They must also be aware of the dynamics of scientific research, the attitudes of scientists to controversy and uncertainty, the importance and limitations of scientific models, the ways in which a scientific assertion advances from hypothesis to accepted fact, and the sociology of the various branches of the scientific and engineering communities. On the technological side, they need to be aware of the time horizon of technological innovation, the influence of politics and economics on its speed and direction, the sociology and culture of the innovator, the entrepreneur and the venture capitalist, and the concepts of technology assessment and responsible innovation.

Natural scientists who wish to deal with these problems, for their part, need to familiarize themselves with ideas that are derived from the social sciences, law and business, and that are not part of their traditional training: cost–benefit analysis; induced innovation, opportunity costs, and the roles and limitations of markets; the ideas of governance, regimes, norms, and adaptive management; the ways in which science is (or is not) incorporated into public opinion, domestic policy, and international diplomacy; and the political and social influences on popular attitudes toward science and toward expertise of all kinds.

The links and concepts that bind science and technology to world affairs extend well beyond the topics we have discussed in this book. These links and these concepts are equally important as they apply to digital currencies, genetically modified organisms, food safety, pollution, ocean resources, international watersheds, polar regions, outer space, and any number of other issues in which science, technology, and world affairs intersect.

Some of these linkages are already addressed in university departments devoted to science diplomacy, a discipline centered on the role of scientists and engineers as advisers to policymakers and diplomats, as national representatives to foreign countries, and as participants in the planning and management of international facilities for scientific and technological research. This is an important vehicle by which scientists and engineers can broaden their horizons beyond their disciplinary training. Other linkages are addressed by science and technology studies (STS), an academic field that was originally centered on the history, philosophy, anthropology and sociology of science, as well as public attitudes toward science and science communication, but has made efforts to broaden itself to include attention to the influence of the political, economic, and social context on science and technology, and vice versa. A related field of long standing is science, technology and society (also abbreviated as STS), which is defined as the "social, political, legal, economic, and cultural environment that shapes research and invention, supports or inhibits it—and is in turn shaped by evolving science and technology."[4]

This book began as an effort to establish the discipline of Science, Technology, and International Affairs (STIA) as a subject that is sufficiently important and well defined as to merit academic and professional attention in its own right. STIA is an issue- and equity-oriented field focused on the multifaceted interaction of science and technology with the broad context of international politics, economics, business, law, psychology, culture, and ethics that we have here called the Survival Nexus. It incorporates the insights derived from its sister disciplines: the two STSs and science diplomacy.

As we have seen in the different chapters in this book, STIA unites and synthesizes crosscutting concepts of the natural and social sciences that have been developed separately for the study of energy, environment, nuclear issues, space, information technology, cybersecurity, and the many other topics that have important dimensions of science and technology and that span national boundaries. STIA studies the impact of new knowledge and technology on the international system; sets forth the international response to the opportunities, risks, and issues to which they give rise; explains the effect of international affairs on scientific research or technological innovation; and explores the issues of ethics and equity that result from new scientific knowledge and new technology.

STIA deserves recognition on its own, so that students and experts trained in the scientific and technological dimension of one of these fields can pass easily into the same dimension of another field, and so that students and experts in the natural and social sciences can become familiar with each other's ideas easily and conveniently. It is important enough to be made part of the curriculum in high schools and colleges.[5]

The Questions We Have Been Asking

All of us, whether natural or social scientists, teachers, ambassadors, or interested citizens, need to approach the scientific and technological dimension of policy issues with a set of questions in mind. These are the questions we have been asking throughout our discussion of the issues addressed in this book and constitute its unifying themes. They vary somewhat, depending on whether we are addressing a question of science, of technology, or of innovation.

For scientific issues, we ask: How strong is the evidence for a given assertion? Is the information misleading or being distorted by sources of funding or other special interests? Are there important outstanding areas of insufficient understanding, and how long is it likely to take to resolve them? Are these areas receiving sufficient attention, and if not, why not? How long is it likely to take to translate scientific findings into practical application? Are there public attitudes and values that must be taken into account in the funding, planning and execution of research, and the application of its results?

A different set of political, ethical, and operational questions applies to any effort to apply technology to a global problem: Are there partial or complete technological fixes to the issue? Who will make the decisions, and by what criteria? Who will gain the benefits of whatever we do, and who will bear the costs and risks? How will existing institutions and policies affect the impact of a proposed action? What should be done to defend the interests of people who bear the costs and risks of technologies from which they do not benefit? How are we to balance the costs and risks of one action against those of another? Are there social, religious or cultural differences that must be taken into account?

On the political side, how can we collaborate with politically isolated countries like Iran and North Korea, for example, on the control of infectious diseases that can readily spread to the rest of the world? How do we collaborate with hostile countries to work out the technical challenges involved in possible arms control agreements on cyberweapons and hypersonic missiles? On the commercial side, how do we collaborate on technological innovations that are critically needed for global health or environmental sustainability with countries with which we are competing politically, commercially, and technologically?

If the issue concerns an innovation that does not yet exist, a different set of questions comes into play: What will be the capabilities and limitations of the technology under consideration, and how soon is it likely to be available? Will there be a commercial market for a proposed product, and if not, is there a social, environmental, or security justification for subsidies or other incentives, or for public-spirited intervention by a government, an NGO or a philanthropy? Are there policy or legal obstacles to a scale-up? Will the innovation threaten the interests of someone who is profiting from the existing situation, and if so, how can this opposition be managed? Are there likely to be social, ethical, or philosophical ramifications to the technology? Do attitudes to these ramifications vary by culture or political philosophy? Will there be negative effects, and is there anything that can be done to anticipate and to mitigate or avoid them?

Norms for Applying Science and Technology to Global Problems

In a world in which numerous entities—from governments to individuals—can take action with worldwide ramifications, it is essential that the development and application of science and technology to global problems be based on widely accepted norms. For some technologies, this may mean strengthening of existing regimes, and restoring public understanding and acceptance of the norms that underlie them, which may have eroded with time. Other technologies discussed in this book—hypersonic missiles, cyberweapons, artificial intelligence, autonomous weapons, and gene drivers—raise new issues that call for new norms whose development and acceptance will require research, discussion, and negotiation. When fully developed and articulated, these norms should be internationally agreed, cross-culturally acceptable guides to the behavior that is expected, not only of governments, but also of the many other entities that influence the Survival Nexus, including businesses, NGOs, community organizations, and individuals. They should be clear enough to be able to be applied to specific situations.

As a procedural matter, these norms might begin as recommendations and guidelines from "norm entrepreneurs": expert groups, researchers, governments, nongovernmental or international organizations, private companies, religious organizations, or perhaps even individuals.[6] If all goes well, different versions of these proposed norms could coalesce into principles that guide a common strategy for addressing the problem. These in time could work their way into more formal instruments: national and local legislation and regulations, and perhaps agreements binding on governments and their citizens.

We here propose a tentative set of general norms and principles to guide the application of science and technology to global problems in a sustainable and equitable direction that is grounded in facts. These broad norms would underlie the measures taken to address individual problems. As a first effort and as a stimulus to discussion, we propose the following norms and principles for addressing the issues involved in the Survival Nexus.

Respect for facts and evolving knowledge

The Survival Nexus is complex and constantly changing. It can best be addressed by the principles of adaptive management. Decision makers must acknowledge and take into account the consensus of natural and social scientists, including the quality of the evidence and any uncertainties. Having decided on an action or a policy, they need to monitor and evaluate the immediate results and longer-term consequences of their actions and feed the resulting lessons and insights back into future decisions and plans, educating and informing the public as they do so. Decision makers also must encourage research into the scientific facts, as well as the social, economic, political, and environmental aspects of the problems they are trying to solve. They must support research, development, and deployment of relevant technology.

Cooperation and collaboration

Global issues like pandemics, climate change, and avoidance of nuclear war require global responses, even among countries with fundamentally different political and cultural values and even amid intense geopolitical conflict and economic and technological competition. This means negotiation in good faith on measures to deal with global problems, as well as international collaboration on basic and pre-competitive research on the scientific and technological dimensions of these problems. Such collaboration should include research on the design, implementation, and refurbishing of regimes and norms for limiting the risk and for maximizing and sharing the benefits of new technology.

Avoiding harm and minimizing risk

The future of events in the Survival Nexus is inherently and unavoidably uncertain. Predictive models in science and assessments of new technology typically depend on incomplete data regarding the past and unpredictable human actions in the future. To the extent that the consequences of their actions can be predicted, decision makers should avoid harmful practices. As a practical implementation of the precautionary principle, they should examine alternatives and seek out and adopt the least risky way to accomplish their objectives. They should consider the benefits, costs, and risks of both action and inaction. Advance preparations should be made to deal with pandemics and foreseeable natural disasters. At the very least, the development and deployment of destabilizing technology—technology that is clearly likely to get out of control with potentially catastrophic results—must be avoided.

Equity, sustainability, and inclusivity

The development and application of technology should embody principles of fairness, both to people living now and to future generations, so that everyone benefits from the application of science and technology.[7] This means that decision makers should consider not only the overall balance of benefits, costs, and risks, but also who gains the benefits and who bears the costs and risks, both now and in the future. Here the Rawls criterion of equity is a good guide:[8] What decision would you make if you did not know in advance who you would be and when and where you were to live when you were affected by the decision being contemplated? At a minimum, the application of science and technology should not trample on anyone's dignity and human rights, and future generations should not bear all the risks of present actions from which we gain all the benefits.

Beyond this, specific measures should be taken to ensure that low-income and disadvantaged people have access to the benefits of improved technology (COVID-19 vaccines are a contemporary example) and not suffer disproportionately from their costs. Inclusivity in decision-making—i.e., involving those who will be affected in the planning and implementation of an action or policy—is important, both to facilitate the best result and as a way to ensure that those affected by a decision will be heard and that their interests will be taken into account.

Adding to knowledge and developing technology

Specific, constructive interventions are needed all over the world to address global scientific and technological issues. For example, researchers all over the world should be supported to contribute to the understanding of the causes and effects of climate change, and to participate in the development and testing of such innovative orphan technologies as vaccines for the diseases of low-income people. Conversely, proposals for scientific research on global problems should not be rejected because funders want to avoid learning about a global issue that they do not wish to face or understand. Nor should research institutions devoted to such research be abolished, or research that is already underway defunded, for such reasons.

Accountability

The Survival Nexus involves not only governments, businesses, and international organizations, but also a wide variety of NGOs, foundations, research institutes, universities, community groups, and religious organizations, each with its own principles, objectives, and governance and leadership structure. It also involves individuals. The world is richer for the resulting diversity in points of view. At the same time, each organization and even individuals should be held accountable for their actions and statements in this area. This means that they are expected to be transparent about what they are planning, what they are doing and why they are doing it, and to give an honest account of what happened after they did it.[9] They should expect consequences for not abiding by this and other norms: loss of reputation and pressures from other countries, communities, or businesses; boycotts or other commercial measures; or other political or economic penalties.

Ethics are Fundamental

The questions raised by the advances in science and technology we have discussed in this book are fundamentally issues of practical ethics: How much effort are we willing to put in today to deal with long-term problems whose costs may fall on someone other than ourselves—decreasing the chance of accidental nuclear war, preparing for pandemics before they begin, and assuring a livable world for our descendants? How do we weigh the relative risks and benefits involved in the application of new technologies for altering Earth's climate or for making permanent changes in the human genome? What compromises are we willing to make with authoritarian regimes to assure a unified worldwide Internet? To what extent are we willing to devote resources to health, food, and employment in low-income countries, whether or not these issues will come to haunt us in the form of pandemics, migration, and social unrest?

These are critical, Earth-in-the-balance questions with implications not only for governments, but also for businesses, schools, advocacy organizations, religious institutions, and, indeed, for everyone—since it is public attitudes toward the broader ethical issues and public willingness to pay economic and political costs and to put up with inconveniences that will in large part determine the policies of more formal institutions. If only for this reason, these issues need a more prominent place, not only in political and religious discussion, but also in the mass media and the arts.

The broad issues raised in this book are shared by great variety of technologies that otherwise seem totally unrelated. The facts underlying these issues and their ramifications are not going away even if political and business leaders would prefer to ignore or to suppress them. Natural and social scientists seek to understand these facts as best they can and to set them forth as clearly as they can. Here businesses and universities have particularly important roles to play. From the point of view of immediate profitability, both valuable and dangerous technologies constitute business opportunities. Whether and how they are used, however, can determine in the long run the health of our grandchildren, of our democracy, and of our planet. Universities have a special responsibility because of their role in shaping how people think about these multidisciplinary problems, a task that is currently complicated by the intellectual distance between scientific and technological disciplines, on the one hand, and the social sciences and humanities, on the other.

Advances in science and technology result from the activities of human beings; they are not unstoppable juggernauts. True, science and technology each have their own inner dynamic—but they are also very much subject to human influence. Social media has exploded in response to human desire for communication in ways far beyond the original ideas of the pioneering inventors of information technology. The successful regimes for arms control and for the defense of the stratospheric ozone layer show that people can, if they so choose, slow the advance of technology and guide it into desirable directions.

The issues that we have been discussing in this book all involve not only science and technology, but also, equally importantly, the links that connect science and technology with politics, economics, business, law, psychology, communications, ethics, and culture. These links do not receive the attention they deserve. They are key to the sustainability of our planet and to the nature of the world we will live in and that we will leave to our children and grandchildren. They are critical to the future of human life and civilization. That's why we call them the Survival Nexus.

Acknowledgments

Writing a book on a sprawling subject like science, technology, and world affairs requires more than a little help from my friends. I want first to thank Carole Sargent, Georgetown University's Director of the Office of Scholarly Publications, who educated me on the practicalities of the publishing world; my literary agent, Don Fehr of TriDent Media, who taught me the basics of writing for a non-academic audience; and Juliette Leader, from whose expert copyediting I have greatly benefited.

Friends and colleagues gave selflessly of their time to orient me on their specialties and save me from needless blunders. Here I want to thank Bernhard Liese and Emily Mendenhall for their advice on global health; Marianne Haug and Joanna Lewis on energy; Robert Gallucci and Kathryn Olesko on nuclear issues; Samuel Visner and Jason Matheny on cybersecurity; Carl Dahlman on competitiveness; William Bonvillian on manufacturing; Marilyn Perrin on molecular biology; and Jonathan Wadsworth, Jan Low, and Alex McCalla on agricultural research. I am, of course, responsible for the errors that remain.

I also want to thank the Georgetown University School of Foreign Service and its program in Science, Technology, and International Affairs (STIA) for research and library support, and the American Association for the Advancement of Science for allowing me to be a Visiting Scholar.

I first started this book in 2010, although I could not concentrate on it until well after my retirement. During all of this time, I have greatly benefited from excellent undergraduate research assistants, both for this book and for previous publications, and want to thank them all here. They are, in alphabetical order: Agree Ahmed, Alexandra Allegra, Kevin Anchukaitis, Libby Bassini, Kate Broderick, Jeffrey Caso, Luis Costa, Gabriel Gorre, Eric Hopper, Matt Kallman, Corrinne Kisner, Gregory Kollmer, Juliette Leader, Jeffrey Martin, Kevin Meurer, Alexander Olesker, Kristen Skilman, Mimi Smith, Sarah Wappner, and Colleen Wood.

Finally, I thank my wife, Edith Brown Weiss, for her constant support, her invaluable advice and help throughout, and much, much more.

Notes

Chapter 1

1. Vernon W. Ruttan, *Is War Necessary for Economic Growth?: Military Procurement and Technology Development* (New York: Oxford University Press, 2006).
2. Basic research seeks to uncover new information about the natural world. It may be motivated by potential application (basic applied research—this is not an oxymoron) or by the curiosity of the researcher.
3. To a philosopher or an anthropologist, the word "technology" may encompass all of the features and conveniences of modern life, typically as a prelude to a discussion (often negative) of the overall direction in which the world is headed. That's not our emphasis in this book.

Chapter 2

1. Carmen J. Giunta, "Thomas Midgley, Jr., and the Invention of Chlorofluorocarbon Refrigerants: It Ain't Necessarily So," *Bulletin for the History of Chemistry 31*, no. 2 (2006): 69.
2. The ups and downs of early research on stratospheric ozone are chronicled in Naomi Oreskes and Erik M. Conway, *Merchants of Doubt* (New York: Bloomsbury Press, 2010).
3. A stronger version of the precautionary principle states that precautionary action *should* be taken in response to "threats of harm to human health or the environment even if some cause-and-effect relationships are not fully established scientifically" (Wingspread Statement on the Precautionary Principle (1998), https://www.iatp.org/sites/default/files/Wingspread_Statement_on_the_Precautionary_Prin.htm). When applied to the application of a new technology, or to a project like a bridge or a dam, the precautionary principle becomes a statement of "look before you leap" and is useful in blocking harmful proposals. On the other hand, critics argue that it can lead in practice to excessive caution: "under no circumstances try anything new." Søren Holm and John Harris, "Precautionary Principle Stifles Discovery," *Nature 400*, no. 6743 (July 1999): 398.
4. R. T. Watson, M. J. Prather, and M. J. Kurylo, "*Present State of Knowledge of the Upper Atmosphere 1988: An Assessment Report*" (Washington, DC: National Aeronautics and Space Administration, June 1, 1988).

5. World Meteorological Organization, *Executive Summary: Scientific Assessment of Ozone Depletion: 2018* (Geneva: Global Ozone Research and Monitoring Project, 2018), https://www.esrl.noaa.gov/csd/assessments/ozone/2018/executivesummary/.
6. M. Rigby et al., "Increase in CFC-11 Emissions from Eastern China Based on Atmospheric Observations," *Nature* 569, no. 7757 (May 2019): 546–50.
7. Megan Lickley et al., "Quantifying Contributions of Chlorofluorocarbon Banks to Emissions and Impacts on the Ozone Layer and Climate," *Nature Communications 11*, no. 1 (March 17, 2020): 1380.
8. Oreskes and Conway, *Merchants of Doubt, chapter 4.*

Chapter 3

1. IPCC, "Summary for Policymakers," in V. Masson-Delmotte et al., eds., *Global Warming of 1.5°C: An IPCC Special Report* (Geneva: World Meteorological Organization, 2018); IPCC, *Special Report on the Ocean and Cryosphere in a Changing Climate*, eds. H. O. Pörtner et al. (Geneva: World Meteorological Organization, 2019); Matthew W. Jones et al., Climate Change Increases the Risk of Wildfires (Norwich, UK: ScienceBrief, January 14, 2020).
2. In this chapter and elsewhere in the book, we shall use the term "climate disruption" interchangeably with "climate change," and the term "global warming" to refer to the rise in the Earth's average temperature.
3. A brief explanation of the science of climate change is found in National Academy of Sciences and Royal Society, *Climate Change: Evidence and Causes: Update 2020* (Washington, DC: National Academies Press, 2020). A concise summary is found in Robert Socolow, "Contending with Climate Change: The Next 25 Years," *Bulletin of the Atomic Scientists* 76, no. 6 (December 2020): 294–301.
4. Nicholas R. Golledge et al., "Global Environmental Consequences of Twenty-First-Century Ice-Sheet Melt," *Nature 566*, no. 7742 (February 2019): 65–72; Scott A. Kulp and Benjamin H. Strauss, "New Elevation Data Triple Estimates of Global Vulnerability to Sea-Level Rise and Coastal Flooding," *Nature Communications 10*, no. 1 (October 29, 2019): 4844. An older assessment predicted much more serious flooding. Barbara Neumann et al., "Future Coastal Population Growth and Exposure to Sea-Level Rise and Coastal Flooding - A Global Assessment," *PLOS ONE 10*, no. 3 (March 11, 2015): e0118571. Golledge et al., "Global Environmental Consequences of Twenty-First-Century Ice-Sheet Melt." *Nature* 566, no. 7742 (February 2019): 65–72.
5. Alice Franek et al., Water for a Sustainable World: Facts and Figures (Perugia, Italy: United Nations World Water Assessment Programme, 2015).
6. IPCC, *Special Report on the Ocean and Cryosphere*, 223.
7. Carbon dioxide is not the only greenhouse gas. Others include methane (CH_4), nitrous oxide (N_2O), and the various CFCs, hydrochlorofluorocarbons (HCFCs) and hydrofluorocarbons (HFCs). Scientists have devised the portmanteau concept of "CO_2 equivalent" or "global warming potential" in order to aggregate the effects of

the different greenhouse gases, each of which has a different capacity to radiate heat back to Earth, and each of which remains in the atmosphere for a different length of time (i.e., has a different residence time) before it decomposes or is washed out of the atmosphere by rain. Because methane has a shorter residence time in the atmosphere than carbon dioxide, reducing methane emissions could slow global warming faster than a comparable reduction in CO_2 emissions.

8. Benjamin P. Horton et al., "Mapping Sea-Level Change in Time, Space, and Probability," *Annual Review of Environment and Resources 43*, no. 1 (October 17, 2018): 481–521.
9. IPBES, The Global Assessment Report on Biodiversity and Ecosystem Services. Summary for Policymakers (Bonn: Intergovernmental Science-Policy Platform on Biodiversity and Ecosystem Services Secretariat, 2019). In addition to climate change, habitat loss, pollution, and exploitation also contribute to these extinctions.
10. The oceanography here is somewhat complicated. Western Europe owes its relatively mild climate in large part to global oceanic currents called the thermohaline circulation, which operate along with the Gulf Stream to bring warm equatorial waters northward to Western Europe. This current depends on heavy, cold, salty water that sinks to the bottom of the ocean off the coast of Greenland, sucking warm water northward from the equator to take its place, increasing its salinity and density by evaporation along the way. Large quantities of fresh water from melting ice from Greenland would dilute the salty water and make it less dense, so that it would no longer sink to the bottom or would do so more slowly. This would reduce the flow of warm water from the equator to the neighborhood of Western Europe. Since the temperature of the ocean has a big effect on the climate of nearby land, this would make Western Europe a lot colder than it is today.
11. S. Sherwood et al., "An Assessment of Earth's Climate Sensitivity Using Multiple Lines of Evidence," *Reviews of Geophysics* (July 2020): e2019RG000678.
12. John Holdren, quoted in David Malakoff, "Let's Call It 'Climate Disruption,' White House Science Adviser Suggests (Again)," *Science* (May 2, 2014), https://www.sciencemag.org/news/2014/05/lets-call-it-climate-disruption-white-house-science-adviser-suggests-again.
13. Matthew E. Kahn et al., "Long-Term Macroeconomic Effects of Climate Change: A Cross-Country Analysis," Working Paper (Washington, DC: National Bureau of Economic Research, August 2019).
14. S. Sherwood et al., "An Assessment of Earth's Climate Sensitivity." This more recent estimate gives a range that is narrower than the prediction of 1.4–5.8°C (2.5–10.4°F) used in the IPCC fifth assessment report and cited in many subsequent publications. IPCC, *Climate Change 2013: The Physical Science Basis. Working Group I Contribution to the Fifth Assessment Report of the Intergovernmental Panel on Climate Change*, ed. T. F. Stocker et al. (Cambridge: Cambridge University Press, 2013).
15. Canty & Associates, "Compare Cities," Weatherbase, 2020, http://www.weatherbase.com/compare.php3?first=593809&second=523281, quoting data from the National Climatic Data Center.

16. Detlef P. van Vuuren et al., "The Shared Socio-Economic Pathways: Trajectories for Human Development and Global Environmental Change," *Global Environmental Change 42* (January 1, 2017): 148–52; summarized in Zeke Hausfather, "Explainer: How 'Shared Socioeconomic Pathways' Explore Future Climate Change," *CarbonBrief*, April 19, 2018, https://www.carbonbrief.org/explainer-how-shared-socioeconomic-pathways-explore-future-climate-change.
17. Robert Henson, Warming World, Impacts by Degree (Washington, DC: National Academies of Science, 2011).
18. Bette L. Otto-Bliesner et al., "How Warm Was the Last Interglacial? New Model–Data Comparisons," *Philosophical Transactions of the Royal Society A: Mathematical, Physical and Engineering Sciences 371*, no. 2001 (October 28, 2013), doi.org/10.1098/rsta.2013.0097.
19. IPCC, *Global Warming of 1.5°C*; IPCC, *Climate Change 2014: Synthesis Report. Contribution of Working Groups I, II and III to the Fifth Assessment Report of the Intergovernmental Panel on Climate Change*, ed. Core Writing Team, R. K. Pachauri, and L. A. Meyer (Cambridge: Cambridge University Press, 2014); Potsdam Institute for Climate Impact Research and Climate Analytics, Turn Down the Heat: Why a 4C° Warmer World Must Be Avoided (Washington, DC: World Bank, 2012). A detailed table, based on a thorough literature search, is found in Robert McSweeney, "The Impacts of Climate Change at 1.5C, 2C and Beyond," *CarbonBrief*, October 4, 2018, https://interactive.carbonbrief.org/impacts-climate-change-one-point-five-degrees-two-degrees/. A good visual, based on the numbers in the 2014 IPCC report, is found at David Roberts, "This Graphic Explains Why 2 Degrees of Global Warming Will Be Way Worse than 1.5," *Vox*, January 19, 2018, https://www.vox.com/energy-and-environment/2018/1/19/16908402/global-warming-2-degrees-climate-change.
20. Susan Solomon et al., "Irreversible Climate Change Due to Carbon Dioxide Emissions," *Proceedings of the National Academy of Sciences 106*, no. 6 (February 10, 2009): 1704–9.
21. IPCC, *Global Warming of 1.5°C*, distilled in Kelly Levin, "8 Things You Need to Know About the IPCC 1.5˚C Report," World Resources Institute, October 7, 2018, https://www.wri.org/blog/2018/10/8-things-you-need-know-about-ipcc-15-c-report. See also Henson, *Warming World*, and IPCC, *Climate Change 2014*, 1140. IPCC, *Global Warming of 1.5°C*, 178.
22. IPCC, *Global Warming of 1.5°*, 178.
23. Mark A. Chandler, "The Climate of the Pliocene: Simulating Earth's Last Great Warm Period," NASA Goddard Institute for Space Studies, April 1997, https://www.giss.nasa.gov/research/features/199704_pliocene/.
24. Michael Sigmond, John C. Fyfe, and Neil C. Swart, "Ice-Free Arctic Projections Under the Paris Agreement," *Nature Climate Change 8*, no. 5 (May 2018): 404–8.
25. Potsdam Institute, *Turn Down the Heat.*
26. Timothy Bralower and David Bice, "Ancient Climate Events: Paleocene Eocene Thermal Maximum," Pennsylvania State University, accessed April 4, 2020, https://www.e-education.psu.edu/earth103/node/639.

27. Henrike Brecht et al., "Sea-Level Rise and Storm Surges: High Stakes for a Small Number of Developing Countries," *The Journal of Environment & Development 21*, no. 1 (March 1, 2012): 120–38, cited in Potsdam Institute, *Turn Down the Heat.* The countries are Mozambique, Venezuela, Madagascar, Mexico, India, Bangladesh, Indonesia, Vietnam, and the Philippines.
28. Mark Lynas, *Six Degrees: Our Future on a Hotter Planet* (London: Fourth Estate, 2007), presents a detailed, apocalyptic prediction of an almost inhabitable Earth at 5°C and 6°C of global warming. See also Will Steffen et al., "Trajectories of the Earth System in the Anthropocene," *Proceedings of the National Academy of Sciences 115*, no. 33 (August 14, 2018): 8252–59.
29. William D. Nordhaus, Can We Control Carbon Dioxide? (Laxenburg, Austria: International Institute for Applied Systems Analysis, June 1975), cited in Brian Palmer, "How Much of an Increase in Temperature Can We Tolerate?," *Pacific Standard*, April 24, 2015, https://psmag.com/environment/the-climate-change-debate-is-about-rising-temperatures.
30. More precisely, scientists can estimate the amount of CO_2 that would give rise to a certain probability—typically, one-third, one-half, and two-thirds—of a given change in average temperature. See, for example IPCC, *Climate Change 2014*, Table 2.2.
31. The IPCC 1.5°C report estimates that limiting total net carbon dioxide emissions to 550–770 gigatons will provide a two-thirds probability of limiting global warming to 1.5°C. The prediction varies, depending on exactly what average temperature is being predicted and on whether all greenhouse gases and all sources are being taken into consideration. A gigaton is a billion (thousand million) tons.
32. For comparison, the total carbon dioxide stored in all fossil fuels yet unburned amounts to some 80,000 gigatons, or about 2000 years of emissions at the current rate. (H. H. Rogner et al., "Aggregate Fossil Energy Occurrences," in United Nations Development Program, *World Energy Assessment: Energy and the Challenge of Sustainability* (New York:, 2000), Table 5.7, p. 149, quoted in Robert Socolow, "Climate Change and Destiny Studies: Creating Our Near and Far Futures," *Bulletin of the Atomic Scientists 71*, no. 6 (2015): 18–28.
33. S. Pacala and R. Socolow, "Stabilization Wedges: Solving the Climate Problem for the Next 50 Years with Current Technologies," *Science 305*, no. 5686 (August 13, 2004): 968–72.
34. IPCC, *Global Warming of 1.5°C*, and Henson, *Warming World.*
35. An excellent account of the early history of research on climate change and its impact on policymakers is found in Spencer Weart, "The Discovery of Global Warming - A History," American Institute of Physics, 2009, https://history.aip.org/climate/index.htm.
36. IPCC, "Overview," in *Climate Change: The 1990 and 1992 IPCC Assessments* (Geneva: World Meteorological Organization, 1992), https://www.ipcc.ch/site/assets/uploads/2018/05/ipcc_90_92_assessments_far_overview.pdf..
37. Gro Harlem Brundtland, *Report of the World Commission on Environment and Development: Our Common Future*, UN General Assembly A/42/427 (Geneva: United Nations, 1987). See also Edith Brown-Weiss, *In Fairness to Future*

Generations: International Law, Common Patrimony, and Intergenerational Equity (Tokyo: United Nations University, 1989).

38. There is also a "strong" version of the precautionary principle: that action to protect the environment *should* be taken *even in the absence* of firm scientific evidence (my italics). The problem with this formulation is that in the absence of understanding, action intended to protect the environment might do more harm than good. For more extensive discussion, see Charles Weiss, "Can There Be Science-Based Precaution?," *Environmental Research Letters 1*, no. 1 (October 2006), 014003, and the references therein.
39. United Nations, "United Nations Framework Convention on Climate Change," UN Doc. FCCC/INFORMAL/84, 1992, https://unfccc.int/resource/docs/convkp/conveng.pdf. See Article 3, section 3.
40. Naomi Oreskes and Erik M. Conway, *Merchants of Doubt* (New York: Bloomsbury Press, 2009), chapter 6.
41. J. M. Turner and Andrew C. Isenberg, *The Republican Reversal* (Cambridge, MA: Harvard University Press, 2018).
42. In the case of Exxon, for many years the leading supporter of the climate denialists, business decisions were reinforced by the confidence of a chief executive officer, a Ph.D. chemical engineer, that his technical judgment was at least as good as that of the IPCC. Steve Coll, *Private Empire: ExxonMobil and American Power* (New York: Penguin Books, 2013).
43. Thomas S. Kuhn, *The Structure of Scientific Revolutions* (Chicago: University of Chicago Press, 1962).
44. Union of Concerned Scientists, "Debunking Misinformation About Stolen Climate Emails," August 25, 2011, https://www.ucsusa.org/resources/debunking-misinformation-about-stolen-climate-emails.
45. Joel Primack and Frank von Hippel, *Advice and Dissent: Scientists in the Political Arena* (New York: Basic Books, 1974).
46. IPCC, Understanding Climate Change: 22 Years of IPCC Assessment (Geneva: IPCC Secretariat, 2010).
47. Michael D. Mastrandrea et al., Guidance Note for Lead Authors of the IPCC Fifth Assessment Report on Consistent Treatment of Uncertainties (Geneva: IPCC, 2010). https://www.ipcc.ch/site/assets/uploads/2017/08/AR5_Uncertainty_Guidance_Note.pdf.
48. Justin Gillis, "Climate Panel Cites Near Certainty on Warming," *The New York Times*, August 19, 2013, https://www.nytimes.com/2013/08/20/science/earth/extremely-likely-that-human-activity-is-driving-climate-change-panel-finds.html.
49. European Commission, "Kyoto 1st Commitment Period (2008–12)," EU Climate Action, November 23, 2016, https://ec.europa.eu/clima/policies/strategies/progress/kyoto_1_en.
50. UNFCCC Conference of the Parties, "Adoption of the Paris Agreement," UN Doc. FCCC/CP/2015/L.9/Rev/1, December 12, 2015, https://unfccc.int/resource/docs/2015/cop21/eng/l09r01.pdf.

51. The website of the Global Covenant of Mayors for Climate and Energy is at https://www.globalcovenantofmayors.org/about.
52. European Parliament and Council of Europe, *Regulation (EU) 2019/2088 on Sustainability-Related Disclosures in the Financial Services Sector*, EUR-Lex, November 27, 2019, https://eur-lex.europa.eu/eli/reg/2019/2088/oj.
53. Céline Ramstein, Goran Dominioni, and Sanaz Ettehad, State and Trends of Carbon Pricing: 2019 (Washington, DC: World Bank Group, 2019) and World Bank, "Carbon Pricing Dashboard," April 1, 2020, https://carbonpricingdashboard.worldbank.org/about.
54. John W. Byrd, Elizabeth S. Cooperman, and Kent Hickman, "Capital Budgeting and Climate Change: Does Corporate Internal Carbon Pricing Reduce CO_2 Emissions?," SSRN Scholarly Paper (Rochester, NY: Social Science Research Network, April 14, 2020), doi.org/10.2139/ssrn.3575769.
55. Carbon Pricing Leadership Council, Report of the High-Level Commission on Carbon Prices (Washington, DC: World Bank, May 2017) https://static1.squarespace.com/static/54ff9c5ce4b0a53decccfb4c/t/59b7f2409f8dce5316811916/1505227332748/CarbonPricing_FullReport.pdf..
56. Nicolette Bartlett, Hannah Cushing, and Sara Law, *Putting a Price on Carbon: Integrating Climate Risk into Business Planning*, CDP, October 2017, https://b8f65cb373b1b7b15feb-c70d8ead6ced550b4d987d7c03fcdd1d.ssl.cf3.rackcdn.com/cms/reports/documents/000/002/738/original/Putting-a-price-on-carbon-CDP-Report-2017.pdf?1507739326.
57. William B. Bonvillian and Charles Weiss, *Technological Innovation in Legacy Sectors* (New York: Oxford University Press, 2015), p. 71.
58. See the Climate Action 100+ website at https://climateaction100.wordpress.com/about-us/.
59. Allen A. Fawcett et al., "Can Paris Pledges Avert Severe Climate Change?," *Science 350*, no. 6265 (November 26, 2015): 1168–69.
60. Darby, Megan, and Isabelle Gerretsen, "Which Countries Have a Net Zero Climate Goal?," *Climate Home News*, June 14, 2019, https://www.climatechangenews.com/2019/06/14/countries-net-zero-climate-goal/.

Chapter 4

1. The implications of climate change for U.S. foreign policy are addressed in John Podesta and Todd Stern, "A Foreign Policy for the Climate," *Foreign Affairs* 99 (3) (May/June 2020): 39–46.
2. Anteneh G. Dagnachew et al., *Insight into Energy Scenarios: A Comparison of Key Transition Indicators of 2 °C Scenarios* (The Hague: PBL Netherlands Environmental Assessment Agency, September 2019). A scenario by which to reach a target of 1.5°C is found at Johan Röckstrom et al., "A Roadmap for Rapid Decarbonization," *Science 355*, no. 6331 (2017): 1269–71.

3. Critics have called attention to the possibility of "technological lock-in," whereby the dominant market position of mature technologies like corn-based ethanol, silicon-based photovoltaics, and lithium-ion batteries, often achieved with the help of government subsidies or purchasing programs, makes it difficult for more advanced technology to enter the market. See Varun Sivaram, "Unlocking Clean Energy," *Issues in Science and Technology* 33, no. 2 (2017): 31–40.
4. Charles Weiss and William Bonvillian, *Structuring an Energy Technology Revolution* (Cambridge, MA: MIT Press, 2014), 70–71. Peter L. Singer and William B. Bonvillian, "innovation orchards—Helping Start-ups Scale," (Washington: Innovation Technology and Innovation Foundation, March 2017), pp. 7–14, http://www2.itif.org/2017-innovation-orchards.pdf?_ga=2.243967621.1083592602.1619646987-70827581.1619551935.
5. A thorough treatment of the role of technological innovation in climate mitigation, and the necessary policies and likely time horizons for their development and deployment, is found in IEA, *Energy Technology Perspectives 2017: Catalyzing Energy Technology Transformation* (Paris: International Energy Agency, 2017).
6. John Helveston and Jonas Nahm, "China's Key Role in Scaling Low-Carbon Energy Technologies," *Science 366*, no. 6467 (November 15, 2019): 794–96. China is also vigorously exporting coal-burning equipment, but this is a separate story.
7. The Andlinger Center for Energy and Environment at Princeton University has published a series of readable "distillates" that review the current state of many of the technologies that are required for energy production and that do not produce CO_2 emissions. These are available at https://acee.princeton.edu/distillates/.
8. Umair Irfan, "The Best Case for and Against a Fracking Ban," *Vox*, September 12, 2019, https://www.vox.com/energy-and-environment/2019/9/12/20857196/democratic-debate-climate-change-fracking-ban.
9. IEA, *World Energy Outlook 2019* (Paris: International Energy Agency, November 2019).
10. IRENA, Global Energy Transformation: The REmap Transition Pathway (Abu Dhabi: International Renewable Energy Agency, 2019), p. 63.
11. Global CCS Institute, The Global Status of CCS 2018 (Melbourne: Global CCS Institute, 2018).
12. Brad Page, Guloren Turan, and Alex Zapantis, Global Status of CCS 2019: Targeting Climate Change (Melbourne: Global Carbon Capture and Storage Institute, 2019).
13. This is greater than the solar share of world energy production because solar generating capacity isn't used when the sun doesn't shine. IRENA, Future of Solar Photovoltaic: Deployment, Investment, Technology, Grid Integration and Socio-Economic Aspects (Abu Dhabi: International Renewable Energy Agency, November 2019).
14. Levi Golston et al., Wind Power: An Energy Technology Distillate (Princeton, NJ: Andlinger Center for Energy and the Environment, March 2019).
15. IRENA, Future of Wind: Deployment, Investment, Technology, Grid Integration and Socio-Economic Aspects (Abu Dhabi: International Renewable Energy Agency, October 2019).

16. Robert Socolow et al., Grid-Scale Electricity Storage (Implications for Renewable Energy): An Energy Technology Distillate (Princeton, NJ: Andlinger Center for Energy and the Environment, June 2014).
17. Varun Sivaram, *Taming the Sun: Innovations to Harness Solar Energy and Power the Planet* (Cambridge, MA: MIT Press, 2018).
18. Valerie Masson-Delmotte et al., eds., *Climate Change and Land: IPCC Special Report on Climate Change, Desertification, Land Degradation, Sustainable Land Management, Food Security, and Greenhouse Gas Fluxes in Terrestrial Ecosystems - Summary for Policymakers* (Geneva: Intergovernmental Panel on Climate Change, 2020), p. 24.
19. Vaclav Smil, *Power Density: A Key to Understanding Energy Sources and Uses* (Cambridge, MA: MIT Press, 2016).
20. Sui-Lee Wee, "Thousands Being Moved from China's Three Gorges - Again," *Reuters*, August 22, 2012, https://www.reuters.com/article/us-china-threegorges-idUSBRE87L0ZW20120822.
21. Jefferson W. Tester et al., *Sustainable Energy: Choosing Among Options* (Cambridge, MA: The MIT Press, 2005); Idaho National Laboratory, *The Future of Geothermal Energy: Impact of Enhanced Geothermal Systems (EGS) on the United States in the 21st Century* (Idaho Falls, ID: MIT, 2006).
22. Nuclear *fusion* is likely to have advantages over nuclear fission in safety, radioactive pollution, and resistance to the proliferation of nuclear weapons should it come to pass, but it is not yet at the proof-of-concept stage and is unlikely to be implemented at commercial scale before the middle of the twenty-first century. The difference between fission and fusion is briefly explained in Annex A to chapter 5.
23. MIT Energy Initiative, *The Future of Nuclear Energy in a Carbon-Constrained World: An Interdisciplinary MIT Study* (Cambridge, MA: MIT Energy Initiative 2018). A high-density source produces a lot of energy in a small land area.
24. Briefly, there are three approved approaches to the decommissioning of nuclear power plants: removing all contaminated materials and storing them for about seven years separately from the disused plant; containing and monitoring the plant for sixty years until the radioactivity has decayed to safe levels; and permanently encasing (entombing) the contaminated materials in concrete, as was done to the reactor that exploded at Chernobyl. See EIA, "Decommissioning Nuclear Reactors Is a Long-Term and Costly Process," *Energy Information Agency*, November 17, 2017, http://www.eia.gov/todayinenergy/detail.php?id=33792.
25. United Nations Scientific Committee on the Effects of Atomic Radiation, Sources, Effects and Risks of Ionizing Radiation: Report to the General Assembly with Scientific Annexes 2013 (New York: United Nations, 2014).
26. Frank N. von Hippel and Michael Schoeppner, "Reducing the Danger from Fires in Spent Fuel Pools," *Science & Global Security 24*, no. 3 (2016): 141–73. See especially Annex A.
27. M. Granger Morgan et al., "US Nuclear Power: The Vanishing Low-Carbon Wedge," *Proceedings of the National Academy of Sciences 115*, no. 28 (July 10, 2018): 7184–89.
28. Chapter 11 of Working Group III in IPCC, *Climate Change 2014.*

29. No-till farming will be affected by evidence that the leading weed killer, glyphosate (Roundup™), a Bayer/Monsanto product, is linked to cancer in humans and animals. (See, for example, Luoping Zhang et al., "Exposure to Glyphosate-Based Herbicides and Risk for Non-Hodgkin Lymphoma: A Meta-Analysis and Supporting Evidence," *Mutation Research 781* (September 2019): 186–206.) More than forty thousand lawsuits have been filed, and Bayer has set aside over $10 billion to cover existing and future settlements. Glyphosate has been banned in several countries and localities but is still being produced and sold. Approved chemical replacements are probably ten years off. New technologies of automated, plant-by-plant weeding, using GPS, artificial intelligence and robotics, may eventually replace herbicides in advanced farming systems.
30. Chapters 8, 9, 10, and 12 of Working Group III in IPCC, *Climate Change 2014*.
31. IEA, *Energy Technology Perspectives 2016: Towards Sustainable Urban Energy Systems* (Paris: International Energy Agency, 2016), 3.
32. Chapter 9 of Working Group III in IPCC, *Climate Change 2014*.
33. Chapter 8 of Working Group III in IPCC, *Climate Change 2014*.
34. Chapter 7 of Working Group III in IPCC, *Climate Change 2014*.
35. Thibaud Voïta, "The Power of China's Energy Efficiency Policies" (Paris: Institut Français des Relations Internationales, September 2018).
36. ISO, "ISO 50001: Energy Management Systems—Requirements with Guidance for Use," International Standards Organization, 2011, https://www.iso.org/obp/ui#iso:std:iso:50001:ed-1:v1:en.
37. See Working Group II in IPCC, *Climate Change 2014*, which delves into these vulnerabilities.
38. National Research Council, *Climate Intervention: Carbon Dioxide Removal and Reliable Sequestration* (Washington, DC: National Academies Press, 2015). National Academies of Sciences, Engineering, and Medicine. *Reflecting Sunlight: Recommendations for Solar Geoengineering Research and Research Governance*. Washington, DC: The National Academies Press (2021).
39. Matthias J. Pickl, "The Renewable Energy Strategies of Oil Majors—From Oil to Energy?," *Energy Strategy Reviews 26* (November 1, 2019): 100370.
40. Edith Brown Weiss, *In Fairness to Future Generations: International Law, Common Patrimony, and Intergenerational Equity* (Tokyo: United Nations University, 1989).

Chapter 5

1. Paul Scharre, *Army of None: Autonomous Weapons and the Future of War* (New York: W.W. Norton, 2018), 1.
2. Matthew Bunn, Anthony Wier, and John P. Holdren, *Controlling Nuclear Warheads and Materials: A Report Card and Action Plan* (Washington, DC: Nuclear Threat Initiative, March 2003), 15–16. A dramatic recreation of a hypothetical nuclear attack on Washington, D.C., can be seen at https://www.youtube.com/watch?v=SUuOskX3z7U.

3. The atomic bombs that exploded over Hiroshima and Nagasaki had the explosive power of 15,000 and 18,000–20,000 tons (kilotons) of TNT, respectively. For comparison, the fertilizer bomb that destroyed the Murrah Office Building in Oklahoma City in 1995 had the force of 1.5 tons, or 0.0015 kilotons of TNT.
4. Alex Wellerstain, "Nukemap," 2020. An 800-kiloton bomb would destroy all life within a radius of ten miles, an area that extends across the island of Manhattan and into the middle of Staten Island, the edge of the Bronx, eastern Queens, and well into New Jersey. https://nuclearsecrecy.com/nukemap. Jonathan Schell, "Nuclear Holocaust," in Charles W. Kegley, Jr. and Eugene R. Wittkopf, eds., *The Nuclear Reader: Strategy, Weapons, War* (New York: St. Martin's Press, 1985), 268–69.
5. A strategic strike is one "designed or trained to strike an enemy at the sources of its military, economic, or political power," as opposed to a tactical strike employed at the battlefront. Merriam-Webster online dictionary, https://www.merriam-webster.com/dictionary/strategic.
6. Joseph Cirincione, *Bomb Scare: The History and Future of Nuclear Weapons* (New York: Columbia University Press, 2007), 87.
7. For authoritative material on arms control and the nuclear threat, see the publications of the Arms Control Association, the spring 2020 issue of *Daedalus*, and the Elders policy paper—Mary Robinson et al., "*Nuclear Non-Proliferation and Disarmament,*" Policy paper (London: The Elders Foundation, November 2019).
8. The origin of the technology for nuclear weapons and nuclear power in the discovery of nuclear fission by basic scientists was the inspiration for the pipeline model of innovation that underlies much of U.S. science policy. According to this model, technological innovation and the economic growth that it makes possible are driven by inventions derived from basic research aimed at increasing knowledge of science and engineering, so the best government policy for the stimulation of innovation lies in public support to basic research. See Vannevar Bush, *Science: The Endless Frontier* (Washington, DC: Government Printing Office, 1945), and William B. Bonvillian and Charles Weiss, *Technological Innovation in Legacy Sectors* (New York: Oxford University Press, 2015), pp. 24–30.
9. Leo Szilard, who drafted the letter and delivered it to Einstein, got lost while he was looking for Einstein's summer home on Long Island so that he could get Einstein's signature, and was about to give up and return home when he was directed to it by a local youngster, a dramatic example of the world-historical effect of random events. Robert Jungk, *Brighter than a Thousand Suns: A Personal History of the Atomic Scientists* (Boston: Houghton Mifflin Harcourt, 1958), 84.
10. For accounts of the discovery of nuclear fission and the nuclear chain reaction, see Richard Rhodes, *The Making of the Atomic Bomb* (New York: Simon & Schuster, 1986), 271; Jungk, *Brighter than a Thousand Suns*; and Ruth Lewin Sime, "Science and Politics: The Discovery of Nuclear Fission 75 Years Ago," *Annalen Der Physik 526*, no. 3–4 (2014): A27–31.
11. Ralph E. Lapp, "The Einstein Letter That Started It All," *The New York Times*, August 2, 1964, http://www.nytimes.com/1964/08/02/the-einstein-letter-that-started-it-all.html.

12. J. Robert Oppenheimer, "Physics in the Contemporary World," *Bulletin of the Atomic Scientists* 4, no. 3 (1948): 65–68, 85–86. Oppenheimer later said that ". . . he did not regret his work at Los Alamos . . . When he said that physicists had 'known sin,' the sin he had in mind was not murder, but pride." Ray Monk, *Robert Oppenheimer: A Life Inside the Center* (New York: Anchor Books, 2014), 685.
13. James Franck et al., "The Franck Report: A Report to the Secretary of War, June 1945," Federation of American Scientists, June 1945, https://fas.org/sgp/eprint/franck.html. The exact text of Franck's letter is as follows: "[A] demonstration of the new weapon might best be made, before the eyes of representatives of all the United Nations, on the desert or a barren island. The best possible atmosphere for the achievement of an international agreement could be achieved if America could say to the world, 'you see what sort of a weapon we had but did not use. We are ready to renounce its use in the future if other nations join us in this renunciation and agree to the establishment of an efficient control.'"
14. Fred Kaplan has reconstructed the bureaucratic history of the triad and shown that it derives from the need to give each of the uniformed services (Army, Navy, and Air Force) a "piece of the action." See Fred Kaplan, *The Bomb: Presidents, Generals, and the Secret History of Nuclear War* (New York: Simon & Schuster, 2020).
15. Joseph Rotblat, *Scientists in the Quest for Peace: A History of the Pugwash Conferences* (Cambridge, MA: MIT Press, 1972); Joseph Rotblat, "The Early Days of Pugwash," *Physics Today* 54, no. 6 (June 2001): 50–55.
16. Edward Kaplan, *To Kill Nations: American Strategy in the Air-Atomic Age and the Rise of Mutually Assured Destruction* (Ithaca, NY: Cornell University Press, 2015), 180–83.
17. Herman Kahn, *On Thermonuclear War* (New Brunswick, NJ: Transaction Publishers, 2007), 433.
18. Kaplan, *The Bomb*.
19. A summary of the history of the launch-on-warning doctrine, along with key documents, is found at National Security Archive, The "Launch on Warning" Nuclear Strategy and Its Insider Critics, (Washington: George Washington University, June 11, 2019), https://nsarchive.gwu.edu/briefing-book/nuclear-vault/2019-06-11/launch-warning-nuclear-strategy-its-insider-critics
20. This incident illustrates the critical importance of command and control—who has authority to order the use of nuclear weapons, and how these weapons are to be protected from theft and from accidental or unauthorized use—especially at times of crisis. The goal is to prevent the accidental launch of a nuclear missile, or launch by a rogue military officer or by an officer in league with a terrorist group. In nearly all nuclear powers, strategic weapons are kept carefully under civilian control, for example, by requiring multiple "fingers on the trigger"—i.e., that more than one person independently type a secret code into the device that controls the launch—or by keeping these weapons disassembled and their parts stored at different locations. (In the latter case, simply assembling the weapons at a time of crisis so as to make them ready for immediate use would constitute a significant escalation of threat and tension.) The security of these measures and of the weapons themselves is a particular point of concern in nuclear powers like Pakistan, which are under the threat of terrorism or civil

unrest, or in situations where fully assembled weapons are located close to a battlefield or on an ally's territory far from the nuclear power's capital.

21. T. V. Paul, Richard J. Harknett, and James J. Wirtz, eds., *The Absolute Weapon Revisited: Nuclear Arms and the Emerging International Order* (Ann Harbor, MI: University of Michigan Press, 1998), 294; James G. Blight, Bruce J. Alyn, and David A. Welch, *Cuba on the Brink: Castro, the Missile Crisis, and the Soviet Collapse* (New York: Rowman & Littlefield Publishers, 2002). See also Robert McNamara's commentary in the movie *The Fog of War* and Robert S. Norris and Hans M. Kristensen, "The Cuban Missile Crisis: A Nuclear Order of Battle, October and November 1962," *Bulletin of the Atomic Scientists* 68, no. 6 (November 1, 2012): 85–91.

22. This technical definition of an international regime is adapted from Stephen Krasner, *International Regimes* (Ithaca, NY: Cornell University Press, 1983), 2.

23. Earthquakes and nuclear explosions produce waves in the Earth with different characteristics that can be detected and distinguished at long distances from their origin by the International Monitoring System, a worldwide, international network of laboratories and seismographic stations. A nuclear explosion originates from a point and sends out pressure waves (P waves) inside the Earth in all directions. These waves are known as compression waves because they alternately compress and expand the rock or other material through which they pass, the way sound waves act on the air. These compressions and expansions take place in the direction the P wave travels. An earthquake, by contrast, originates from sideways or up-and-down movement along a linear geological fault, not from a single point. It therefore predominantly sends out transverse shear waves (S waves) perpendicular to the direction in which they travel, analogous to the waves that are created when one snaps a rope or whip. (Nuclear explosions also produce S waves, but unlike those from earthquakes, they are smaller than the P waves). Seismologists tell whether a given seismograph reading (a signal) is caused by an earthquake or an explosion by comparing the strength of the P and S waves—a high S/P ratio means an earthquake—and by comparing a given signal to those received in the past from previous comparable explosions and earthquakes. It was once feared that cheaters could muffle (decouple) the seismic waves from a small nuclear explosion by conducting it in a huge cavity in an underground salt deposit, thus concealing it from outside observers. Researchers showed that this would be hard to keep secret because it would be possible only with low-yield explosions in a few, well-known locations that could easily be inspected if such inspections were permitted. See Lawrence Livermore National Security, "*Sleuthing Seismic Signals*," Science & Technology Review: Deciphering Seismic Data (March 2009), 4–12 https://str.llnl.gov/content/pages/past-issues-pdfs/2009.03.pdf.

24. The IAEA verifies that nonnuclear weapons countries party to the NPT do not have "undeclared nuclear material or activities" through inspectors' site visits to nuclear facilities, complemented by information from satellites and foreign intelligence services. Pierre Goldschmidt, "Strengthened Safeguards," *IAEA Bulletin 34*, no. 4 (2001), https://www.iaea.org/publications/magazines/bulletin/43-4/strengthened-safeguards. These visits verify that the nuclear materials on the facility's internal records and the equipment observed by the inspectors correspond to those reported

to the IAEA, and that the reported material is actually present. The inspectors also look for traces of telltale isotopes that are inevitably left in nearby soil, which are detectable even long after extensive efforts at concealment. In "bulk" facilities that produce large quantities of plutonium, the IAEA examines containment measures (fences, sensors, dogs, etc.) that guard against smuggling even relatively small amounts of nuclear material.

25. Cirincione, *Bomb Scare*, 24–26; Jungk, *Brighter than a Thousand Suns.*
26. Robert Pool, *Beyond Engineering: How Society Shapes Technology* (New York: Oxford University Press, 1997).
27. South Korea and Taiwan abandoned their consideration of nuclear programs under U.S. pressure and with the assurance that they would be protected by the American nuclear umbrella. Brazil and Argentina ended their programs by mutual agreement. South Africa ended its program just before the fall of the White apartheid regime, in part to prevent a government controlled by Black Africans from having nuclear weapons. Sweden and Switzerland also abandoned their plans for nuclear weapons, in part for budgetary reasons, and signed the NPT.
28. Graham Allison, "*What Happened to the Soviet Superpower's Nuclear Arsenal? Clues for the Nuclear Security Summit*" (Cambridge, MA: Belfer Center for Science and International Affairs, March 2012), 19.
29. Kingston Reif, "New START at a Glance," Arms Control Association, January 2020, https://www.armscontrol.org/factsheets/NewSTART. See also Anders Åslund, "How Small Is the Soviet National Income?," in Henry S. Rowen and Charles Wolf Jr., eds., *The Impoverished Superpower: Perestroika and the Soviet Military Burden* (San Francisco: Institute for Contemporary Studies, 1990), 49; Cirincione, *Bomb Scare*, 97.
30. Department of State. New START Treaty. https://www.state.gov/new-start/
31. There is a critical difference between the nuclear umbrella—the commitment to defend allies against nuclear attack—and *extended deterrence*, the commitment to defend allies against all attack, nuclear or conventional, as if it were an attack on the United States itself. The public statement of U.S. nuclear doctrine is ambiguous on this distinction. US Air Force, "Air Force Doctrine Publication 3-72: Nuclear Operations," Curtis LeMay Center, May 19, 2015, updated 18 December 2020. https://www.doctrine.af.mil/Portals/61/documents/Annex_3-72/3-72-D12-NUKE-OPS-Extended-Deterrence.pdf.
32. Joseph A. Christoff, Weapons of Mass Destruction: Additional Russian Cooperation Needed to Facilitate U.S. Efforts to Improve Security at Russian Sites (Washington, DC: Government Accountability Office, March 2003).
33. Director of National Intelligence, Annual Report to Congress on the Safety and Security of Russian Nuclear Facilities and Military Forces (Washington, DC: National Intelligence Council, 2011).
34. Ed Pilkington and Martin Pengally, "'Let It Be an Arms Race': Donald Trump Appears to Double Down on Nuclear Expansion," *The Guardian*, December 24, 2016, http://www.theguardian.com/us-news/2016/dec/23/donald-trump-nuclear-weapons-arms-race.
35. Bernard Brodie, ed., *The Absolute Weapon: Atomic Power and World Order* (Freeport, NY: Harcourt Brace Jovanovich, 1946); Kenneth N. Waltz *The Spread of Nuclear*

Weapons: More May Be Better, Adelphi Papers, No. 171 (London: International Institute for Strategic Studies, 1981).

36. Anne I. Harrington, Eliza Gheorghe, and Anya Loukianova Fink, "What Arguments Motivate Citizens to Demand Nuclear Disarmament?," *Bulletin of the Atomic Scientists 73*, no. 4 (July 4, 2017): 255–63.
37. The full text of the treaty is available at ICAN. *Treaty on the Prohibition of Nuclear Weapons*. New York: United Nations, 2017. https://www.icanw.org/full_text_of_the_treaty.
38. George P. Shultz, William J. Perry, Henry A. Kissinger, and Sam Nunn, "A World Free of Nuclear Weapons," *Wall Street Journal*, January 4, 2007; George P. Shultz, William J. Perry, Henry A. Kissinger and Sam Nunn, "Toward a Nuclear-Free World," *Wall Street Journal*, January 16, 2008, https://www.wsj.com/articles/SB116787515251566636.
39. International Atomic Energy Agency, "Agreed Framework of 21 October 1994 Between the United States of America and the Democratic People's Republic of Korea," Information Circular 457, November 2, 1994, https://web.archive.org/web/20031217175315/http://www.iaea.org/Publications/Documents/Infcircs/Others/infcirc457.pdf.
40. The full text of the Iran nuclear deal, known formally as the Joint Comprehensive Plan of Action, can be found at http://apps.washingtonpost.com/g/documents/world/full-text-of-the-iran-nuclear-deal/1651.
41. The nuclear doctrine of a country is a statement of the principles that govern the role of nuclear weapons in its overall political and military strategy: the choice of weapons, the delivery system, the pattern of deployment, and the system of command and control. They address such questions as: Under what circumstances might nuclear weapons actually be used? To avoid losing an important regional conflict? To defend a key ally? To defend its country's own soil? To assure the continuation of its government in office? Against what kind of target might they be used? Who is authorized to order their use? Might the country ever be the first to use nuclear weapons in a given conflict? The importance of a nuclear doctrine lies in the fact that it is a guideline for quick decision; a crisis is not the best time to ask these basic questions for the first time.
42. Samuel F. Wells, Jr., "The Origins of Massive Retaliation," *Political Science Quarterly 96*, no. 1 (1981): 31–52.
43. Nikolai N. Sokov, "Why Russia Calls a Limited Nuclear Strike 'De-Escalation,'" *Bulletin of the Atomic Scientists 13* (March 2014) https://thebulletin.org/2014/03/why-russia-calls-a-limited-nuclear-strike-de-escalation/.
44. Alan Robock, "Nuclear Winter," *WIREs Climate Change 1* (May/June 2010): 418–27.
45. Brigitte Lebens Nacos, *Terrorism and Counterterrorism*, 4th ed. (Boston: Pearson Longman, 2012), 156.
46. Rolf Mowatt-Larssen, "*Islam and the Bomb: Religious Justification for and Against Nuclear Weapons*" (Cambridge, MA: Belfer Center for Science and International Affairs, 2011).
47. Michael Levi, *On Nuclear Terrorism* (Cambridge, MA: Harvard University Press, 2007), 72.

48. Greg Mello, Andrew Lichterman, and William Weida, "The Stockpile Stewardship Charade," *Issues in Science and Technology 15*, no. 3 (Spring 1999). As a technical matter, computer simulations have become so sophisticated that explosive testing of nuclear weapons is thought no longer to be necessary. Even so, simulations do not provide the "feel" of a real explosion. The government's description of the nuclear stewardship program can be found at https://annual.llnl.gov/annual06/pdfs/stewardship.pdf.
49. "A Hero at Home, a Villain Abroad," *The Economist*, June 19, 2008, https://www.economist.com/international/2008/06/19/a-hero-at-home-a-villain-abroad.
50. The details of the arrangement by which this was effected are summarized in Nuclear Threat Initiative, "India: Nuclear Weapons Program," November 2019, http://www.nti.org/learn/countries/india/nuclear).
51. We are here discussing fully autonomous weapons, which raise the relevant practical and ethical issues in their purest form. There is also a much broader category of semiautonomous weapon, for example, armed drones that are capable of surveying a situation and selecting a target but cannot fire or bomb without approval from a human operator. One might also conceive of an intermediate level of autonomy, in which an armed drone is authorized to engage any target within a well-defined area known to be hostile, a situation analogous to the free-fire zones of the Vietnam War. In either case, human supervision would be more difficult in a hostile environment in which communication between human and machine may be jammed. Scharre, *Army of None.*
52. A detailed summary of national programs to develop autonomous weapons is found in Andrew Feickert et al., U.S. Ground Forces Robotics and Autonomous Systems (RAS) and Artificial Intelligence: Considerations for Congress, *R45392* (Washington, DC: Congressional Research Service, November 20, 2018).
53. Michael T. Klare, "Autonomous Weapons Systems and the Laws of War," *Arms Control Today*, March 2019, https://www.armscontrol.org/act/2019-03/features/autonomous-weapons-systems-laws-war.
54. Samuel Bendett, "The Rise of Russia's Hi-Tech Military," *Fletcher Security Review 6*, no. 1 (2019), https://sites.tufts.edu/fletcherrussia/files/2019/06/The-Rise-of-Russia%E2%80%99s-Hi-Tech-Military-Samuel-Bendett.pdf.
55. We leave aside discussions of the possibility of robots that are programmed to constantly improve themselves, and robots that are more intelligent than humans (artificial general intelligence), on the assumption that both of these dystopic nightmares are thought to be far in the future.
56. Scharre, *Army of None,* 180; Nicole Kobie, "To Cripple AI, Hackers Are Turning Data Against Itself," *Wired*, September 11, 2018, https://www.wired.co.uk/article/artificial-intelligence-hacking-machine-learning-adversarial.
57. The assertion that an action is intrinsically forbidden under all circumstances, no matter whether its consequences are good or bad, is a statement of deontological ethics. This contrasts with consequentialist ethics, in which an ethical proposition is judged on whether or not it is thought to lead to a favorable outcome.

58. "Endorsers," The Campaign to Stop Killer Robots, 2019, https://www.stopkillerrobots.org/endorsers/.
59. Scharre, *Army of None*, 331, provides a comprehensive table of these historic efforts and of their successes (for example, the prohibition of biological warfare), but mostly of their failures.
60. Scharre, *Army of None*, ch. 16 and references therein.
61. Nita Bhalla, "Nations Dawdle on Agreeing Rules to Control 'Killer Robots' in Future Wars," *Reuters*, January 17, 2020, https://www.reuters.com/article/us-global-rights-killer-robots-idUSKBN1ZG151.

Annex A

1. Much of this annex is based on unpublished notes prepared for classroom use by the late Allan Krass.
2. Plutonium could, in principle, be used in a gun-type bomb but only with considerable loss of efficiency. This is because of the ^{240}Pu impurity, which produces so many "spontaneous" neutrons—neutrons not produced by fission—that a critical mass, once formed, would fly apart before the bomb could explode with optimum effect.
3. A dirty bomb is not a nuclear device at all but, rather, a chemical explosive surrounded by radioactive material, which it spreads when it explodes. The immediate casualties from such an explosion are due mostly to the chemical blast, but the resulting radioactive contamination can deny the use of an area until the completion of expensive cleanup operations that may take months or years. They are therefore often called "weapons of mass *disrup*tion," rather than *destruct*ion.
4. A nanosecond is a billionth of a second, or about the time it takes light to go a foot in the air or in a vacuum.
5. These forms of radiation should be distinguished from the radioactive *material* that *emits* the radiation.
6. Scientists may never be able to settle the controversy over whether low doses of radiation can cause cancer, or whether natural mechanisms can repair small amounts of radiation damage before it causes mutations or malignancies. Existing research methods would require an impractically large number of laboratory animals to settle this issue.
7. Charles Ferguson, *Nuclear Energy: What Everyone Needs to Know* (Oxford: Oxford University Press, 2011), 107.
8. Still a third technology is the calutron, a sort of industrial scale mass spectrometer.
9. World Nuclear Association, "Uranium Enrichment," 2016. http://www.world-nuclear.org/information-library/nuclear-fuel-cycle/conversion-enrichment-and-fabrication/uranium-enrichment.aspx.
10. The equations for these reactions look scary but are actually simple to understand if you remember that $^{238}U_{92}$ has 92 protons and 146 neutrons, and that a change in the number of protons means a change in the name of the element and also in its chemical

characteristics. This means that the equation for adding a neutron to the nucleus of ^{238}U is $^{238}U_{92} + {}^{1}n \rightarrow {}^{239}U_{92}$. The equation for the transformation of a neutron in the ^{238}U nucleus into a proton and an electron is then $^{239}U_{92} \rightarrow {}^{239}Np_{93}$ + e, and that for a neutron in the ^{239}Np being similarly transformed is $^{239}Np_{93} \rightarrow {}^{239}Pu_{94}$ + e, with "e" being the symbol for an electron. (Pu is the chemical symbol for the element plutonium; Np is the symbol for neptunium. When you add a proton to a nucleus, you change the element and the elemental symbol, in this case from Pu to Np.)

Chapter 6

1. The story of the Peruvian cholera epidemic is engagingly recounted in Ulysses B. Panisset, *International Health Statecraft: Foreign Policy and Public Health in Peru's Cholera Epidemic* (Lanham, MD: University Press of America, 2000).
2. The World Health Organization website defines epidemiology as "the study of the distribution and determinants of health-related states or events (including disease), and the application of this study to the control of diseases and other health problems."
3. Laurie Garrett, *The Coming Plague: Newly Emerging Diseases in a World Out of Balance* (New York: Penguin, 1995).
4. Wenjun Ma, Robert E Kahn, and Juergen A Richt, "The Pig as a Mixing Vessel for Influenza Viruses: Human and Veterinary Implications," *Journal of Molecular and Genetic Medicine* 3, no. 1 (November 27, 2008): 158–66.
5. Emily K. Shuman, "Global Climate Change and Infectious Diseases," *New England Journal of Medicine 362*, no. 12 (March 25, 2010): 1061–63.
6. World Health Organization, *Malaria, Global Health Observatory Data*, 2018,; World Health Organization, "10 Facts About Schistosomiasis," December 2017, www.who.int/gho/malaria/en/.
7. K. Singh and S. Mehta, "The Clinical Development Process for a Novel Preventive Vaccine: An Overview," *Journal of Postgraduate Medicine 62*, no. 1 (2016): 4–11. Governments are aware of these and other pressures, and they may assume the risk of liability connected to the manufacture of essential vaccines or offer a guaranteed market or exclusive intellectual property rights to any company that develops an effective new vaccine for diseases for which none had previously existed, as was done for COVID-19.
8. World Health Organization, WHO Global Health: Expenditure Atlas (Geneva: WHO, September 2014).
9. John-Arne Røttingen et al., "Mapping of Available Health Research and Development Data: What's There, What's Missing, and What Role Is There for a Global Observatory?," *Lancet 382*, no. 9900 (October 12, 2013): 1286–307.
10. Linda P. Fried et al., "Global Health Is Public Health," *Lancet 375*, no. 9714 (February 13, 2010): 535–37.
11. The concept of intermestic issues is developed in Victor D. Cha, "Globalization and the Study of International Security," *Journal of Peace Research 37*, no. 3 (2000): 391–403.

12. Anya Sarang et al., "Prisons as a Source of Tuberculosis in Russia," *International Journal of Prisoner Health 12* (March 14, 2016): 45–56.
13. Jeffrey D. Klausner et al., "Correlates of Gonococcal Infection and of Antimicrobial-Resistant Neisseria Gonorrhoeae Among Female Sex Workers, Republic of the Philippines, 1996–1997," *The Journal of Infectious Diseases 179*, no. 3 (March 1, 1999): 729–33.
14. Paul M. Sharp and Beatrice H. Hahn, "Origins of HIV and the AIDS Pandemic," *Cold Spring Harbor Perspectives in Medicine 1*, no. 1 (September 1, 2011): a006841. In some West African countries, people were forced to go deeper into the forest in search of game (bushmeat), thereby increasing their exposure to zoonoses, because their government had cut off traditional sources of protein by selling fishing rights to overseas fleets in the "exclusive economic zone" within 200 miles of the coastline.
15. Almudena Marí Saéz et al., "Investigating the Zoonotic Origin of the West African Ebola Epidemic," *EMBO Molecular Medicine 7*, no. 1 (January 1, 2015): 17–23; Angellar Manguvo and Benford Mafuvadze, "The Impact of Traditional and Religious Practices on the Spread of Ebola in West Africa: Time for a Strategic Shift," *The Pan African Medical Journal 22*, no. S1 (October 10, 2015), doi.org/10.11694/pamj.supp.2015.22.1.6190.
16. Ashok Moloo, "River Blindness: Shifting from Prevention to Surveillance and Elimination," *World Health Organization*, November 12, 2017, http://www.who.int/neglected_diseases/news/shifting_from_prevention_to_surveillance_elimination/en/.
17. Ernst B Haas, "Function," *Beyond the Nation-State: Functionalism and International Organization* (Stanford, CA: Stanford University Press, 1964), 27–28. Haas continues, "[Functionalists] would concentrate on commonly experienced needs initially, expecting the circle of the noncontroversial to expand at the expense of the political, as practical cooperation became coterminous with the totality of interstate relations. At that point, a true world community will have arisen." From Haas's lips to God's ear.
18. International collaboration has also had its downside, as when United Nations peacekeepers sent to keep order after the 2010 earthquake in Haiti caused a cholera epidemic by poor sanitation practices that contaminated downstream drinking water.
19. Leslie George Norman, Road Traffic Accidents: Epidemiology, Control, and Prevention (Geneva: World Health Organization, 1962).
20. This byplay was also a prominent feature of the history of the WHO's predecessor organizations.
21. Anita Hardon, "Consumers Versus Producers: Power Play Behind the Scenes," in Najmi Kenjo et al., eds., *Drugs Policy in Developing Countries* (London: Zed Books, 1992), pp. 48–64
22. Selam Gebrekidan et al., "In Hunt for Virus Source, WHO Let China Take Charge," *The New York Times*, 3 November 2020, A1.
23. R. Weissman, "A Long, Strange TRIPS: The Pharmaceutical Industry Drive to Harmonize Global Intellectual Property Rules, and the Remaining WTO Legal Alternatives Available to Third World Countries," *University of Pennsylvania Journal of International Economic Law 17*, no. 4 (1996): 1069–125.

24. Gil Walt, "WHO Under Stress: Implications for Health Policy," *Health Policy 24*, no. 2 (1993): 125–44. The United States was the only country to vote against a proposed WHO marketing code for infant formula in 1981, and in 2018 it brought strong pressure that forced sponsors to withdraw an innocuous World Health Assembly resolution encouraging breastfeeding of infants, in both cases in order to protect the interests of manufacturers of baby formula. Tobacco and asbestos firms have also undertaken campaigns against WHO initiatives aimed at the use of their products (see chapter 7).
25. Matthew M. Heaton, "Health and Medicine in Colonial Society," in Martin S. Shanguhyia and Toyin Falola, eds., *The Palgrave Handbook of African Colonial and Postcolonial History* (New York: Palgrave Macmillan, 2018), 303–18; Jeremy Greene et al., "Colonial Medicine and Its Legacies," in Paul Farmer et al., eds., *Reimagining Global Health: An Introduction* (Berkeley, CA: University of California Press, 2013), 33–73.
26. Fexinidazole, a nontoxic oral cure for African sleeping sickness, was announced in 2018 and is undergoing final testing.
27. See, for example, Helen Tilley, "Medicine, Empires, and Ethics in Colonial Africa," *AMA Journal of Ethics 18*, no. 7 (July 1, 2016): 743–53. A broad history of global health is found in Ann-Emmanuelle Birn, Yogan Pillay, and Timothy H. Holtz, *Textbook of International Health: Global Health in a Dynamic World*, 3rd ed. (New York: Oxford University Press, 2009), ch. 2.
28. Greene, "Colonial Medicine and Its Legacies."
29. Randall M. Packard, *A History of Global Health: Interventions into the Lives of Other Peoples* (Baltimore: Johns Hopkins Press, 2016), 152–80.
30. Randolph Quaye, *Balancing Public and Private Health Care Systems: The Sub-Saharan African Experience* (Lanham, MD: University Press of America, 2010), 2–4.
31. United Nations, "Constitution of the World Health Organization," *United Nations Treaty Series 14* (July 22, 1946): 186.
32. See, for example, the UN Covenant on Economic, Cultural and Social Rights, the UN Convention on the Rights of the Child, and the UN Convention on the Rights of Persons with Disabilities.
33. Lawrence Gostin and Devi Srindha, "Global Health and the Law," *The New England Journal of Medicine 370*, no. 18 (2014): 1732.
34. Marcelo Cueto, "The Origins of Primary Health Care and Selective Primary Health Care," *American Journal of Public Health 94*, no. 11 (2004): 1864–74.
35. J. A. Walsh and K. S. Warren, "Selective Primary Health Care: An Interim Strategy for Disease Control in Developing Countries," *The New England Journal of Medicine 301*, no. 18 (November 1, 1979): 967–74. See chapter 10 for additional examples of appropriate technology in health.
36. Olivier Fontaine, Paul Garner, and M. K. Bahn, "Oral Rehydration Therapy: The Simple Solution for Saving Lives," *British Medical Journal* 334, (January, 2007), S14..
37. Joy E. Lawn, "The Child Survival Revolution: What Next?," *The Lancet 384*, no. 9947 (September 13, 2014): 931–33; World Health Organization, The World Health Report 2002 - Reducing Risks, Promoting Healthy Life (Geneva: WHO, 2002).

38. Raymond B. Fosdick, *The Story of the Rockefeller Foundation* (New York: Harper and Brothers, 1952), 41. Many other foundations have made and continue to make important contributions to global health. On the corporate side, Merck and Dupont contributed Ivormectin and fine-mesh filters to the fights against river blindness and Guinea worm (dracontiasis), two debilitating parasitic diseases, respectively.
39. Irfan Nooruddin and Joel W. Simmons, "The Politics of Hard Choices: IMF Programs and Government Spending," *International Organization 60*, no. 4 (2006): 1001–33.
40. "A Plague of Finance," *The Economist*, September 29, 2001, https://www.economist.com/special-report/2001/09/29/a-plague-of-finance. The *Human Development Report* was developed by the UN Development Fund and UNICEF to call attention to the non-economic dimensions of development, in contrast to the *World Development Report* published by the World Bank.
41. World Health Organization, "Voluntary Contributions," April 19, 2018, http://www.who.int/about/finances-accountability/funding/voluntary-contributions/en/.
42. Barton Gellman, "AIDS Is Declared Threat to Security," *Washington Post*, April 30, 2000, https://www.washingtonpost.com/archive/politics/2000/04/30/aids-is-declared-threat-to-security/c5e976e4-3fe8-411b-9734-ca44f3130b41/.
43. As originally defined by WHO, the concept of global health security refers to infectious diseases that "endanger the collective health of populations living across geographical regions and international boundaries". (World Health Organization, "Health Security," http://www.who.int/health-security/en/). The idea was later stretched to include all aspects of global health from maternal health to salty foods. For purposes of analysis as opposed to those of fundraising, health security threats are now defined as diseases that spread fast, are little understood, have no cure or treatment, have high rates of mortality or morbidity (sickness), or induce visceral fears, in order to distinguish them from health emergencies, risks, and concerns. By this definition, COVID-19, which has already caused the deaths of more U.S. citizens than were killed in World War II, clearly qualifies as a security threat. Clare Wenham, "The Oversecuritization of Global Health: Changing the Terms of Debate," *International Affairs 95*, no. 5 (September 1, 2019): 1093–110.
44. World Bank, World Development Report: Investing in Health (New York: Oxford University Press, 1993).
45. Paul E. Farmer, B. Nizeye, S. Stulac, S. Keshavjee, "Structural Violence and Clinical Medicine," *PLoS Medicine 3*, no. 10 (2006): e449.
46. World Health Organization, quoted in <Thomas Gaulkin, "Pandemic Failure or Convenient Scapegoat: How Did WHO Get Here?," *Bulletin of the Atomic Scientists*, July 9, 2020. https://thebulletin.org/2020/07/pandemic-failure-or-convenient-scapegoat-how-did-who-get-here/
47. David Gootnick, Trends in U.S. Spending for Global HIV/AIDS and Other Health Assistance in Fiscal Years 2001–2008 (Washington, DC: US Government Accountability Office, October 2010).
48. Nathan C. Lo, A. Lowe, and E. Bendavid, "Abstinence Funding Was Not Associated with Reductions in HIV Risk Behavior in Sub-Saharan Africa," *Health Affairs 35*, no. 5 (2016): 856–63.

49. In another example of the imposition of the concerns of high-income countries on aid recipients, the U.S. government has intermittently imposed a policy, known to its critics as a "gag rule," that prohibits NGOs that receive its global health funds in aid-recipient countries from providing abortions and from referring women for abortions, whether or not they are legal in the country, whether or not U.S. money is directly involved, and regardless of the other health services the NGO may be providing. Barbara B. Crane, Nils Daulaire, and Alex C. Ezeh, "Reproductive Health in Culture Wars Crossfire," *Science 358*, no. 6360 (October 13, 2017): 175–76, and references therein.
50. GAVI's primary support is to immunization against diphtheria, tetanus, and pertussis (whooping cough). It also supports vaccination against pneumococcal meningitis, rotavirus (a major cause of diarrhea), meningitis, human papillomavirus (a cause of cervical and penile cancer), typhoid fever, polio, measles, yellow fever, cholera, and Japanese encephalitis.
51. Brigg Reilley, Michel Van Herp, Dan Sermand, and Nicoletta Dentico, "SARS and Carlo Urbani," *New England Journal of Medicine 348* (2003): 1951–52.
52. David P. Fidler, *SARS, Governance and the Globalization of Disease* (New York: Palgrave Macmillan, 2004).
53. World Health Organization, *International Health Regulations*, 3rd ed. (Geneva: WHO, 2005).
54. World Health Organization, "Report of the Ebola Interim Assessment Panel," July 2015, http://www.who.int/csr/resources/publications/ebola/report-by-panel.pdf; Laurie Garrett, "Ebola's Lessons: How the WHO Mishandled the Crisis," *Foreign Affairs 94*, no, 5 (2017): 80–107.
55. Punam Chuhan-Pole and Angwaf Manka, "*Yes, Africa Can: Success Stories from a Dynamic Continent*" (Washington, DC: World Bank, 2011), 433–43.
56. The sustainable development goals are "the blueprint to achieve a better and more sustainable future for all [by the year 2030]. They address the global challenges we face, including those related to poverty, inequality, climate, environmental degradation, prosperity, and peace and justice." The full text of the SDGs is available at https://www.un.org/sustainabledevelopment/sustainable-development-goals/.
57. Felicia Marie Knaul et al., "The Quest for Universal Health Coverage: Achieving Social Protection for All in Mexico," *Lancet 380*, no. 9849 (October 6, 2012): 1259–79.
58. Octavio Gómez-Dantés, Michael R. Reich, and Francisco Garrido-Latorre, "Political Economy of Pursuing the Expansion of Social Protection in Health in Mexico," *Health Systems & Reform 1*, no. 3 (2015): 207–16; Julio Frenk, "Bridging the Divide: Global Lessons from Evidence-Based Health Policy in Mexico," *Lancet 368*, no. 9539 (2006): 954–61.
59. Julio Frenk et al., "Comprehensive Reform to Improve Health System Performance in Mexico," *Lancet 368*, no. 9546 (October 28, 2006): 1524–34.
60. Jaime Sepúlveda et al., "Improvement of Child Survival in Mexico: The Diagonal Approach," *Lancet 368*, no. 9551 (December 2, 2006): 2017–27.
61. Frenk's concessions to the Philip Morris Company in connection with a tobacco control law blocked the possibility of his election to be director-general of the WHO,

although they were probably unavoidable given Mexican political realities, while his support for distributing the morning-after pill through government clinics drew strong opposition from the Catholic Church. He left the ministry in 2006 to pursue a distinguished career in the United States. Alejandra Xanic von Bertrab, "Part II: A Troubled Model for Reform in Mexico," Center for Public Integrity, November 15, 2010, https://publicintegrity.org/health/part-ii-a-troubled-model-for-reform-in-mexico/.

62. Daniel Cotlear et al., Going Universal: How 24 Developing Countries Are Implementing Universal Health Coverage from the Bottom Up (Washington, DC: World Bank, 2015), 112.
63. The evidence for asymptomatic human-to-human transmission at early stages of the COVID-19 pandemic was circumstantial but nevertheless convincing. First of all, the virus load in the nose of people infected with the virus was the same whether or not the person showed symptoms. Second, the infection spread quickly in ski lodges, religious gatherings, cruise ships, and in a U.S. Navy aircraft carrier, the *USS Theodore Roosevelt*—too quickly for the few people there who were showing symptoms to have done all the spreading. Subsequent research found more direct physical evidence for human-to-human transmission, as, for example, patients who had clearly caught the disease from other patients, and virus-laden aerosols in closed indoor systems in hospitals and restaurants. Some 60% of COVID-19 transmission was eventually shown to come from asymptomatic carriers.
64. Chris Buckley et al, "How Coronavirus Escaped from China's Grasp," *The* New York Times, December 19, 2020, A1 updated 3 February 2021. https://www.nytimes.com/2020/12/30/world/asia/china-coronavirus.html.
65. There are many published timelines of the early Chinese response to the COVID-19 epidemic. Two I found useful are: Zaheer Allam, "The First 50 days of COVID-19: A Detailed Chronological Timeline and Extensive Review of Literature Documenting the Pandemic," In *Surveying the Covid-19 Pandemic and Its Implications: Urban Health, Data Technology and Political Economy* (Elsevier Public Health Emergency Collection), 1–7, Elsevier, July 24, 2020, https://doi.org/10.1016/B978-0-12-824313-8.00001-2; and Susan V. Lawrence, "*COVID-19 and China: A Chronology of Events (December 2019–January 2020)*" (Washington: Congressional Research Service, May 13, 2020), https://fas.org/sgp/crs/row/R46354.pdf.
66. Jamie P. Horsely, "*Let's End the COVID-19 Blame Game: Reconsidering China's Role in the Pandemic*" (Washington: Brookings, 19 August 2020). https://www.brookings.edu/blog/order-from-chaos/2020/08/19/lets-end-the-covid-19-blame-game-reconsidering-chinas-role-in-the-pandemic/.
67. Raymond Zhong et al, "No 'Negative' News: How China Censored the Coronavirus," *The* New York Times, 19 December 2020, https://www.nytimes.com/2020/12/19/technology/china-coronavirus-censorship.html.
68. The fact that the Wuhan Institute of Virology, located just a few miles from the wild-life market, has long been studying viruses carried by bats has led to charges that the spread of coronavirus was the result of an accident or a breach in safety procedures at that laboratory. (Nicholas Wade, "The Origin of COVID: Did People or Nature

Open Pandora's Box at Wuhan?" *Bulletin of the Atomic Scientists*, 5 May 2021, https://thebulletin.org/2021/05/the-origin-of-covid-did-people-or-nature-open-pandoras-box-at-wuhan and references therein.) Leading international virologists have argued that this is unlikely, and Shi Zhengli, the leading bat researcher at the Institute, insists that she carefully reviewed her records and found no record of any such lapse or escape. Jane Qiu, "How China's 'Bat Woman' Hunted Down Viruses from SARS to the New Coronavirus," *Scientific American*, June 1, 2020. https://www.scientificamerican.com/index.cfm/_api/render/file/?method=inline&fileID=E1FDF8DE-9E22-4CE5-AD8B2E4682F52A86. More data is required to resolve this controversy.

69. Chinese and Russian authorities took advantage of worldwide eagerness for access to vaccines by releasing vaccines for widespread dissemination while large-scale ("phase III") tests for safety and effectiveness are still incomplete, with the intention of gaining friends through the use of "soft" (nonmilitary) power and at the same time (in the case of the Chinese) establishing themselves as important players on the world biotechnology stage. (Emily Rauhala et al., "Chinese and Russian Vaccines Remain Unproven—but Desperate Countries Plan to Use Them Anyway," *Washington Post*, December 18, 2020, https://www.washingtonpost.com/world/vaccines-china-russia-safety/2020/12/18/8c87ba54-3637-11eb-9699-00d311f13d2d_story.html).

70. Two false alarms from the research community have made some people reluctant to vaccinate themselves or their children, a phenomenon known as vaccine hesitancy. This phenomenon has destroyed herd immunity to measles and other contagious diseases in several communities and created unnecessary epidemics. Contrary to scare propaganda based on fraudulent research, there is no reason to believe that vaccines are related to autism. Nor is the amount of mercury that is present in many vaccines large enough to do harm, and even that small amount is being phased out. In the specific case of COVID-19, vaccine hesitancy has also been due to fear of long- or short-term side effects of the vaccine, to distrust of government and of the scientific community, and to skepticism as to the seriousness of the disease and the pandemic. Experts are confident that the benefits of the vaccine far outweigh the risks.

Annex B

1. Fischetti, Mark, and Veronica Falconieri Hays. "Inside the Coronavirus: What Scientists Know About the Inner Workings of the Pathogen That Has Infected the World." *Scientific American* 323, no. 1 (July 2020): 32–37.
2. Akiko Iwasaki and Patrick Wong, "The Immune Havoc of COVID-19," *Scientific American* (January 2021): 35–41.
3. Iwasaki, "Immune Havoc."
4. This step gained some notoriety because it is inhibited by hydroxychloroquine, a malaria remedy that was repeatedly recommended by President Trump and his allies in the United States and in foreign countries but was found in clinical tests not to have sufficient benefits to justify its harmful side effects.

5. Nicholas P. Jewell, Joseph A. Lewnard, and Britta L. Jewell, "Caution Warranted: Using the Institute for Health Metrics and Evaluation Model for Predicting the Course of the COVID-19 Pandemic," *Annals of Internal Medicine 173*, no. 3 (April 14, 2020): 226–27.
6. This section is a summary of "The Good and Bad Campaigns," in Randall Packard, *A History of Global Health: Interventions into the Lives of Other Peoples* (Baltimore: Johns Hopkins Press, 2016), chapter 8..
7. José A. Nájera, Matiana González-Silva, and Pedro L. Alonso, "Some Lessons for the Future from the Global Malaria Eradication Programme (1955–1969)," *PLOS Medicine 8*, no. 1 (January 25, 2011): e1000412.
8. World Health Organization, "The Top 10 Causes of Death," *Media Centre*, May 24, 2018, https://www.who.int/news-room/fact-sheets/detail/the-top-10-causes-of-death; Jennifer R. Herricks et al., "The Global Burden of Disease Study 2013: What Does It Mean for the NTDs?," *PLOS Neglected Tropical Diseases 11*, no. 8 (August 3, 2017): e0005424.

Chapter 7

1. Dean Jamison, "Global Health 2035: A World Converging Within a Generation," *The Lancet 382*, no. 9908 (December 13, 2013): 1898–955.
2. The global median age is 29.6 years. Of the 7.3 billion people on Earth, 12% are over sixty and 26% are under fifteen years of age and will be looking for jobs in a few years—a point to which we shall return when we consider the problems of competitiveness and employment in chapter 11. 60% of the world's people live in Asia and 37% in China and India. See United Nations, Department of Economic and Social Affairs, Population Division, "The World Population Prospects: 2015 Revision," Working Paper ESA/P/WP.241 (New York: United Nations, July 2015), www.un.org/en/development/desa/publications/world-population-prospects-2015-revision.html; United Nations Population Division, "Population Ages 0–14 (% of Total)," World Bank, 2017, data.worldbank.org/indicator/SP.POP.0014.TO.ZS?end=2016&start=2016; World Health Organization, "Ageing and Health," *Media Centre*, February 5, 2018, www.who.int/mediacentre/factsheets/fs404/en/.; United Nations. *ESCAP Population Data Sheet*. 2016. https://www.unescap.org/sites/default/d8files/knowledge-products/SPPS%20PS%20data%20sheet%202016%20v15-2.pdf.
3. United Nations, "The World Population Prospects."
4. "The Global Burden of Disease Study 2015: From Big Data to Meaningful Change," *The Lancet 388*, no. 10053 (October 8, 2016): 1447–850.
5. The WHO website defines epidemiology as "the study of the distribution and determinants of health-related states or events (including disease), and the application of this study to the control of diseases and other health problems."
6. World Health Organization. "Social Determinants of Health." 2021. https://www.who.int/health-topics/social-determinants-of-health#tab=tab_1 The WHO list includes "income and social protection, education, unemployment and job insecurity, working life conditions, food insecurity, housing, basic amenities and the

environment, early childhood development, social inclusion and non-discrimination, structural conflict, [and] access to affordable health services of decent quality".

7. See Emily Mendenhall et al., "Non-Communicable Disease Syndemics: Poverty, Depression, and Diabetes Among Low-Income Populations," *Lancet 389*, no. 10072 (04 2017): 951–63. This number of *The Lancet* is devoted to syndemics.
8. World Health Organization, "NCD Mortality and Morbidity," August 24, 2018, www.who.int/gho/ncd/mortality_morbidity/en/.
9. The formal title of the case is "Thailand—Restrictions on Importation of and Internal Taxes on Cigarettes," *General Agreements on Tariffs and Trade (*GATT) Basic Instruments and Selected Documents *(BISD) S37* (1991): 200. The Thai government could, in principle, have discouraged smoking without violating WTO rules by putting a tax on all cigarettes, both domestic and foreign, but that isn't what they did.
10. United Nations, "WHO Framework Convention on Tobacco Control," *United Nations Treaty Series, 2302*, no. 41032 (May 21, 2003), https://treaties.un.org/Pages/showDetails.aspx?objid=0800000280055410&clang=_en.
11. Marie-Lyne Grenier, "The Impact of Brain Drain in Underserved Countries: Implementing a Global Ethical Leadership Approach," *International Journal of Migration, Health and Social Care 11*, no. 3 (January 21, 2015): 218–22.
12. The industry had a legitimate worry that its profitable markets in the United States and Europe could have been undercut by the re-export of pills sold cheaply in Africa, but it could have protected its position by selling these medicines to low-income countries in odd colors or other distinctive formulations that would have been difficult to re-export.
13. William W. Fisher and Cyrill P. Rigamonti, *The South Africa AIDS Controversy: A Case Study in Patent Law and Policy, Harvard Law School*, February 10, 2005, boris.unibe.ch/69881/.
14. Taken as a whole, traditional medicines are neither fakes nor panaceas. They are probably especially effective for common, readily recognizable ailments like parasites, diarrhea, and fungus infections. (Otherwise, no one would patronize traditional healers!) Many other traditional medicines are basically placebos or less effective or unstandardized versions of well-known pharmaceutical principles. Some (like various supposed aphrodisiacs made from animal parts and the beets and garlic that the South African government once promoted as a cure for HIV/AIDS) are outright fakes. But some, like those listed in the main text, have made important contributions to Western allopathic medicine.
15. Charles Weiss and Thomas Eisner, "Partnerships for Bioprospecting," *Technology in Society 20* (1998): 481–98.
16. Joe Jackson, *The Thief at the End of the World: Rubber, Power, and the Seeds of Empire* (New York: Viking Press, 2008). The export of these seeds was not illegal under Brazilian law, but the license for export was likely obtained under questionable circumstances.
17. The texts of the CBD and of the follow-on Nagoya Protocol on Access to Genetic Resources and the Fair and Equitable Sharing of Benefits Arising from Their Utilization are available at f and https://www.cbd.int/abs/doc/protocol/nagoya-protocol-en.pdf,

respectively. See also "U.N. Biodiversity Summit Yields Welcome and Unexpected Progress," *Science* 330, no. 6605 (November 5, 2010): 742–43.

18. K. Divakaran Prathapan et al., "When the Cure Kills—CBD Limits Biodiversity Research," *Science 360*, no. 6396 (June 29, 2018): 1405–6.
19. Weiss and Eisner, "Partnerships for Bioprospecting."
20. Ambarish Mukherjee et al., "Modernization of Ayurveda: A Brief Overview of Indian Initiatives," *Natural Product Communications 9*, no. 2 (February 1, 2014). Research on Chinese traditional medicine has not found clear evidence for its effectiveness. See Institute of Complementary and Alternative Medicine, "Traditional Chinese Medicine: In Depth," *National Center for Complementary and Integrative Health*, 2012, https://nccih.nih.gov/health/whatiscam/chinesemed.htm. Chinese government policy is available at People's Republic of China State Council Information Office, "Traditional Medicine in China," December 6, 2016, http://english.gov.cn/archive/white_paper/2016/12/06/content_281475509333700.htm.
21. Peter Gelling, "Indonesia Defiant on Refusal to Share Bird Flu Samples," *The New York Times*, March 25, 2007, www.nytimes.com/2007/03/26/world/asia/26cnd-flu.html. The free exchange of genetic material related to basic food crops has made possible major advances in technology for small-scale farmers in low-income countries, as we discuss in chapter 10.
22. Anthony Robbins, "The CIA's Vaccination Ruse," *Journal of Public Health Policy 33*, no. 4 (2012): 387–89.
23. Emmanuel Kwaku Akyeampong, Allan G. Hill, and Arthur Kleinman, eds., *The Culture of Mental Illness and Psychiatric Practice in Africa* (Bloomington, IN: Indiana University Press, 2015); World Health Organization and Calouste Gulbenkian Foundation, *Social Determinants of Mental Health* (Geneva: WHO, 2014).
24. Murray Feshbach, *Russia's Health and Demographic Crises: Policy Implications and Consequences* (Washington, DC: Chemical and Biological Arms Control Institute, 2003).
25. Hans Zinsser, *Rats, Lice and History* (New York: Little, Brown, 1935).
26. *Isaiah* 37:36.
27. Peter Sarris, "The Justinianic Plague: Origins and Effects," *Continuity and Change 17*, no. 2 (2002): 169–82.
28. Charles C. Mann, *1491: New Revelations of the Americas Before Columbus* (New York: Knopf, 2005).
29. John Robert McNeill, *Mosquito Empires: Ecology and War in the Greater Caribbean, 1620–1914* (New York: Cambridge University Press, 2010).
30. Robert J. T. Joy, "Malaria in American Troops in the South and Southwest Pacific in World War II," *Medical History 43*, no. 2 (1999): 192–207. The U.S. military finally succeeded in controlling the disease by a combination of protective long clothing, destroying the breeding sites of mosquitos, and "bug bombs"—aerosol containers that dispersed pyrethrum insecticide. (DDT reached the troops in the field too late to have a decisive effect.)
31. Mark Wheelis, "Biological Warfare at the 1346 Siege of Caffa," *Emerging Infectious Diseases 8*, no. 9 (2002): 971–75.

32. Whether Amherst's proposal ever got beyond the discussion stage is subject to dispute, but one of his compatriots, a Captain Ecuyer, had definitely used this tactic earlier. Elizabeth A. Fenn, "Biological Warfare in Eighteenth-Century North America: Beyond Jeffery Amherst," *The Journal of American History 86*, no. 4 (2000): 1552–80.
33. Ken Alibek and Stephen Handelman, *Biohazard: The Chilling True Story of the Largest Covert Biological Weapons Program in the World—Told from the Inside by the Man Who Ran It* (New York: Random House, 1999).
34. Hyun-Kyung Kim, Elizabeth Philipp, and Hattie Chung, *North Korea's Biological Weapons Program: The Known and Unknown* (Cambridge, MA: Harvard University Kennedy School, 2017).
35. Sonia Ben Ouagrham-Gormley, "Potemkin or Real? North Korea's Biological Weapons Program," *Bulletin of the Atomic Scientists*, July 18, 2017, https://thebulletin.org/2017/07/potemkin-or-real-north-koreas-biological-weapons-program/.
36. The formal name of the 1925 treaty is "Protocol for the Prohibition of the Use in War of Asphyxiating, Poisonous or Other Gases, and of Bacteriological Methods of Warfare (Geneva Protocol)." Its text may be found at https://unoda-web.s3-accelerate.amazonaws.com/wp-content/uploads/assets/WMD/Bio/pdf/Status_Protocol.pdf.
37. United Nations, "Convention on the Prohibition of the Development, Production and Stockpiling of Bacteriological (Biological) and Toxin Weapons and on their Destruction," *United Nations Treaty Series 1015*, no. 14860 (April 10, 1972): 163.
38. Joseph S. Nye, *Soft Power the Means to Success in World Politics* (New York: PublicAffairs, 2005), 7.
39. Bill Gates, "Here's My Plan to Improve Our World—and How You Can Help," *Wired*, November 12, 2013, www.wired.com/2013/11/bill-gates-wired-essay/.
40. Lisa Stiffler, "Bill Gates Sr. at 90: A Giant Impact on Technology, Philanthropy and the Seattle Region," *GeekWire*, November 26, 2015, www.geekwire.com/2015/life-and-times-of-bill-gates-sr-a-giant-impact-on-technology-philanthropy-and-the-seattle-region/.
41. The document was Jeffrey D. Sachs, *Macroeconomics and Health: Investing in Health for Economic Development* (Geneva: WHO, 2001).
42. Mark Curtis, "*Gated Development: Is the Gates Foundation Always a Force for Good?*" 2nd ed. (London: Global Justice Now, June 2016).
43. Sandi Doughton, "Not Many Speak Their Mind to Gates Foundation," *The Seattle Times*, August 2, 2008, https://www.seattletimes.com/seattle-news/not-many-speak-their-mind-to-gates-foundation/.
44. The foundation's strategy vis-à-vis malaria is set forth on its website at "Malaria Strategy Overview," *Bill & Melinda Gates Foundation*, 2018, www.gatesfoundation.org/What-We-Do/Global-Health/Malaria.
45. The full WHO definition is "the activities required, both proactive and reactive, to minimize vulnerability to acute public health events that endanger the collective health of populations living across geographical regions and international boundaries" (World Health Organization, "Health Security.")
46. Clare Wenham, "The Oversecuritization of Global Health: Changing the Terms of Debate," *International Affairs 95*, no. 5 (September 1, 2019): 1093–110.

Chapter 8

1. In both Manila and Egypt, once the protests had begun, separate communities of sportsmen, fashionistas, etc., exploited their own long-standing word-of-mouth channels of communication independent of the Internet, a logical necessity in view of the fact that only about a quarter of the Egyptian population participated in the Net. Merlyna Lim, "Clicks, Cabs, and Coffee Houses: Social Media and Oppositional Movements in Egypt, 2004–2011," *Journal of Communication 62*, no. 2 (2012): 231–48.
2. Later studies showed that only the initial middle-class protests in Manila were organized with the aid of Internet communications, and that later demonstrations of both protesters and Estrada supporters around the country relied on more traditional means of communication. Manuel Castells et al., "Electronic Communication and Socio-Political Mobilization: A New Form of Civil Society," in Helmut K. Anheier, Mary Kaldor, and Marlies Glasius, eds., *Global Civil Society 2005/6* (London: SAGE Publications, 2006), 266–85.
3. David R. Johnson and David Post, "Law and Borders: The Rise of Law in Cyberspace," *Stanford Law Review 48*, no. 5 (1996): 1367–402.
4. "State of Privacy: Egypt," Privacy International, January 26, 2019, http://privacyinternational.org/state-privacy/1001/state-privacy-egypt.
5. Evgeny Morozov, *The Net Delusion: The Dark Side of Internet Freedom* (New York: PublicAffairs, 2011).
6. Xiao Qiang, "President XI's Surveillance State," *Journal of Democracy 30*, no. 1 (2019): 53–67; Marco Rubio and Chris Smith, 2018 Annual Report: Xinjiang (Washington, DC: Congressional-Executive Commission on China, 2018), 9; Rory Cellan-Jones, "China to Track Cars in Terrorism Crackdown," *BBC News*, February 21, 2017, https://www.bbc.com/news/world-asia-china-39038364.
7. Charles Weiss, "How Do Science and Technology Affect International Affairs?," *Minerva 53* (2015): 411–30.
8. Practical quantum computers are probably more than a decade in the future but may someday make it possible to break the codes that are now used to keep messages private. On the other hand, by then it may be possible to develop cryptographic methods that even quantum computers cannot overcome, with unpredictable consequences for the balance between cryptographers who write codes and those who break them.
9. For general histories of the Internet, see Janet Abatte, *Inventing the Internet* (Cambridge, MA: MIT Press, 1991), and Johnny Ryan, *A History of the Internet and the Digital Future* (London: Reaktion Books, 2010).
10. The agency was later renamed the Defense Advanced Projects Research Agency (DARPA).
11. Ryan, *A History of the Internet*, 25–26; and Katie Hafner and Matthew Lyon, *Where the Wizards Stay Up Late: The Origins of the Internet* (New York: Touchstone, 1996), 12–13.
12. Officially, assigning IP addresses was the job of a nonprofit organization called the Internet Assigned Numbers Authority (IANA), of which Postel was director.

13. Fred Kaplan, *Dark Territory: The Secret History of Cyber War* (New York: Simon & Schuster, 2016), 9.
14. Two subtler simplifications in the early Net, made to avoid delays in deployment, hampered later efforts to improve Internet security. First, hypertext software allowed any web page to link to any other web page but did not provide a way to inform the second web page (the one to which the first one had linked) that this linkage had been established. This later made it more complicated and time-consuming to trace a malicious attack back to its originator. It also made it almost impossible to stop the spread of intellectual property without compensation to its owner. (See Mike Holderness, "It Could All Be So Much Better [Interview with Ted Nelson]," *New Scientist*, 191, 54–55 (September 16, 2006). Second, any changes to a web page erased previous versions, making it impossible to reconstruct the original version if the revision was destroyed or made inaccessible. This complicates efforts to defend against ransomware, a criminal technique in which victims lose access to their own unbacked-up data unless they pay for its release.
15. The transcript of Gore's interview on CNN's *Late Edition* in 1999 shows that he said, "during my service in the United States Congress, I took the initiative in creating the Internet." See Ryan, *A History of the Internet*, 95, for a more detailed explanation of the controversy.
16. International Telecommunication Union, "Individuals Using the Internet (% of Population)," *World Bank*, October 2, 2019, https://data.worldbank.org/indicator/IT.NET.USER.ZS.
17. For a striking example, see Nicholas Negroponte, *Being Digital* (New York: Vintage Books, 1995), 229–38.
18. James Curran, Natalie Fenton, and Des Freedman, *Misunderstanding the Internet*, 2nd ed. (London: Routledge, 2016).
19. Deborah Spar, *Ruling the Waves: Cycles of Discovery, Chaos, and Wealth from the Compass to the Internet* (New York: Harcourt, 2001).
20. International Telecommunications Union, "Individuals Using the Internet"; United Nations Population Division, "Population, Total," World Bank, October 16, 2019, https://data.worldbank.org/indicator/SP.POP.TOTL.
21. Pippa Norris, *Digital Divide: Civic Engagement, Information Poverty, and the Internet Worldwide* (New York: Cambridge University Press, 2001).
22. International Telecommunications Union, "Individuals Using the Internet."
23. Christopher Wray, "Transcript: The Threat Posed by the Chinese Government and the Chinese Communist Party to the Economic and National Security of the United States," *Hudson Institute*, July 7, 2020, https://www.hudson.org/research/16201-transcript-the-threat-posed-by-the-chinese-government-and-the-chinese-communist-party-to-the-economic-and-national-security-of-the-united-states.
24. Rebecca A. Fannin, *Tech Titans of China* (Boston: Nicholas Brealey, 2019); David E. Sanger, *The Perfect Weapon: War, Sabotage and Fear in the Cyber Age* (New York: Crown Publishing, 2018), 108.

25. Stuart A. Thompson and Charlie Werzel, "Twelve Million Phones, One Dataset, Zero Privacy," *The* New York Times, 19 December 2019, https://www.nytimes.com/interactive/2019/12/19/opinion/location-tracking-cell-phone.html.
26. Fan Liang, Vishnupriya Das, Nadiya Kostyuk, and Muzammil M. Hussain, "Constructing a Data-Driven Society: China's Social Credit System as a State Surveillance Infrastructure," *Policy & Internet 10*, no. 4 (2018): 415–53. Although full-scale implementation of the social credit system was once planned for 2020, it is not clear how effective local efforts at pilot implementation have been (Elizabeth C. Economy, *The Third Revolution: Xi Jinping and the New Chinese State* (New York: Oxford University Press, 2018)).
27. Raymond Zhong, "China's Cryptocurrency Plan Has a Powerful Partner: Big Brother," *The New York Times*, October 18, 2019. https://www.nytimes.com/2019/10/18/technology/china-cryptocurrency-facebook-libra.html.. Facebook once planned a similar cryptocurrency named Libra. This would have provided for user anonymity.
28. John Perry Barlow, "A Declaration of the Independence of Cyberspace," *Electronic Frontier Foundation*, February 8, 1996, https://www.eff.org/cyberspace-independence.
29. Economy, *The Third Revolution*; Randy James, "A Brief History Of Chinese Internet Censorship," *Time*, March 18, 2009, http://content.time.com/time/world/article/0,8599,1885961,00.html.
30. Adam Segal, "When China Rules the Internet," *Foreign Affairs 85*, no. 5 (2018): 10–18.
31. Freedom House, "Freedom of the Net 2020: China."
32. Economy, *The Third Revolution*.
33. Raymond Zhong et al, "No 'Negative News': How China Censored the Coronavirus," *The* New York Times, 19 December 2020, https://www.nytimes.com/2020/12/19/technology/china-coronavirus-censorship.html; Freedom House, "Freedom of the Net 2020: China."
34. Jonah Force Hill, "The Growth of Data Localization Post-Snowden," Lawfare Research Paper Series, 2, No. 3. July 21, 2014, 1–41. https://assets.documentcloud.org/documents/7276302/THE-GROWTH-OF-DATA-LOCALIZATION-POST-SNOWDEN.pdf.
35. For a concise summary, see Rachel F. Fefer and Kristin Archick, *E.U. Data Protection Rules and U.S. Implications* (Washington, DC: Congressional Research Service, 2019), https://crsreports.congress.gov/product/pdf/IF/IF10896..
36. Martin A. Weiss and Kristin Archick, *U.S.-EU Data Privacy: From Safe Harbor to Privacy Shield*, CRS 7–5700 (Washington, DC: Congressional Research Service, 2016). https://epic.org/crs/R44257.pdf..
37. World Summit on the Information Society, *Tunis Agenda for the Information Society*, WSIS 05/TUNIS/DOC/6(Rev. 1)-E (2005), November 18, 2005, https://www.itu.int/net/wsis/docs2/tunis/off/6rev1.html).
38. Joel Snyder, Konstantinos Komaitis, and Andrei Robachevsky, "*The History of IANA: An Extended Timeline with Citations and Commentary*" (Reston, VA: Internet Society, January 2017); and K. Raustiala, "Governing the Internet," *American Journal of International Law 110* (2016): 491–503.

39. Nigel Inkster, *China's Cyber Power* (Washington, DC: Routledge and the International Institute for Strategic Studies, 2018).
40. A ban on exports of U.S.-designed semiconductors could, in principle, put a crimp on Chinese surveillance technology, but video cameras and other equipment of high quality are increasingly designed and manufactured in China. Further increases in self-sufficiency are a major goal of Chinese industrial and security policy.
41. Edith Brown Weiss, *Establishing Norms in a Kaleidoscopic World, General Course on Public International Law* 396 (Leiden: Brill/Nijhoff, 2018), chapter 11.

Chapter 9

1. For general treatments of cybersecurity and cyberwar, see Fred Kaplan, *Dark Territory: The Secret History of Cyber War* (New York: Simon & Schuster, 2016); Bruce Schneier, *Click Here to Kill Everybody: Security and Survival in a Hyperconnected World* (New York: W. W. Norton, 2018); and David Sanger, *The Perfect Weapon: War, Sabotage and Fear in the Cyber Age* (New York: Crown Publishing, 2018).
2. Sanger, *The Perfect Weapon*, 154.
3. A worm is like a virus but can spread from computer to computer on its own without needing a host program. A zero-day vulnerability in a piece of software is a flaw that is unknown to those responsible for fixing it until it is exploited or revealed by a hacker. The specialized computers attacked by Stuxnet software controlled the speed of the centrifuges that separate uranium hexafluoride gas made from uranium-238 ($^{238}UF_6$) from the chemically identical gas made from the fissionable isotope uranium-235 ($^{235}UF_6$) (see chapter 5 and Annex A.)
4. There were actually two versions of the Stuxnet worm; the description here is of the second version. It is drawn from Ralph Langner, "To Kill a Centrifuge" (Arlington, VA: Langner Group, 2013).
5. David Albright, Paul Brannan, and Christina Walrond, Did Stuxnet Take Out 1,000 Centrifuges at the Natanz Enrichment Plant? (Washington, DC: Institute for Science and International Security, December 22, 2010), https://isis-online.org/isis-reports/detail/did-stuxnet-take-out-1000-centrifuges-at-the-natanz-enrichment-plant/; and Marie Baezner and Patrice Robin, Hotspot Analysis: Stuxnet (Zurich: ETH CyberDefense Project, 2017), https://css.ethz.ch/content/dam/ethz/special-interest/gess/cis/center-for-securities-studies/pdfs/Cyber-Reports-2017-04.pdf.
6. Sanger, *The Perfect Weapon*, 43.
7. Kaplan, *Dark Territory*, 67–71. A red team in this context is a team authorized to attack the cybersecurity of an organization, using techniques that would be used by an adversary.
8. Kaplan, *Dark Territory*, 86.
9. Lionel D. Alford, Jr., "Cyber Warfare: Protecting Military Systems," *Acquisition Review Quarterly*, 99 (Spring 2000): 100–120, at 103.

10. Christopher Wray, "Transcript: The Threat Posed by the Chinese Government and the Chinese Communist Party to the Economic and National Security of the United States," *Hudson Institute*, July 7, 2020, https://www.hudson.org/research/16201-transcript-the-threat-posed-by-the-chinese-government-and-the-chinese-communist-party-to-the-economic-and-national-security-of-the-united-states.
11. This account of the Snowden leak is taken from Kaplan, *Dark Territory*, and James R. Clapper, *Facts and Fears: Hard Truths from a Life in Intelligence* (New York: Viking, 2018).
12. A Web crawler is an automated program that methodically searches for and downloads copies of web pages of a specified type.
13. Paul Lewis, "Snowden documents Show NSA Gathering 5bn Cell Phone Records Daily," *The Guardian*, December 5, 2013, https://www.theguardian.com/world/2013/dec/04/nsa-storing-cell-phone-records-daily-snowden.
14. The program was reformed in 2015 so that most of the records are now held by the telecommunications companies and are accessible to the NSA only with a warrant issued by the secret Foreign Intelligence Surveillance (FISA) court.
15. Under current Supreme Court precedent, information is protected by the Fourth Amendment to the U.S. Constitution only if there is a "reasonable expectation of privacy," a concept that is subject to various interpretations and varies with time and technology. This doctrine would mean that the expectation of privacy is lost when information is turned over to a third party during its normal course of business and is thus knowingly exposed to the public, as when calling parties disclose a telephone number to the telephone company when they make a call. Courts have differed over how and whether to apply this "third-party disclosure doctrine" now that huge quantities of data can be aggregated and analyzed. A recent Supreme Court decision involving cellphone location data (Carpenter v. United States, 138 S. Ct. 2206, 2215–216 (2018)) indicates that courts are beginning to reconsider this doctrine in light of the revolution in information and communications.See Timothy Geverd, "Bulk Telephony Metadata Collection and the Fourth Amendment: The Case For Revisiting the Third-Party Disclosure Doctrine in the Digital Age," *The John Marshall Journal of Information Technology & Privacy Law 31*, no. 2 (January 1, 2014): 191–235; and Charles Weiss, "The Coming Technology of Knowledge Discovery: A Final Blow to Privacy Protection?," *University of Illinois Journal of Law, Technology and Policy 2004*, no. 2 (2004): 253–81.
16. A back door bypasses the security mechanism of a computer or software system in order to allow a hacker to gain high-level access without being detected.
17. Shazeda Ahmed, Natasha E. Bajema, Samuel Bendett, et al., *AI, China, Russia, and the Global Order: Technological, Political, Global, and Creative Perspectives*, Strategic Multilayer Assessment Periodic Publication (Boston, MA: Defense Technical Information Center, December 1, 2018), part I, https://apps.dtic.mil/docs/citations/AD1066673.
18. Quoted in Adam Segal, "When China Rules the Internet," *Foreign Affairs 85*, no. 5 (2018): 10–18.

19. Ellen Nakashima, "U.S. Pushes Hard for a Ban on Huawei in Europe, but the Firm's 5G Prices Are Nearly Irresistible," *Washington Post*, May 29, 2019, https://www.washingtonpost.com/world/national-security/for-huawei-the-5g-play-is-in-europe--and-the-us-is-pushing-hard-for-a-ban-there/2019/05/28/582a8ff6-78d4-11e9-b7ae-390de4259661_story.html; Ryan McMorrow, "Huawei a Key Beneficiary of China Subsidies That US Wants Ended," *Phys.org*, May 30, 2019, https://phys.org/news/2019-05-huawei-key-beneficiary-china-subsidies.html.
20. Article 7 of the 2017 Chinese National Intelligence Law mandates that "any organization or citizen shall support, assist and cooperate with the state intelligence work." This law has been interpreted to mean that Huawei would be obliged to spy on its users if requested by the Chinese government, though Huawei strongly disputes this claim.
21. Alina Polyakova and Chris Meserole, *Exporting Digital Authoritarianism, Democracy & Disorder policy brief, Brookings Institution*, August 26, 2019, https://www.brookings.edu/wp-content/uploads/2019/08/FP_20190826_digital_authoritarianism_polyakova_meserole.pdf.
22. Valery Gerasimov, "The Value of Science Is in the Foresight: New Challenges Demand Rethinking the Forms and Methods of Carrying Out Combat Operations," *Military Review* 96, no. 1 (January/February 2016): 23–29. https://www.armyupress.army.mil/Portals/7/military-review/Archives/English/MilitaryReview_20160228_art008.pdf. See also Charles K. Bartles, Getting Gerasimov Right, *Military Review* 96, no. 1 (January/February 2016): 30-38. ; Sarah P. White, Understanding Cyberwarfare: Lessons from the Russia-Georgia War (West Point, NY: Modern War Institute, 2018), https://mwi.usma.edu/wp-content/uploads/2018/03/Understanding-Cyberwarfare.pdf.
23. For the declared U.S. cybersecurity policy, see Department of Defense, Cyber Strategy (Arlington, VA: US Department of Defense, 2018), https://media.defense.gov/2018/Sep/18/2002041658/-1/-1/1/CYBER_STRATEGY_SUMMARY_FINAL.PDF. For a clearer explanation, see Nina Kollars and Jacqueline Schneider, "Defending Forward: The 2018 Cyber Strategy Is Here," *War on the Rocks,* September 20, 2018, https://warontherocks.com/2018/09/defending-forward-the-2018-cyber-strategy-is-here/.
24. Diego A. Martin and Jacob N. Shapiro, *Trends in Online Foreign Influence Efforts* (Princeton, NJ: ESOC Publications, 2019).
25. Clapper, *Facts and Fears.*
26. Clapper, *Facts and Fears*, 114.
27. Tim Weiner, *Legacy of Ashes: The History of the CIA* (New York: Anchor Books, 2008), 346.
28. In this context, a troll is a person who posts inflammatory material on an online community to provoke discord. A bot is an automated program that imitates the action of a person in order to spread a propaganda message online. Bots are also used in DDoS attacks and for many other purposes, both legitimate and illegitimate.

29. One of the most widely shared of these strange theories held that Hillary Clinton was running a (nonexistent) pedophile ring out of the (nonexistent) basement of an (existent) Virginia pizzeria. This tale was recycled during the 2020 election campaign.
30. Clapper, *Facts and Fears*, ch. 11.
31. Clapper, *Facts and Fears*, 354.
32. Peter Sanger, Nicole Perlroth, and Julian E. Barnes, "As Understanding of Russian Hacking Grows, So Does Alarm," *The New York Times,* January 2, 2021, updated January 5, 2021, https://www.nytimes.com/2021/01/02/us/politics/russian-hacking-government.html. A supply chain hack penetrates a computer system by attacking a vulnerable outside partner or provider that has access to that system.
33. David E. Sanger and Nicole Perlroth, "FireEye, a Top Cybersecurity Firm, Says It Was Hacked by a Nation-State," *The New York Times*, December 8, 2020, updated February 6, 2021, https://www.nytimes.com/2020/12/08/technology/fireeye-hacked-russians.html.
34. B. Mazanes, *The Evolution of Cyberwar: International Norms for Emerging Technology Weapons* (Lincoln: University of Nebraska Press, 2015); Edith Brown Weiss, *Establishing Norms in a Kaleidoscopic World*. General Course on Public International Law 396 (Leiden: Brill/Nijhoff, 2018).
35. M. Schmitt, ed., *Tallinn Manual 2.0 on the International Law Applicable to Cyber Operations*, NATO Cooperative Cyber Defense Center of Excellence (Cambridge: Cambridge University Press, 2017).
36. Dan Ephron and Yuval Shany, "A Rule Book on the Shelf? Tallinn Manual 2.0 on Cyberoperations and Subsequent State Practice," *American Journal of International Law 112*, no. 4 (2019): 583–657. Obama's threat did not apply to election interference in the form of propaganda or disinformation. In a separate incident, China agreed to curtail its program of extensive digital industrial espionage as part of the price for a summit meeting with the United States in 2015. Sanger, *The Perfect Weapon*, 121.
37. United Nations Secretary-General, *Group of Governmental Experts on Developments in the Field of Information and Telecommunications in the Context of International Security*, A/70/174 (New York: UN, July 22, 2015).
38. Anders Henriksen, "The End of the Road for the UN GGE Process: The Future Regulation of Cyberspace," *Journal of Cybersecurity* 5, no. 1 (January 1, 2019). The present status of the discussions is summarized in the Chair's Summary of the Third Substantive Session, held March 8–12, 2021, of the Open-Ended Working Group on Developments in the Field of Information and Telecommunications in the Context of International Security (Conference Room Paper, UN General Assembly). https://front.un-arm.org/wp-content/uploads/2021/03/Chairs-Summary-A-AC.290-2021-CRP.3-technical-reissue.pdf
39. The Code of Conduct was proposed in 2011 and revised in 2015. The text of the 2015 revision, together with a line-by-line analysis of the changes made by its sponsors to the original, is found at Sarah McKune, "An Analysis of the International Code of Conduct for Information Security," Citizen Lab Report, Munk School of Global Affairs, University of Toronto, 2015. **https://openeffect.ca/code-conduct/**. In this context, a State is a country.

40. Brad Smith, "Transcript: The Need for a Digital Geneva Convention," *Keynote Address, RSA Conference*, February 14, 2017, San Francisco.
41. Office of the Press Secretary, "President Xi Jinping's State Visit to the United States," White House Archives, September 25, 2015, https://obamawhitehouse.archives.gov/the-press-office/2015/09/25/fact-sheet-president-xi-jinpings-state-visit-united-states.
42. Customary international law consists of rules that governments obey in the belief that they are legally obligated to do so, despite the absence of a formal agreement.

Chapter 10

1. By contrast, the sophisticated electromechanical leg provided to a U.S. soldier wounded in Iraq or Afghanistan is so good that it gives "handicapped" runners an advantage over their "able-bodied" competitors—but it does not look like a natural leg and foot, and it costs about $15,000.
2. Yet another frugal technology for countries cursed with land mines is the use of trained giant Mozambique rats as mine detectors. Unlike dogs, these rats are too light to set off antipersonnel mines, and their noses are better mine detectors than the best man-made instruments of analytical chemistry. Timothy L. Edwards et al., "Giant African Pouched Rats (Cricetomys Gambianus) That Work on Tilled Soil Accurately Detect Land Mines," *Journal of Applied Behavior Analysis 48*, no. 3 (2015): 696–700.
3. Michael DeGusta, "Are Smart Phones Spreading Faster than Any Technology in Human History?," *MIT Technology Review*, May 9, 2012, https://www.technologyreview.com/s/427787/are-smart-phones-spreading-faster-than-any-technology-in-human-history/.
4. Robert J. Saunders, Jeremy J. Warford, and Bjorn Wellenius, *Telecommunications and Economic Development* (Baltimore, MD: Johns Hopkins University Press, 1994).
5. Donald Maitland, *The Missing Link: Report of the Independent Commission for World-Wide Telecommunications Development* (Geneva: International Telecommunications Union, 1985). See also Saunders, *Telecommunications and Economic Development.*
6. M-PESA is owned by Vodaphone and Safaricom, British and Kenyan telecommunications firms, respectively. Researchers from Gamos, a British consulting firm, and from the Commonwealth Telecommunications Organization, an intergovernmental organization, collaborated with Vodaphone in the conception and launch of M-PESA. See Irene Yuan Sun, *The Next Factory of the World: How Chinese Investment Is Reshaping Africa* (Cambridge, MA: Harvard Business Press, 2017).
7. Frugal innovations share some features with "disruptive" innovations, which are defined as products initially aimed at special requirements of the neglected, low-cost end of a market, and which later improve to the point that they eat away at the market for the more expensive, well-established products and even replace them. Low-cost airlines are the classic example. See Clayton Christensen, *The Innovator's Dilemma: When New Technologies Cause Great Firms to Fail* (Cambridge, MA: Harvard Business School Press, 1997). Unlike most frugal innovations, disruptive innovations as defined by Christensen are profitable throughout this process and do not require subsidy.

8. E. F. Schumacher, *Small Is Beautiful: Economics as if People Mattered* (London: Blond & Briggs, 1973).
9. George McRobie, *Small Is Possible* (New York: Harper and Row, 1981).
10. Raphael Kaplinsky, *The Economies of Small: Appropriate Technology in a Changing World* (London: Intermediate Technology Press, 1990).
11. Daniel M. Kammen, "Cookstoves for the Developing World," *Scientific American 273*, no. 1 (July 1, 1995): 72–75. doi.org/10.1038/scientificamerican0795-72.
12. Sunna Malla and Govina R. Timilsina, "Household Cooking Fuel Choice and Adoption of Improved Cookstoves in Developing Countries: A Review," Policy Research Working Paper No. 6903 (Washington, DC: World Bank, 2014).
13. Meena Khandelwal et al., "Why Have Improved Cook-stove Initiatives in India Failed?" *World Development 92* (April 1, 2017):13–27.
14. Not having benefited from sex education in school, Muruganantham hadn't even known about the existence of menstruation until he was in his thirties. Once he realized its importance and set his mind to doing something about it, he faced so much difficulty in getting women to discuss so intimate a matter that he was forced to try out new pad designs by simulating menstruation on himself! He did this by attaching a plastic tube to a soccer-ball bladder that squirted goat blood when it was pressed.
15. Yudhijit Bhattacharjee, "How an Indian Innovator Reverse-Engineered the Making of Sanitary Pads," *The New York Times*, November 10, 2016, https://www.nytimes.com/interactive/2016/11/13/magazine/design-issue-sanitary-pads-india.html.
16. C. K. Prahalad, *The Fortune at the Bottom of the Pyramid: Eradicating Poverty Through Profits* (Upper Saddle River, NJ: Wharton School Publishing, 2005).
17. Alessandra N. Bazzano et al., "The Life Course Implications of Ready to Use Therapeutic Food for Children in Low-Income Countries," *International Journal of Environmental Research and Public Health 14*, no. 4 (April 2017): 403; CMAM, "Key Data," PlumpyField Network, 2018, http://www.plumpyfield.com/about/key-data.
18. Subir Roy, "Tata Nano—Why Did the People Not Want the People's Car?," *The Wire*, July 17, 2018, https://thewire.in/business/tata-nano-why-did-the-people-not-want-the-peoples-car.
19. Berk Özler, "One Laptop Per Child Is Not Improving Reading or Math. But, Are We Learning Enough from These Evaluations?," *World Bank Blogs, Development Impact (blog)*, June 14, 2012, https://blogs.worldbank.org/impactevaluations/one-laptop-per-child-is-not-improving-reading-or-math-but-are-we-learning-enough-from-these-evaluati.
20. The Equalize Health website is at https://equalizehealth.org.
21. Neonatal jaundice (yellowed skin) is due to the accumulation of toxic bilirubin, derived from the breakdown of hemoglobin from worn out red blood cells, faster than the baby's liver can detoxify it. Blue-green light energizes an oxidation reaction whose product can be readily excreted. Chiara Greco et al., "Neonatal Jaundice in Low- and Middle-Income Countries: Lessons and Future Directions from the 2015 Don Ostrow Trieste Yellow Retreat," *Neonatology 110*, no. 3 (2016): 172–80.
22. Nicholas Bakalar, "For Neonatal Jaundice, a New Option in Resource-Poor Nations," *The New York Times*, September 26, 2011, https://www.nytimes.com/2011/09/27/health/27jaundice.html.

23. This is the BMVSS Foundation. The letters stand for Bhagwan Mahaveer Viklang Sahayata Samit, which means "Bhagwan Mahaveer Disabled Assistance Committee."
24. The evaluating firm, ImpactMatters, relied upon D-REV's own monitoring data to make this calculation and did not make its own field evaluation, which would have been prohibitively expensive. ImpactMatters, *D-Rev Impact Audit: Newborn Health Program and Mobility Program (2013–2015)* (New York: ImpactMatters, September 26, 2017), 2..
25. McKinsey & Company, "Driving Down the Cost of High-Quality Care: Lessons from the Aravind Eye Care System," *Health International 11* (2011): 1–27.
26. An animal reservoir was recently discovered in dogs and some African wild animals, dimming hopes for total eradication.
27. Li Liu et al., "Global, Regional, and National Causes of Child Mortality: An Updated Systematic Analysis for 2010 with Time Trends since 2000," *Lancet 379*, no. 9832 (June 9, 2012): 2151–61.
28. The defining feature of the VIP latrine, now copied all over the world, is an offset ventilation pipe that vents smelly gases from the excreta pile directly into the air, bypassing the cabin where people do their business. A screen atop the pipe blocks flies and other insects from getting into the cabin via the pipe. If flies do get in, perhaps because users leave the door open, adult flies that eventually hatch are attracted up the pipe by the daylight, are blocked by the screen, and eventually die. Charles G. Gunnerson, "Sanitation Systems for Developing Countries," in Charles Weiss and Nicolas Jequier, eds., *Technology, Finance and Development* (Lexington, MA: D.C. Heath & Co., 1984).
29. Umut Türker, "Alternative Sewerage Solution: Condominial Method and Its Application," *Physics and Chemistry of the Earth, Parts A/B/C*, special issue, Water in the Mediterranean Basin, *36*, no. 5 (January 1, 2011): 179–86.
30. Felicia Wu and William P. Butz, "The Green Revolution," in *The Future of Genetically Modified Crops* (Santa Monica, CA: RAND Corporation, 2004), 29.
31. William Paddock and Paul Paddock, *Famine 1975! America's Decision: Who Will Survive?* (Boston, MA: Little-Brown, 1967); Paul Ehrlich, *The Population Bomb* (New York: Sierra Club-Ballantine, 1969).
32. For a summary of the social and environmental consequences of the Green Revolution, see Gordon Conway, *One Billion Hungry* (Ithaca, NY: Cornell University Press, 2012), 44–62.
33. The authoritative account of the early history of the CGIAR is Warren Baum, *Partners Against Hunger: The Consultative Group on International Agricultural Research* (Washington, DC: World Bank, 1986).
34. The Washington Consensus also led to experiments that required farmers to pay for the advice of extension workers, by analogy to the requirement that patients make copayments for health services. These experiments did not go well in either health or agriculture. William M. Rivera and John W. Cary, "Privatizing Agricultural Extension," in Burton E. Swanson, Robert P. Bentz, and Andrew J. Sofranko, eds., *Improving Agricultural Extension: A Reference Manual* (Rome: Food and Agricultural Organization, 1997).
35. In order not to divert too much money from agricultural research and to avoid conflict with the FAO, ISNAR was mandated to do *research on how best to run* an effective national agricultural research and extension system. This required pioneering

multidisciplinary research, for which a satisfactory intellectual framework did not exist. ISNAR was unable to develop such a framework. It therefore spent most of its resources providing technical assistance and advice to existing NARSs, contrary to its official mandate. Donors gradually starved it of resources and finally closed it down in 2004, merging its remaining research efforts with those of the International Food Policy Research Institute (IFPRI), another CGIAR-supported center, where they languished and eventually ceased. This has forced crop breeders in the remaining CGIAR-supported centers to divert their own limited resources in order to spend more time with NARSs so as to assure the adaptation and diffusion of the varieties they develop.

36. The donors have set up a Global Crop Diversity Trust. However, this has yet to reach the size needed to fund the CGIAR gene banks in perpetuity, so they are still financed out of regular CGIAR funds. The Crop Trust manages the international "last resort" doomsday seed vault deep in the permafrost of Svaalbard, Norway.
37. This section is based on interviews with CGIAR scientists, and on Selcuk Ozgediz, *The CGIAR at 40: Institutional Evolution of the World's Premier Agricultural Research Network* (Washington, DC: CGIAR Fund, 2011).
38. Prabhu L. Pingali, "Green Revolution: Impacts, Limits, and the Path Ahead," *Proceedings of the National Academy of Sciences 109*, no. 31 (July 31, 2012): 12302–8.
39. This project was identified as a priority by researchers, who sought and received dedicated funding from the Gates Foundation.
40. William B. Bonvillian and Charles Weiss, *Technological Innovation in Legacy Sectors* (New York: Oxford University Press, 2014).
41. The DigitalGreen website is at https://www.digitalgreen.org/.
42. Taken at face value, the term "appropriate technology" suffers from a tautology: If a method works and was widely adopted, it must have been appropriate, no? Besides, technology can be both sophisticated and appropriate to the problem at hand. Space technology, for example, is needed to quickly assess damage from a tsunami or other natural disaster.
43. These events, sometimes dubbed "ideathons," often center around a specific challenge presented by a local NGO (e.g., the Harvard Law School 2015 ideathon on sanitation in Ghana), a focus area (e.g., the University of Toronto 2016 ideathon around global health), or even the UN Sustainable Development Goals (e.g., the Urban Innovation Challenge held in 2019 in Seoul).

Chapter 11

1. Peter Dicken, *Global Shift: Mapping the Changing Contours of the World Economy*, 7th ed. (New York: Guilford Press, 2015), 54.
2. Suzanne Berger, *How We Compete* (Cambridge, MA: MIT Press, 2006).
3. Raphael Kaplinsky, *Globalization, Poverty and Inequality* (Cambridge, MA: Polity Press, 2005).

4. Stephen Nellis, "Apple's IPhone X Has Higher Margin than IPhone 8: Analysis," *Reuters*, November 6, 2017, https://www.reuters.com/article/us-apple-iphone-idUSKBN1D62RZ.
5. Dicken, *Global Shift*.
6. Cheryl Katz, "Piling Up: How China's Ban on Importing Waste Has Stalled Global Recycling," *Yale Environment 360 (blog)*, March 7, 2019, https://e360.yale.edu/features/piling-up-how-chinas-ban-on-importing-waste-has-stalled-global-recycling.
7. William B. Bonvillian and Peter L. Singer, *Advanced Manufacturing: The New American Innovation Policies* (Cambridge, MA: MIT Press, 2017); William B. Bonvillian and Charles Weiss, *Technological Innovation in Legacy Sectors* (New York: Oxford University Press, 2014).
8. Karl-Benedikt Frey and Michael A. Osborne, "The Future of Employment: How Susceptible Are Jobs to Computerization?" *Technology Forecasting and Social Change 114* (January 2014): 254–80.
9. These speculations go back at least to John Maynard Keynes, "Economic Possibilities for Our Grandchildren," in *Essays in Persuasion*, 358–73 (New York: W. W. Norton, [1930] 1963). More recent predictions along the same lines are Bill Joy, "Why the Future Doesn't Need Us?," *Wired*, April 1, 2000, https://www.wired.com/2000/04/joy-2/; and Ray Kurzweil, *The Singularity Is Near* (New York: Penguin Books, 2006).
10. Josh Bivens, *Updated Employment Multipliers for the US Economy, Economic Policy Institute*, January 23, 2019, https://www.epi.org/publication/updated-employment-multipliers-for-the-u-s-economy/.
11. Martin Neil Baily and Barry P. Bosworth, "US Manufacturing: Understanding Its Past and Its Potential Future," *Journal of Economic Perspectives 28*, no. 1 (2014): 3–26.
12. Baily and Bosworth, "US Manufacturing."
13. David Autor et al., The Work of the Future: Shaping Technology and Institutions - Fall 2019 Report (Cambridge, MA: MIT Work of the Future, 2019).
14. Anthony P. Carnevale et. al, Good Jobs Are Back: College Graduates Are First in Line (Washington, DC; Georgetown Center for Education and the Workforce, 2015).
15. Antoine van Agtmael and F. Bakker, *The Smartest Places on Earth: Why Rustbelts Are the Emerging Hotspots of Global Innovation* (New York: PublicAffairs, 2018); Enrico Moretti, *The New Geography of Jobs* (New York: Harcourt-Brace, 2012); Mark Muro et al., Digitalization and the American Work Force (Washington, DC: Brookings Institution, 2017).
16. Erik Brynjolfsson and Andrew McAfee, *The Second Machine Age: Work, Progress and Prosperity in a Time of Brilliant Technologies* (New York: W.W. Norton, 2014); David Autor, "Why Are There Still So Many Jobs?," *Journal of Economic Perspectives 29*, no. 3 (2015): 3–30.
17. Autor, "Why Are There Still So Many Jobs?"
18. Autor, "The Work of the Future."
19. To be sure, not all service jobs are low skill and low wage. High-skilled service jobs in finance and insurance pay very well.
20. Up-to-date country-level economic data are available from the OECD at https://data.oecd.org/korea.htm.

21. Hyung-Sup Choi, "Technology for Development: The South Korean Experience," in Jairam Ramesh and Charles Weiss, eds., *Mobilizing Technology for World Development* (New York: Praeger Publishers, 1979), 130–39. http://documents.worldbank.org/curated/en/463821467989554789/pdf/UNN251000Mobil0or0world0development.pdf..
22. Alice H. Amsden, *Asia's Next Giant: South Korea and Late Industrialization* (New York: Oxford University Press, 1992).
23. Hong Kong is now part of the People's Republic of China (PRC), under the principle of "one country, two systems," which recognized the legitimacy of Hong Kong's capitalist economy. This status is now under stress. The PRC considers Taiwan to be a province.
24. Steven Schlossstein, *Asia's New Little Dragons: The Dynamic Emergence of Indonesia, Thailand, and Malaysia* (Chicago: Contemporary Books, 1991).
25. Mariana Mazzucatto, *The Entrepreneurial State* (New York: Public Affairs, 2014).
26. U.S. venture capital companies are receptive to business plans that promise rewards on a three- to seven-year timescale but are less receptive to proposals in fields like energy, which generally carry a ten- to twelve-year time horizon. See Bonvillian and Weiss, *Technological Innovation in Legacy Sectors*. Chinese banks and government agencies have moved to take advantage of some of these opportunities. In biotechnology, another United States strength, strong intellectual property protection and certification of successful innovations by the Food and Drug Administration help U.S. firms attract long-term private investment.
27. European Commission, *Turning Europe into a True Innovation Union*, MEMO/10/473, October 6, 2010, https://ec.europa.eu/commission/presscorner/detail/en/MEMO_10_473.
28. Bonvillian and Weiss, *Technological Innovation in Legacy Sectors*, 146–49.
29. Bonvillian and Singer, *Advanced Manufacturing*.
30. Ezra F. Vogel, *Japan as Number One: Lessons for America* (New York: Harper and Row, 1985).
31. Jack Womack et al., *The Machine That Changed the World* (New York: Rawson Associates, 1990).
32. Bonvillian and Singer, *Advanced Manufacturing*; Bonvillian and Weiss, *Technological Innovation in Legacy Sectors*. The manufacturing institutes established by the Obama Administration, and continued thereafter with bipartisan Congressional support, are an effort to restore some of these important supports to domestic manufacturing.
33. Charles Wessner, ed., *21st Century Manufacturing: The Role of the Manufacturing Extension Partnership Program* (Washington, DC: National Academies Press, 2013), appendix 2; Bonvillian and Weiss, *Technological Innovation in Legacy Sectors*, 145–48.
34. Robert D. Putnam, *Bowling Alone: The Collapse and Revival of American Community* (New York: Simon & Schuster, 2000).
35. David H. Autor, "Work of the Past, Work of the Future," *AEA Papers and Proceedings 109* (May 2019): 1–32.
36. Autor, "The Work of the Future."

37. Strictly speaking, there is no such thing as maintaining competitiveness, since almost anyone can be competitive if their wages are low enough. Properly stated, the objective is to maintain a high level of productivity that justifies a high level of wages, even in the face of international competition.
38. Dutch disease has many dimensions. Foreign exchange revenues from natural resource exports (or from foreign aid) tend to enrich and concentrate political power in a relatively few people and to create a relatively small number of well-paying jobs, worsening income inequality. Even in countries with honest and effective government, these revenues put money under the control of government officials, who tend to prefer a few large public investments (rather than many small ones) because they are easier to manage. In countries with less effective governments, these big projects invite waste and corruption. If the military of the country is weak, these resources and revenues invite rebellion or foreign invasion. The macroeconomic dimension of Dutch Disease is more complicated. Natural resource exports tend to raise the value of local currency to the point where local agricultural or industrial products cannot compete with imports. High wages in the natural resources sector make it difficult for other branches of the economy to compete for labor. In the economists' shorthand, the exchange rates of these countries and the terms of trade between the natural resource sector and other parts of the economy are set by the extractive natural resource industry too high for agricultural or industrial exports to compete. The effect on foreign exchange works as follows. Unless a country sequesters its large receipts of foreign exchange in a "sovereign wealth fund," or uses them to encourage imports, preferably of capital equipment needed for industrialization or infrastructure, an influx of foreign currency raises the demand for local currency to pay for local goods and services more than it raises the demand for imports. This increases the value of local currency and makes it harder to export anything other than natural resources.
39. Michael L. Ross, "The Political Economy of the Resource Curse," *World Politics 51*, no. 2 (January 1999): 297–322.
40. Dani Rodrik, "Premature Deindustrialization," *Journal of Economic Growth 21*, no. 1 (2016): 1–33.
41. Garry Chua Kenn, Education, Skills, and Decent Work in Low and Middle Income Countries: Trends and Results from an Adult Skills Survey, UNESCO, 2016, https://unesdoc.unesco.org/ark:/48223/pf0000245622.
42. United Nations Department of Economic and Social Affairs, Population Division, World Population Prospects 2019: Highlights (New York: United Nations, 2019).
43. Carl Dahlman, *World Under Pressure: How China and India Are Influencing the Global Economy and Environment* (Palo Alto, CA: Stanford University Press, 2012).
44. Elizabeth C. Economy, "The Great Leap Backward? The Costs of China's Environmental Crisis," *Foreign Affairs 86*, no. 5 (2007): 38–59.
45. Jonas Nahm and Edward S. Steinfeld, "Scale-Up Nation: China's Specialization in Innovative Manufacturing," *World Development 54* (February 2014): 288–300.
46. For Chinese wind technology, see Joanna I. Lewis, *Green Innovation in China* (New York: Columbia University Press, 2010); for links between innovation and manufacturing and a compact summary of the Chinese innovation system, see Bonvillian and Weiss, *Technological Innovation in Legacy Sectors*, 151–61.

47. The BRI is Xi Jinping's signature foreign policy and economic development project: a sprawling, centrally managed program that invests heavily in loans for infrastructure projects across nearly seventy countries. Many of the infrastructure programs it has financed would not pass the cost/benefit tests that are normally required to justify such large loans. As a result, some of the governments of the countries with BRI projects have defaulted on their loans, making many observers skeptical of the BRI's promises of economic development. In 2019, the Chinese central government rebooted the program in order to improve its image ("China's Reboot of the Belt and Road Initiative," *The Economist Intelligence Unit*, April 29, 2019, https://country.eiu.com/article.aspx?articleid=597952843&Country=China&topic=Politics). Like Japanese investments before them, Chinese natural resource-oriented investments have been criticized for creating environmental damage, for using exclusively Chinese labor and firms, for mistreating Chinese workers abroad, and for the racist attitudes of their managers and workers (Aaron Halegua and Jerome A. Cohen, "The Forgotten Victims of China's Belt and Road Initiative," *Washington Post*, April 23, 2019, https://www.washingtonpost.com/opinions/2019/04/23/forgotten-victims-chinas-belt-road-initiative/). These problems may fade as the Chinese, like the Japanese before them, gain more experience. The Asian Infrastructure Investment Bank (AIIB), in which China plays a leading role, has a better record in all these areas (Eduard Fernandez, "What the Belt and Road Initiative Can Learn from the AIIB," *Infrastructure Investor*, February 27, 2019, https://www.infrastructureinvestor.com/what-the-belt-and-road-can-learn-from-the-aiib/).
48. Dieter Ernst, "Advanced Manufacturing and China's Future for Jobs," in Eva Paus, ed., *Confronting Dystopia: The New Technological Revolution and the Future of Work* (Ithaca, NY: Cornell University Press, 2018).
49. State Council of China, "Made in China 2025," July 7, 2015, http://www.cittadellascienza.it/cina/wp-content/uploads/2017/02/IoT-ONE-Made-in-China-2025.pdf. The full list of technologies includes: next generation information technology, high-end numerical control machinery and robotics, aerospace and aviation equipment, maritime engineering equipment and high-tech maritime vessel manufacturing, advanced rail equipment, energy-saving and new energy vehicles, electrical equipment, new materials, biomedicine and high-performance medical devices, and agricultural machinery and equipment.
50. Douglas B. Fuller, *Paper Tigers, Hidden Dragons: Firms and the Political Economy of China's Technological Development* (New York: Oxford University Press, 2019); Anton Malkin, Beyond "Forced" Technology Transfers: Analysis of and Recommendations on Intangible Economy Governance in China (Waterloo, ON: Center for International Governance Innovation, March 2020).
51. Nigel Inkster, *China's Cyber Power* (Washington, DC: Routledge and the International Institute for Strategic Studies, 2018), 68–69; EY Greater China, "China Is Poised to Win the 5G Race. Are You Up to Speed?," *EY*, June 13, 2018, https://www.ey.com/en_cn/tmt/china-is-poised-to-win-the-5g-race-are-you-up-to-speed.
52. Technical standards are voluntary agreements among producers that set forth detailed technical specifications that ensure that products, services, and systems are

safe, consistent, and reliable, and that they are mutually compatible and interoperable with other technical systems. See ITU Telecommunication Standardization Bureau, Understanding Patents, Competition & Standardization in an Interconnected World (Geneva: International Telecommunications Union, 2014). De facto technical standards may be set by the fact that one brand dominates a particular market, as when the need for compatibility with Microsoft Windows set the standard for software when desktop computers were dominant. Alternatively, technical standards may be set by a governmental or nongovernmental organization (for example, the Technical Inspection Association (*Technischer Überwachungsverein,* (TuV), in Germany, or the American Society of Mechanical Engineers (ASME) in the United States), or by negotiations among competing firms and their governments, overseen by the ISO or some other international organization.

53. Hilary McGeachy, "*US-China Technology Competition: Impacting a Rules-Based Order,*" United States Studies Centre at the University of Sydney, May 2, 2019, https://www.ussc.edu.au/analysis/us-china-technology-competition-impacting-a-rules-based-order.
54. Andrew Kipnis and Shanfeng Li, "Is Chinese Education Underfunded?," *The China Quarterly, 202* (2010): 327–43; and Li Hui and Ben Blanchard, "China Taps Tech Training to Tackle Labor Market Mismatch," *Reuters*, June 7, 2014, https://www.reuters.com/article/us-china-labour-education idUSKBN0EJ01A20140608.
55. Amy Qin, "Fraud Scandals Sap China's Dream of Becoming a Science Superpower," *The New York Times*, October 13, 2017, https://www.nytimes.com/2017/10/13/world/asia/china-science-fraud-scandals.html.
56. China Power Team, "*Is China a Global Leader in Research and Development?,*" Center for Strategic and International Studies, March 12, 2020, https://chinapower.csis.org/china-research-and-development-rnd/.
57. Elizabeth Economy, *The Third Revolution: Xi Jinping and the New Chinese State* (New York: Oxford University Press, 2018).
58. Richard McGregor, "How the State Runs Business in China," *The Guardian*, July 25, 2019, https://www.theguardian.com/world/2019/jul/25/china-business-xi-jinping-communist-party-state-private-enterprise-huawei.
59. Barbara Lippert and Volker Perthes, Strategic Rivalry Between the United States and China: Causes, Trajectories, and Implications for Europe (Berlin: Stiftung Wissenschaft und Politik - German Institute for International and Security Affairs, April 2020).
60. Cecilia Kang and Alan Rappeport, "Trump Blocks Broadcom's Bid for Qualcomm," *The New York Times*, March 12, 2018, https://www.nytimes.com/2018/03/12/technology/trump-broadcom-qualcomm-merger.html; Steve Lohr, "United States Moves to Ban Huawei from Government Contracts," *The New York Times*, August 7, 2019, https://www.nytimes.com/2019/08/07/business/huawei-us-ban.html.
61. Fabrizio Zilibotti, "Growing and Slowing Down Like China," *Journal of the European Economic Association 15*, no. 5 (October 1, 2017): 943–88.
62. This aphorism is frequently attributed to New York Yankees catcher Yogi Berra and to the Danish physicist Nils Bohr, one of the founding fathers of quantum mechanics. It is also said to be an old Danish proverb.

63. Irene Yuan Sun, *The Next Factory of the World: How Chinese Investment Is Reshaping Africa* (Boston: Harvard Business School Press, 2017).

Chapter 12

1. Chinese official codes of scientific ethics are comparable to those in Europe and the United States, but there are few consequences for violating them, and Chinese scientists generally do not receive training in ethics.
2. Dr. He's experiment altered a gene involved in the synthesis of a protein called CCR5, which is an entry point, or receptor, for the virus that causes HIV/AIDS. His idea was that eliminating this protein would prevent the babies in the experiment from ever contracting the disease. However, he seems to have left unchanged many other genes that also play a part in controlling this protein. Besides, changes in the gene for CCR5 create the risk of future health problems unrelated to HIV/AIDS, including changes in cognition and increased probability of serious complications of influenza. Ed Yong, "The CRISPR Baby Scandal Gets Worse by the Day," *The Atlantic*, December 3, 2018.
3. A comprehensive account of the invention, current and potential applications, and ethical ramifications of gene driver technology is found in Jennifer A. Doudna and Samuel H. Sternberg, *A Crack in Creation: Gene Editing and the Unthinkable Power to Control Evolution* (Boston: Houghton-Mifflin, 2017). Professor Doudna is the inventor of CRISPR-Cas9.
4. CRISPR is an acronym standing for "clustered regularly interspaced short palindromic repeats." Cas9 stands for "CRISPR-associated protein #9," How they work will be explained shortly.
5. In DNA-RNA hybridization, double-stranded DNA separates into single strands, one of which binds to a complementary sequence of nucleotides in a single-stranded RNA.
6. Strictly speaking, curing a genetic disease by editing the gene of a somatic cell *does* affect the gene pool of future generations in the sense that it enables the carrier of the defective gene to survive and reproduce when they would otherwise not have done so. This argument would apply to enabling the carrier of any disease with a genetic component to survive and reproduce, whether or not gene editing was involved.
7. When CRISPR-cas9 edits the DNA of a sperm or egg cell, it inserts a gene for the CRISPR-cas9 gene editing system in addition to whatever other edits it makes. Let's say it was a sperm cell; it would work the same way for an edited egg cell. If the edited sperm cell fertilizes an unedited egg cell, one of the chromosomes of the resulting zygote will include the edited gene, plus a gene for CRISPR-cas9. During fertilization, this chromosome will pair with a chromosome derived from the unedited egg cell that the sperm had fertilized, so that the zygote has one edited and one unedited chromosome. The CRISPR-cas9 gene now springs into action. The zygote turns on the CRISPR-cas9 gene (expresses it) and synthesizes the CRISPR-cas9 gene editor system, which clips out the unedited gene of the egg-derived chromosome and inserts in its place the changes specified by the gene driver, including the gene for CRISPR-Cas9.

Now *both* chromosomes of the zygote have the modified gene or genes and the gene for CRISPR-Cas9. The same process will take place when the organism that eventually develops from the zygote matures and reproduces, ensuring that its descendants, too, will pass the new, edited gene on to *their* descendants, whether or not it conveys to them any advantageous characteristics.

8. Scientists are working on a "reversal drive" and other possible measures that may be able to erase or counteract the changes wrought by a gene driver gone awry.
9. For example, these diseases can be addressed by creating a number of fertilized embryos in a cell culture outside of the mother and choosing an embryo that is not afflicted with the undesirable gene for implantation in the womb to create a pregnancy. The remaining embryos would then be discarded or used for research. This process is known as preimplantation genetic diagnosis or in vitro fertilization and embryo selection.
10. Paul Berg, "Asilomar 1975: DNA Modification Secured," *Nature 455* (September 18, 2008): 290–91.
11. Michael Rogers, "The Pandora's Box Congress," *Rolling Stone 189* (June 19, 1975): 37–40, 42, 74, 77–78, 82.
12. J. Morgan and W. J. Whelan, *Recombinant DNA and Genetic Recombination* (New York: Pergamon Press, 1979).
13. See, for example, Sheldon Krimsky, *Genetic Alchemy: The Social History of the Recombinant DNA Controversy* (Cambridge, MA: MIT Press, 1982).
14. The acceptance of the need for broad public discussion of the risks and benefits of biotechnology contrasts sharply with the standard practice in the field of classical biological control, as, for example, in the campaign against cassava mealybug described in chapter 10. The issues raised by these technologies are similar to those raised by gene drivers but are typically handled entirely by the scientists directly involved, following accepted professional standards with little or no public input. See, for example, Mark Kenis et al., "*Guide to the Classical Biological Control of Insect Pests in Planted and Natural Forests*" (Rome: Food and Agricultural Organization, 2019); and John M. Randall and Mandy Tu. "Biological Control." In *Weed Control Methods Handbook*, edited by Mandy Tu, Callie Hurd, and John M. Randall. Arlington, VA: The Nature Conservancy, 2003, chapter 4, pp. 4.1–4.24. https://www.invasive.org/gist/products/handbook/06.BiologicalControl.pdf.
15. Todd Kuiken, "Governance: Learn from DIY Biologists," *Nature News 531*, no. 7593 (March 10, 2016): 167. The draft codes are available at DIYBio. "Codes". https://diybio.org/codes.
16. For journalistic accounts of the summits, see John Travis, "Germline Editing Dominates DNA Summit," *Science 350*, no. 6266 (December 11, 2015): 1299–1300; and Victor J. Dzau, Marcia McNutt, and Chunli Bai, "Wake-Up Call from Hong Kong," *Science 362*, no. 6420 (December 14, 2018): 1215. For the summit declarations, see National Academies of Sciences, Engineering and Medicine, *International Summit on Human Gene Editing: A Global Discussion* (Washington, DC: National Academies Press, 2015); and National Academies of Sciences, Engineering, and Medicine, *Second International Summit on Human Genome Editing: Continuing the Global Discussion*

(Washington, DC: National Academies Press, 2019). These and other declarations and codes of conduct are summarized in Henry T. Greely, "How Should Science Respond to CRISPR'd Babies?" *Issues in Science and Technology* 35, no. 3 (2019): 32–37.

17. The case for such a "global observatory" is set forth in Sheila Jasanoff and J. Benjamin Hurlbut, "A Global Observatory for Gene Editing," *Nature* 555, no. 7697 (March 2018): 435–37.
18. Intergovernmental discussions that could, in principle, lead to such a treaty are currently stymied by a somewhat arcane disagreement over whether regulations designed to control the international movement of the older generation of organisms genetically engineered through recombinant DNA (called "living modified organisms," or LMOs, in diplomat-speak) should apply to organisms that have been modified by gene editors or gene drivers. The legal distinction is important, since the Cartagena Protocol of the Convention on Biological Diversity (CBD) requires its 172 parties (which do not include the United States) to establish "ways to control" living modified organisms. For the full text, see Cartagena Protocol on Biosafety to the Convention on Biological Diversity (Montreal: Secretariat of the Convention on Biological Diversity, 2000).
19. Robert Paarlberg, *Starved for Science: How Biotechnology Is Being Kept Out of Africa* (Cambridge, MA: Harvard University Press, 2008).
20. "The organizing committee calls for an ongoing international forum to foster broad public dialogue, develop strategies for increasing equitable access to meet the needs of underserved populations, speed the development of regulatory science, provide a clearinghouse for information about governance options, contribute to the development of common regulatory standards, and enhance coordination of research and clinical applications through an international registry of planned and ongoing experiments." National Academies, *Second International Summit on Human Genome Editing*, 8.
21. ETC Group and Heinrich Boll Foundation, *Forcing the Farm: How Gene Drive Organisms Could Entrench Industrial Agriculture and Threaten Food Sovereignty*, October 2018.
22. There has already been such a private initiative on a smaller scale: in 2012, an American businessman dumped iron filings into the Atlantic Ocean to try to stimulate an algal bloom that could absorb carbon dioxide. Henry Fountain, "A Rogue Climate Experiment Outrages Scientists," *The New York Times*, October 18, 2012. https://www.nytimes.com/2012/10/19/science/earth/iron-dumping-experiment-in-pacific-alarms-marine-experts.html.
23. J. G. S. Shepherd et al., *Geoengineering the Climate: Science, Governance and Uncertainty, RS Policy Document 10/09* (London: The Royal Society, 2009).
24. For a quick summary with references, see Brad Allenby "Geoengineering Redivivus," *Elementa: Science of the Anthropocene* 2:000023 (2014).. We discuss CCS technology, also known as CCUS, in more detail in chapter 4.
25. The least dangerous strategy for the implementation of geoengineering would be to use it to prevent overshoot, an approach known as peak shaving. The assumption here is that mitigation measures will eventually cause the increase in the average temperature of the Earth to settle down at the target range of 1.5–2.0°C over preindustrial

values, but that it might overshoot these values temporarily and thereby cause irreversible ecological damage. In this strategy, geoengineering technology would be used as a temporary measure to avert this overshoot.

26. Peter J. Irvine and David W. Keith, "Halving Warming with Stratospheric Aerosol Geoengineering Moderates Policy-Relevant Climate Hazards," *Environmental Research Letters 15*, no. 4 (March 2020): 044011. https://doi.org/10.1088/1748-9326/ab76de..
27. A public good is something that anyone can use, and whose use by one person does not affect its utility to others. (In the jargon, it is "non-rival and non-excludable").
28. *Asilomar International Conference on Climate Intervention Technologies: Conference Report* (Pacific Grove, CA: Climate Institute, 2010). http://climate.org/archive/resources/climate-archives/conferences/asilomar/report.html.
29. National Academy of Sciences, Reflecting Sunlight: Recommendations for Solar Geoengineering Research and Research Governance (Washington: National Academies Press, 2021).

Chapter 13

1. Hans M. Kristensen et al., "Alert Status of Nuclear Weapons," *AIP Conference Proceedings* 1898, no. 1 (November 15, 2017): 020003.
2. These ideas are further developed in Charles Weiss, "How Do Science and Technology Influence International Affairs?," *Minerva 53,* no. 4 December 2015): 411–30.
3. Herman Kahn, *On Thermonuclear War* (New York: Routledge, 2017), 44. Originally published by Princeton University Press in 1960.
4. Stanford University Program in Science, Technology and Society, What is the Study of STS? https://sts.stanford.edu/about/what-study-sts
5. Full disclosure: As I mentioned in the preface to this book, STIA is the name of the program I used to direct at the Georgetown University School of Foreign Service. You can call the subject "Science, Technology, and World Affairs" if you prefer—but STWA is harder to pronounce.
6. Finnemore, International Norm Dynamics.
7. The International Covenant on Economic, Social and Cultural Rights (Article 15 (1)(b) provides that "The States Parties to the present Covenant recognize the right of everyone . . . [t]o enjoy the benefits of scientific progress and its applications." https://www.ohchr.org/EN/ProfessionalInterest/Pages/CESCR.aspx. The Covenant is binding on the 171 countries that are parties. The United States is not a party to the Covenant because many of its provisions are difficult to implement under the U.S. federal system.
8. John Rawls, *A Theory of Justice* (Cambridge, MA: Harvard University Press, 1971).
9. Edith Brown Weiss, *Establishing Norms in a Kaleidoscopic World* (The Hague: Brill/Nijhoff and the Hague Academy of International Law, 2020), chapter 12.

Bibliography

Chapter 1. Introduction: Science, Technology, and Survival

Ruttan, Vernon W. *Is War Necessary for Economic Growth? Military Procurement and Technology Development*. New York: Oxford University Press, 2006.

Chapter 2. Stratospheric Ozone: A Success with Complications

Giunta, Carmen J. "Thomas Midgley, Jr., and the Invention of Chlorofluorocarbon Refrigerants: It Ain't Necessarily So." *Bulletin for the History of Chemistry* 31, no. 2 (2006): 69.

Holm, Søren, and John Harris. "Precautionary Principle Stifles Discovery." *Nature* 400, no. 6743 (July 1999): 398. doi.org/10.1038/22626.

Lickley, Megan, Susan Solomon, Sarah Fletcher, Guus J. M. Velders, John Daniel, Matthew Rigby, Stephen A. Montzka, Lambert J. M. Kuijpers, and Kane Stone. "Quantifying Contributions of Chlorofluorocarbon Banks to Emissions and Impacts on the Ozone Layer and Climate." *Nature Communications* 11, no. 1 (March 17, 2020): 1380. doi.org–10.1038/s41467-020-15162-7.

Oreskes, Naomi, and Erik M. Conway. *Merchants of Doubt*. New York: Bloomsbury Press, 2010.

Rigby, M., S. Park, T. Saito, L. M. Western, A. L. Redington, X. Fang, S. Henne, et al. "Increase in CFC-11 Emissions from Eastern China Based on Atmospheric Observations." *Nature* 569, no. 7757 (May 2019): 546–50. doi.org/10.1038/s41586-019-1193-4.

Tickner, Joel, ed. *Precaution, Environmental Science and Preventive Public Policy*. Washington, DC: Island Press, 2003.

Watson, R. T., M. J. Prather, and M. J. Kurylo. *Present State of Knowledge of the Upper Atmosphere 1988: An Assessment Report*. Washington, DC: National Aeronautics and Space Administration, June 1, 1988. https://ntrs.nasa.gov/search.jsp?R=19880019849.

Wingspread Statement on the Precautionary Principle (1998), https://www.iatp.org/sites/default/files/Wingspread_Statement_on_the_Precautionary_Prin.htm).

World Meteorological Organization. *Executive Summary: Scientific Assessment of Ozone Depletion: 2018*. Geneva: Global Ozone Research and Monitoring Project, 2018. https://www.esrl.noaa.gov/csd/assessments/ozone/2018/executivesummary/.

Chapter 3. Climate Disruption, Not Just Global Warming

Bartlett, Nicolette, Hannah Cushing, and Sara Law. *Putting a Price on Carbon: Integrating Climate Risk into Business Planning*. CDP. October 2017. https://b8f65cb373b1b7b15feb-c70d8ead6ced550b4d987d7c03fcdd1d.ssl.cf3.rackcdn.com/cms/reports/documents/000/002/738/original/Putting-a-price-on-carbon-CDP-Report-2017.pdf?1507739326.

Bonvillian, William B., and Charles Weiss. *Technological Innovation in Legacy Sectors*. New York: Oxford University Press, 2015.

Bralower, Timothy, and David Bice. "Ancient Climate Events: Paleocene Eocene Thermal Maximum." Pennsylvania State University. Accessed April 4, 2020. https://www.e-education.psu.edu/earth103/node/639.

Brecht, Henrike, Susmita Dasgupta, Benoit Laplante, Siobhan Murray, and David Wheeler. "Sea-Level Rise and Storm Surges: High Stakes for a Small Number of Developing Countries." *The Journal of Environment & Development* 21, no. 1 (March 1, 2012): 120–38. doi.org/10.1177/1070496511433601.

Brown Weiss, Edith. *In Fairness to Future Generations: International Law, Common Patrimony, and Intergenerational Equity*. Tokyo: United Nations University, 1989.

Brundtland, Gro Harlem. *Report of the World Commission on Environment and Development: Our Common Future*, UN General Assembly A/42/427. Geneva: United Nations, 1987. http://www.un-documents.net/our-common-future.pdf.

Byrd, John W., Elizabeth S. Cooperman, and Kent Hickman. "Capital Budgeting and Climate Change: Does Corporate Internal Carbon Pricing Reduce CO_2 Emissions?" SSRN Scholarly Paper. Rochester, NY: Social Science Research Network, April 14, 2020. doi.org/10.2139/ssrn.3575769.

Canty & Associates. "Compare Cities." Weatherbase. 2020. http://www.weatherbase.com/compare.php3?first=593809&second=523281.

Carbon Pricing Leadership Council. *Report of the High-Level Commission on Carbon Prices*. Washington, DC: World Bank, May 2017. https://static1.squarespace.com/static/54ff9c5ce4b0a53decccfb4c/t/59b7f2409f8dce5316811916/1505227332748/CarbonPricing_FullReport.pdf.

Chandler, Mark A. "The Climate of the Pliocene: Simulating Earth's Last Great Warm Period." NASA Goddard Institute for Space Studies. April 1997. https://www.giss.nasa.gov/research/features/199704_pliocene/.

Coll, Steve. *Private Empire: ExxonMobil and American Power*. New York: Penguin Books, 2013.

Darby, Megan, and Isabelle Gerretsen. "Which Countries Have a Net Zero Climate Goal?" *Climate Home News*, June 14, 2019. https://www.climatechangenews.com/2019/06/14/countries-net-zero-climate-goal/.

European Commission. "Kyoto 1st Commitment Period (2008–12)." EU Climate Action. November 23, 2016. https://ec.europa.eu/clima/policies/strategies/progress/kyoto_1_en.

European Parliament and Council of Europe. *Regulation (EU) 2019/2088 on Sustainability-related Disclosures in the Financial Services Sector*. PE/87/2019/REV/1. EUR-Lex, November 27, 2019. https://eur-lex.europa.eu/eli/reg/2019/2088/oj.

Fawcett, Allen A., Gokul C. Iyer, Leon E. Clarke, James A. Edmonds, Nathan E. Hultman, Haewon C. McJeon, Joeri Rogelj, et al. "Can Paris Pledges Avert Severe Climate Change?" *Science* 350, no. 6265 (November 26, 2015): 1168–69. doi.org/10.1126/science.aad5761.

Franek, Alice, Engin Koncagul, Richard Connor, and Diwata Hunziker. *Water for a Sustainable World: Facts and Figures*. Perugia, IT: United Nations World Water Assessment Programme, 2015.

Gillis, Justin. "Climate Panel Cites Near Certainty on Warming." *The New York Times*, August 19, 2013. https://www.nytimes.com/2013/08/20/science/earth/extremely-likely-that-human-activity-is-driving-climate-change-panel-finds.html.

Golledge, Nicholas R., Elizabeth D. Keller, Natalya Gomez, Kaitlin A. Naughten, Jorge Bernales, Luke D. Trusel, and Tamsin L. Edwards. "Global Environmental Consequences of Twenty-First-Century Ice-Sheet Melt." *Nature* 566, no. 7742 (February 2019): 65–72. doi.org/10.1038/s41586-019-0889-9.

Hausfather, Zeke. "Explainer: How 'Shared Socioeconomic Pathways' Explore Future Climate Change." CarbonBrief. April 19, 2018. https://www.carbonbrief.org/explainer-how-shared-socioeconomic-pathways-explore-future-climate-change.

Henson, Robert. *Warming World, Impacts by Degree.* Washington, DC: National Academies of Science, 2011. Summary of National Academy of Sciences, *Climate Stabilization Targets: Emissions, Concentrations, and Impacts over Decades to Millennia* (Washington: National Academies Press, 2011).

Horton, Benjamin P., Robert E. Kopp, Andra J. Garner, Carling C. Hay, Nicole S. Khan, Keven Roy, and Timothy A. Shaw. "Mapping Sea-Level Change in Time, Space, and Probability." *Annual Review of Environment and Resources* 43, no. 1 (October 17, 2018): 481–521. doi.org/10.1146/annurev-environ-102017-025826.

IPBES. *The Global Assessment Report on Biodiversity and Ecosystem Services: Summary for Policymakers.* Bonn: Intergovernmental Science-Policy Platform on Biodiversity and Ecosystem Services Secretariat, 2019. https://ipbes.net/sites/default/files/inline/files/ipbes_global_assessment_report_summary_for_policymakers.pdf.

IPCC. *Climate Change 2013: The Physical Science Basis. Working Group I Contribution to the Fifth Assessment Report of the Intergovernmental Panel on Climate Change.* Edited by T. F. Stocker, D. Qin, G.-K. Plattner, M. Tignor, S. K. Allen, J. Boschung, A. Nauels, Y. Xia, V. Bex, and P. M. Midgley. Cambridge: Cambridge University Press, 2013.

IPCC. *Climate Change 2014: Synthesis Report. Contribution of Working Groups I, II and III to the Fifth Assessment Report of the Intergovernmental Panel on Climate Change.* Edited by Core Writing Team, R. K. Pachauri, and L. A. Meyer (Cambridge: Cambridge University Press, 2014).

IPCC. *Global Warming of 1.5°C: An IPCC Special Report.* Edited by V. Masson-Delmotte, P. Zhai, H.-O. Pörtner, D. Roberts, J. Skea, P. R. Shukla, A. Pirani, et al. Geneva: World Meteorological Organization, 2018.

IPCC. "Overview." In *Climate Change: The 1990 and 1992 IPCC Assessments.* Geneva: World Meteorological Organization, 1992. https://www.ipcc.ch/site/assets/uploads/2018/05/ipcc_90_92_assessments_far_overview.pdf.

IPCC. *Special Report on the Ocean and Cryosphere in a Changing Climate.* Edited by H. O. Pörtner, D. C. Roberts, V. Masson-Delmotte, P. Zhai, E. Poloczanska, K. Mintenbeck, M. Tignor, et al. Geneva: World Meteorological Organization, 2019.

IPCC. "Summary for Policymakers." In *Climate Change and Land: IPCC Special Report on Climate Change, Desertification, Land Degradation, Sustainable Land Management, Food Security, and Greenhouse Gas Fluxes in Terrestrial* Ecosystems. Edited by P. R. Shukla, J. Skea, E. Calvo Buendia, V. Masson-Delmotte, H.-O. Pörtner, D. C. Roberts, P. Zhai, et al. Geneva: World Meteorological Organization, 2020.

IPCC. *Understanding Climate Change: 22 Years of IPCC Assessment.* Geneva: World Meteorological Organization, 2010. https://www.ipcc.ch/site/assets/uploads/2018/04/ipcc-brochure_understanding.pdf.

Jones, Matthew W., Adam Smith, Richard Betts, Josep G. Canadell, I. Colin Prentice, and Corinne Le Quéré. *Climate Change Increases the Risk of Wildfires.* Norwich, UK: ScienceBrief, January 14, 2020. https://sciencebrief.org/briefs/wildfires.

Kahn, Matthew E., Kamiar Mohaddes, Ryan N. C. Ng, M. Hashem Pesaran, Mehdi Raissi, and Jui-Chung Yang. "Long-Term Macroeconomic Effects of Climate Change:

A Cross-Country Analysis." Working Paper. Washington, DC: National Bureau of Economic Research, August 2019. doi.org/10.3386/w26167.

Kuhn, Thomas S. *The Structure of Scientific Revolutions*. Chicago: University of Chicago Press, 1962.

Kulp, Scott A. and Benjamin H. Strauss, "New Elevation Data Triple Estimates of Global Vulnerability to Sea-Level Rise and Coastal Flooding." *Nature Communications* 10, no. 1 (October 29, 2019): 4844.

Levin, Kelly. "8 Things You Need to Know About the IPCC 1.5°C Report." World Resources Institute. October 7, 2018. https://www.wri.org/blog/2018/10/8-things-you-need-know-about-ipcc-15-c-report.

Lynas, Mark. *Six Degrees: Our Future on a Hotter Planet*. London: Fourth Estate, 2007.

Malakoff, David. "Let's Call It 'Climate Disruption,' White House Science Adviser Suggests (Again)." *Science*, May 2, 2014. https://www.sciencemag.org/news/2014/05/lets-call-it-climate-disruption-white-house-science-adviser-suggests-again.

Mastrandrea, Michael D., Katharine J. Mach, Gian-Kasper Plattner, Ottmar Edenhofer, Thomas F. Stocker, Christopher B. Field, Kristie L. Ebi, and Patrick R. Matschoss. "Guidance Note for Lead Authors of the IPCC Fifth Assessment Report on Consistent Treatment of Uncertainties." Geneva: IPCC, 2010. https://www.ipcc.ch/site/assets/uploads/2017/08/AR5_Uncertainty_Guidance_Note.pdf.

McSweeney, Robert. "The Impacts of Climate Change at 1.5C, 2C and Beyond." CarbonBrief. October 4, 2018. https://interactive.carbonbrief.org/impacts-climate-change-one-point-five-degrees-two-degrees/.

National Academy of Sciences and Royal Society. *Climate Change: Evidence and Causes: Update 2020*. Washington, DC: National Academies Press, 2020. doi.org/10.17226/25733.

Neumann, Barbara, Athanasios T. Vafeidis, Juliane Zimmermann, and Robert J. Nicholls. "Future Coastal Population Growth and Exposure to Sea-Level Rise and Coastal Flooding - A Global Assessment." *PLOS ONE* 10, no. 3 (March 11, 2015): e0118571. doi.org/10.1371/journal.pone.0118571.

Nordhaus, William D. "Can We Control Carbon Dioxide?" Laxenburg, Austria: International Institute for Applied Systems Analysis, June 1975.

Oreskes, Naomi, and Erik M. Conway. *Merchants of Doubt*. New York: Bloomsbury Press, 2010.

Otto-Bliesner, Bette L., Nan Rosenbloom, Emma J. Stone, Nicholas P. McKay, Daniel J. Lunt, Esther C. Brady, and Jonathan T. Overpeck. "How Warm Was the Last Interglacial? New Model–Data Comparisons." *Philosophical Transactions of the Royal Society A: Mathematical, Physical and Engineering Sciences* 371, no. 2001 (October 28, 2013). doi.org/10.1098/rsta.2013.0097.

Pacala, S., and R. Socolow. "Stabilization Wedges: Solving the Climate Problem for the Next 50 Years with Current Technologies." *Science* 305, no. 5686 (August 13, 2004): 968–72. doi.org/10.1126/science.1100103.

Palmer, Brian. "How Much of an Increase in Temperature Can We Tolerate?" *Pacific Standard*, April 24, 2015. https://psmag.com/environment/the-climate-change-debate-is-about-rising-temperatures.

Potsdam Institute for Climate Impact Research and Climate Analytics. *Turn Down the Heat: Why a 4C Warmer World Must Be Avoided*. Washington, DC: World Bank, 2012.

Primack, Joel, and Frank von Hippel. *Advice and Dissent: Scientists in the Political Arena*. New York: Basic Books, 1974.

Ramstein, Céline, Goran Dominioni, and Sanaz Ettehad. *State and Trends of Carbon Pricing: 2019*. Washington, DC: World Bank Group, 2019. https://openknowledge.worldbank.org/handle/10986/31755.

Roberts, David. "This Graphic Explains Why 2 Degrees of Global Warming Will Be Way Worse than 1.5." *Vox*, January 19, 2018. https://www.vox.com/energy-and-environment/2018/1/19/16908402/global-warming-2-degrees-climate-change.

Sherwood, S., M. J. Webb, J. D. Annan, K. C. Armour, P. M. Forster, J. C. Hargreaves, and G. Hegerl, et al. "An Assessment of Earth's Climate Sensitivity Using Multiple Lines of Evidence." *Reviews of Geophysics* (July 2020): e2019RG000678. doi.org/10.1029/2019RG000678.

Sigmond, Michael John C. Fyfe, and Neil C. Swart. "Ice-Free Arctic Projections Under the Paris Agreement." *Nature Climate Change* 8, no. 5 (May 2018): 404–8. doi.org/10.1038/s41558-018-0124-y.

Socolow, Robert. "Climate Change and Destiny Studies: Creating Our Near and Far Futures." *Bulletin of the Atomic Scientists* 71, no. 6 (November 1, 2015): 18–28.

Socolow, Robert. "Contending with Climate Change: The Next 25 Years." *Bulletin of the Atomic Scientists* 76, no. 6 (December 2020): 294–301. https://thebulletin.org/premium/2020-12/contending-with-climate-change-the-next-25-years/.

Solomon, Susan, Gian-Kasper Plattner, Reto Knutti, and Pierre Friedlingstein. "Irreversible Climate Change Due to Carbon Dioxide Emissions." *Proceedings of the National Academy of Sciences* 106, no. 6 (February 10, 2009): 1704–9. doi.org/10.1073/pnas.0812721106.

Steffen, Will, Johan Rockström, Katherine Richardson, Timothy M. Lenton, Carl Folke, Diana Liverman, Colin P. Summerhayes, et al. "Trajectories of the Earth System in the Anthropocene." *Proceedings of the National Academy of Sciences* 115, no. 33 (August 14, 2018): 8252–59. doi.org/10.1073/pnas.1810141115.

Turner, James Morton, and Andrew C. Isenberg. *The Republican Reversal: Conservatives and the Environment from Nixon to Trump*. Cambridge, MA: Harvard University Press, 2018.

UNFCCC Conference of the Parties. "Adoption of the Paris Agreement." UN Doc. FCCC/CP/2015/L.9/Rev/1, December 12, 2015. https://unfccc.int/resource/docs/2015/cop21/eng/l09r01.pdf.

Union of Concerned Scientists. "Debunking Misinformation About Stolen Climate Emails." August 25, 2011. https://www.ucsusa.org/resources/debunking-misinformation-about-stolen-climate-emails.*United Nations Framework Convention on Climate Change*. UN Doc. FCCC/INFORMAL/84, 1992. https://unfccc.int/resource/docs/convkp/conveng.pdf.

United Nations Development Program, *World Energy Assessment: Energy and the Challenge of Sustainability* (New York:, 2000).

Van Vuuren, Detlef P. Keywan Riahi, Katherine Calvin, Rob Dellink, Johannes Emmerling, Shinichiro Fujimori, Samir Kc, Elmar Kriegler, and Brian O'Neill. "The Shared Socio-Economic Pathways: Trajectories for Human Development and Global Environmental Change." *Global Environmental Change* 42 (January 1, 2017): 148–52. doi.org/10.1016/j.gloenvcha.2016.10.009.

Weart, Spencer. *The Discovery of Global Warming - A History*. American Institute of Physics, 2009. https://history.aip.org/climate/index.htm.

Weiss, Charles. "Can There Be Science-Based Precaution?" *Environmental Research Letters* 1, no. 1 (October 2006): 014003. doi.org–10.1088/1748-9326/1/1/014003.

World Bank. "Carbon Pricing Dashboard." April 1, 2020. https://carbonpricingdashboard.worldbank.org/about.

Chapter 4. How Do We Address Climate Disruption?

Brown Weiss, Edith. *In Fairness to Future Generations: International Law, Common Patrimony, and Intergenerational Equity*. Tokyo: United Nations University, 1989.

Dagnachew, Anteneh G., Andries F. Hof, Paul L. Lucas, and Detlef P. van Vuuren. *Insight into Energy Scenarios: A Comparison of Key Transition Indicators of 2 °C Scenarios*. The Hague: PBL Netherlands Environmental Assessment Agency, September 2019. https://www.pbl.nl/sites/default/files/downloads/pbl-2019-insight-into-energy-scenarios_3686.pdf.

EIA. "Decommissioning Nuclear Reactors Is a Long-Term and Costly Process." Energy Information Agency. November 17, 2017. http://www.eia.gov/todayinenergy/detail.php?id=33792.

Global CCS Institute. *The Global Status of CCS 2018*. Melbourne: Global CCS Institute, 2018. https://indd.adobe.com/view/2dab1be7-edd0-447d-b020-06242ea2cf3b.

Goldschmidt, Pierre. "Strengthened Safeguards," *IAEA Bulletin 34*, no. 4 (2001). https://www.iaea.org/publications/magazines/bulletin/43-4/strengthened-safeguards

Golston, Levi, Greg Davies, Ryan Edwards, Mark Miller, Mostafa Momen, Tara Nealon, Elie Bou-Zeid, et al. *Wind Power: An Energy Technology Distillate*. Princeton, NJ: Andlinger Center for Energy and the Environment, March 2019. https://acee.princeton.edu/distillates/windpower/.

Helveston, John, and Jonas Nahm. "China's Key Role in Scaling Low-Carbon Energy Technologies." *Science* 366, no. 6467 (November 15, 2019): 794–96. doi.org/10.1126/science.aaz1014.

Idaho National Laboratory. *The Future of Geothermal Energy: Impact of Enhanced Geothermal Systems (EGS) on the United States in the 21st Century*. Idaho Falls, ID: MIT, 2006. http://energy.mit.edu/wp-content/uploads/2006/11/MITEI-The-Future-of-Geothermal-Energy.pdf.

IEA. *Energy Technology Perspectives 2016: Towards Sustainable Urban Energy Systems*. Paris: International Energy Agency, 2016. https://webstore.iea.org/download/summary/1057.

IEA. *Energy Technology Perspectives 2017: Catalyzing Energy Technology Transformation*. Paris: International Energy Agency, 2017. https://webstore.iea.org/energy-technology-perspectives-2017.

IEA. *World Energy Outlook 2019*. Paris: International Energy Agency, 2019. https://webstore.iea.org/world-energy-outlook-2019.

IPCC. *Climate Change 2014: Impacts, Adaptation, and Vulnerability - Working Group II Contribution to the Fifth Assessment Report of the Intergovernmental Panel on Climate Change*. Edited by C. B. Field, V. R. Barros, D. J. Dokken, K. J. Mach, M. D. Mastrandrea, T. E. Bilir, M. Chatterjee, K. L. Ebi, Y. O. Estrada, R. C. Genova, B. Girma, E. S. Kissel, A. N. Levy, S. MacCracken, P. R. Mastrandrea, and L. L.White. Cambridge: Cambridge University Press, 2014.

IPCC. *Climate Change 2014: Mitigation of Climate Change - Working Group III Contribution to the Fifth Assessment Report of the Intergovernmental Panel on Climate Change*. Edited by O. Edenhofer, R. Pichs-Madruga, Y. Sokona, E. Farahani, S. Kadner, K. Seyboth, A. Adler, et al. Cambridge: Cambridge University Press, 2014.

IPCC. *Climate Change and Land: IPCC Special Report on Climate Change, Desertification, Land Degradation, Sustainable Land Management, Food Security, and Greenhouse Gas Fluxes in Terrestrial Ecosystems - Summary for Policymakers*. Edited by P.R. Shukla, J. Skea, E. Calvo Buendia, V. Masson-Delmotte, H.-O. Pörtner, D. C. Roberts, P. Zhai, et al, eds., (Geneva: Intergovernmental Panel on Climate Change, 2020).

IRENA. *Future of Solar Photovoltaic: Deployment, Investment, Technology, Grid Integration and Socio-Economic Aspects*. Abu Dhabi: International Renewable Energy Agency, November 2019. https://www.irena.org/publications/2019/Nov/Future-of-Solar-Photovoltaic.

IRENA. *Future of Wind: Deployment, Investment, Technology, Grid Integration and Socio-Economic Aspects*. Abu Dhabi: International Renewable Energy Agency, October 2019. file:///Users/weissc/Downloads/IRENA_Future_of_wind_2019.pdf.

IRENA. *Global Energy Transformation: The REmap Transition Pathway*. Abu Dhabi: International Renewable Energy Agency, 2019. https://www.irena.org/-/media/Files/IRENA/Agency/Publication/2019/Apr/IRENA_GET_REmap_pathway_2019.pdf.

Irfan, Umair. "The Best Case for and Against a Fracking Ban." *Vox*, October 7, 2020. https://www.vox.com/energy-and-environment/2019/9/12/20857196/democratic-debate-climate-change-fracking-ban.

ISO. "ISO 50001: Energy Management Systems—Requirements with Guidance for Use." International Standards Organization. 2011.

MIT Energy Initiative. *The Future of Nuclear Energy in a Carbon-Constrained World: An Interdisciplinary MIT Study*. Cambridge, MA: MIT, 2018. https://energy.mit.edu/wp-content/uploads/2018/09/The-Future-of-Nuclear-Energy-in-a-Carbon-Constrained-World.pdf.

Morgan, M. Granger, Ahmed Abdulla, Michael J. Ford, and Michael Rath. "US Nuclear Power: The Vanishing Low-Carbon Wedge." *Proceedings of the National Academy of Sciences* 115, no. 28 (July 10, 2018): 7184–89. doi.org/10.1073/pnas.1804655115.

National Academies of Sciences, Engineering, and Medicine. *Reflecting Sunlight: Recommendations for Solar Geoengineering Research and Research Governance*. Washington, DC: National Academies Press (2021).

National Research Council. *Climate Intervention: Carbon Dioxide Removal and Reliable Sequestration*. Washington, DC: National Academies Press, 2015. doi.org/10.17226/18805.

Page, Brad, Guloren Turan, and Alex Zapantis. Global Status of CCS 2019: *Targeting Climate Change*. Melbourne: Global Carbon Capture and Storage Institute, 2019.

Pickl, Matthias J. "The Renewable Energy Strategies of Oil Majors—From Oil to Energy?" *Energy Strategy Reviews* 26 (November 1, 2019): 100370. doi.org/10.1016/j.esr.2019.100370.

Podesta, John, and Todd Stern. "A Foreign Policy for the Climate: How American Leadership Can Avert Catastrophe." *Foreign Affairs* 99, no. 3 (May/June 2020): 39–46.

Rockström, Johan, Owen Gaffney, Joeri Rogelj, Malte Meinshausen, Nebojsa Nakicenovic, and Hans Joachim Schellnhuber. "A Roadmap for Rapid Decarbonization." *Science* 355, no. 6331 (March 24, 2017): 1269–71. doi.org/10.1126/science.aah3443.

Singer, Peter L. and William B. Bonvillian, "Innovation Orchards - Helping Start-ups Scale," (Washington: Innovation Technology and Innovation Foundation, March 2017), pp. 7–14, http://www2.itif.org/2017-innovation-orchards.pdf?_ga=2.243967621.1083592602.1619646987-70827581.1619551935.

Sivaram, Varun. *Taming the Sun: Innovations to Harness Solar Energy and Power the Planet.* Cambridge, MA: MIT Press, 2018.

Sivaram, Varun. "Unlocking Clean Energy." *Issues in Science and Technology* 33, no. 2 (2017): 31–40.

Smil, Vaclav. *Power Density: A Key to Understanding Energy Sources and Uses.* Cambridge, MA: MIT Press, 2016.

Socolow, Robert, Craig Arnold, Greg Davies, Thomas Kreutz, Warren Powell, Michael Schwartz, and Daniel Steingart. *Grid-Scale Electricity Storage - Implications for Renewable Energy: An Energy Technology Distillate.* Princeton, NJ: Andlinger Center for Energy and the Environment, June 2014. https://acee.princeton.edu/wp-content/uploads/2014/06/Andlinger-Energy-Tech-Distillate-Storage.pdf.

Tester, Jefferson W., Elisabeth M. Drake, Michael J. Driscoll, Michael W. Golay, and William A. Peters. *Sustainable Energy: Choosing Among Options.* Cambridge, MA: The MIT Press, 2005.

United Nations Scientific Committee on the Effects of Atomic Radiation. *Sources, Effects and Risks of Ionizing Radiation: Report to the General Assembly with Scientific Annexes 2013.* New York: United Nations, 2014. https://www.unscear.org/docs/publications/2013/UNSCEAR_2013_Report_Vol.I.pdf.

Voïta, Thibaud. *The Power of China's Energy Efficiency Policies.* Paris: Institut Français des Relations Internationales, September 2018. https://www.ifri.org/sites/default/files/atoms/files/voita_power_china_2018.pdf.

von Hippel, Frank N., and Michael Schoeppner. "Reducing the Danger from Fires in Spent Fuel Pools." *Science & Global Security* 24, no. 3 (2016): 141–73.

Wee, Sui-Lee. "Thousands Being Moved from China's Three Gorges - Again." *Reuters,* August 22, 2012. https://www.reuters.com/article/us-china-threegorges-idUSBRE87L0ZW20120822.

Weiss, Charles, and William Bonvillian. *Structuring an Energy Technology Revolution.* Cambridge MA: MIT Press, 2014.

Zhang, Luoping, Iemaan Rana, Rachel M. Shaffer, Emanuela Taioli, and Lianne Sheppard. "Exposure to Glyphosate-Based Herbicides and Risk for Non-Hodgkin Lymphoma: A Meta-Analysis and Supporting Evidence." *Mutation Research* 781 (September 2019): 186–206. doi.org/10.1016/j.mrrev.2019.02.001.

Chapter 5. Nuclear Issues: Civilization in the Balance

Agreed Framework of 21 October 1994 Between the United States of America and the Democratic People's Republic of Korea. International Atomic Energy Agency Information Circular 457, November 2, 1994. https://web.archive.org/web/20031217175315/http://www.iaea.org/Publications/Documents/Infcircs/Others/infcirc457.pdf.

Allison, Graham. "What Happened to the Soviet Superpower's Nuclear Arsenal? Clues for the Nuclear Security Summit." Cambridge, MA: Belfer Center for Science and International Affairs, March 2012. https://www.belfercenter.org/sites/default/files/legacy/files/3%2014%2012%20Final%20What%20Happened%20to%20Soviet%20Arsenals.pdf.

Åslund, Anders. "How Small Is the Soviet National Income?" In *The Impoverished Superpower: Perestroika and the Soviet Military Burden.* Edited by Henry S. Rowen and Charles Wolf Jr. San Francisco: Institute for Contemporary Studies, 1990), pp. 13–61, 288–305.

Bendett, Samuel. "The Rise of Russia's Hi-Tech Military." *Fletcher Security Review* 6, no. 1 (2019). https://sites.tufts.edu/fletcherrussia/files/2019/06/The-Rise-of-Russia%E2%80%99s-Hi-Tech-Military-Samuel-Bendett.pdf.

Bhalla, Nita. "Nations Dawdle on Agreeing Rules to Control 'killer Robots' in Future Wars." *Reuters*, January 17, 2020. https://www.reuters.com/article/us-global-rights-killer-robots-idUSKBN1ZG151.

Blight, James G., Bruce J. Alyn, and David A. Welch. *Cuba on the Brink: Castro, the Missile Crisis, and the Soviet Collapse*. New York: Rowman & Littlefield Publishers, 2002.

Bonvillian, William B., and Charles Weiss. *Technological Innovation in Legacy Sectors*. New York: Oxford University Press, 2015.

Brodie, Bernard, ed. *The Absolute Weapon: Atomic Power and World Order*. Freeport, NY: Harcourt Brace Jovanovich, 1946.

Bunn, Matthew, Anthony Wier, and John P. Holdren. *Controlling Nuclear Warheads and Materials: A Report Card and Action Plan*. Washington, DC: Nuclear Threat Initiative, March 2003.

Bush, Vannevar. *Science: The Endless Frontier*. Washington, DC: Government Printing Office, 1945. https://www.nsf.gov/od/lpa/nsf50/vbush1945.htm#ch3.8.

Christoff, Joseph A. "Weapons of Mass Destruction: Additional Russian Cooperation Needed to Facilitate U.S. Efforts to Improve Security at Russian Sites." Washington, DC: Government Accountability Office, March 2003. https://fas.org/programs/bio/resource/documents/GAO-03482.pdf.

Cirincione, Joseph. *Bomb Scare: The History and Future of Nuclear Weapons*. New York: Columbia University Press, 2007.

Director of National Intelligence. "Annual Report to Congress on the Safety and Security of Russian Nuclear Facilities and Military Forces." Washington, DC: National Intelligence Council, 2011. https://www.dni.gov/files/documents/FOIA/DF-2015-00065.pdf.

"Endorsers." "The Campaign to Stop Killer Robots." 2019. https://www.stopkillerrobots.org/endorsers/.

Feickert, Andrew, Jennifer K. Elsea, Laurence Kapp, and Laurie A. Harris. *U.S. Ground Forces Robotics and Autonomous Systems (RAS) and Artificial Intelligence: Considerations for Congress*. R45392. Washington, DC: Congressional Research Service, November 20, 2018.

Franck, James, D. Hughes, L. Szilard, T. Hogness, E. Rabinowitch, G. Seaborg, and C. J Nickson. *The Franck Report: A Report to the Secretary of War, June 1945*. Federation of American Scientists. June 1945. https://fas.org/sgp/eprint/franck.html.

Goldschmidt, Pierre. "Strengthened Safeguards." *IAEA Bulletin* 34, no. 4 (2001). https://www.iaea.org/publications/magazines/bulletin/43-4/strengthened-safeguards.

Harrington, Anne I., Eliza Gheorghe, and Anya Loukianova Fink. "What Arguments Motivate Citizens to Demand Nuclear Disarmament?" *Bulletin of the Atomic Scientists* 73, no. 4 (July 4, 2017): 255–63. doi.org/10.1080/00963402.2017.1338039.

"A Hero at Home, a Villain Abroad." *The Economist*, June 19, 2008. https://www.economist.com/international/2008/06/19/a- hero-at-home-a-villain-abroad.

Joint Comprehensive Plan of Action. http://apps.washingtonpost.com/g/documents/world/full-text-of-the-iran-nuclear-deal/1651.

Jungk, Robert. *Brighter than a Thousand Suns: A Personal History of the Atomic Scientists*. Boston: Houghton Mifflin Harcourt, 1958.

Kahn, Herman. *On Thermonuclear War*. New Brunswick, NJ: Transaction Publishers, 2007.

Kaplan, Edward. *To Kill Nations: American Strategy in the Air-Atomic Age and the Rise of Mutually Assured Destruction*. Ithaca, NY: Cornell University Press, 2015.

Kaplan, Fred. *The Bomb: Presidents, Generals, and the Secret History of Nuclear War*. New York: Simon & Schuster, 2020.

Klare, Michael T. "Autonomous Weapons Systems and the Laws of War." *Arms Control Today*, March 2019. https://www.armscontrol.org/act/2019-03/features/autonomous-weapons-systems-laws-war.

Kobie, Nicole. "To Cripple AI, Hackers Are Turning Data Against Itself." *Wired*, September 11, 2018. https://www.wired.co.uk/article/artificial-intelligence-hacking-machine-learning-adversarial.

Krasner, Stephen. *International Regimes*. Ithaca, NY: Cornell University Press, 1983.

Lapp, Ralph E. "The Einstein Letter That Started It All." *The New York Times*, August 2, 1964. http://www.nytimes.com/1964/08/02/the-einstein-letter-that-started-it-all.html.

Lawrence Livermore National Security. "Sleuthing Seismic Signals." *Science & Technology Review*, 4–12 (March 2009). https://str.llnl.gov/content/pages/past-issues-pdfs/2009.03.pdf.

Levi, Michael. *On Nuclear Terrorism*. Cambridge, MA: Harvard University Press.

Mello, Greg, Andrew Lichterman, and William Weida. "The Stockpile Stewardship Charade." *Issues in Science and Technology* 15, no. 3 (Spring 1999), 78–85. https://issues.org/mello/

Monk, Ray. *Robert Oppenheimer: A Life Inside the Center*. New York: Anchor Books, 2014.

Mowatt-Larssen, Rolf. "Islam and the Bomb: Religious Justification for and Against Nuclear Weapons." Cambridge, MA: Belfer Center for Science and International Affairs, 2011. https://www.belfercenter.org/sites/default/files/legacy/files/uploads/Islam_and_the_Bomb-Final.pdf

Nacos, Brigitte Lebens. *Terrorism and Counterterrorism*, 4th edition. Boston: Pearson Longman, 2012.

National Security Archive, "The 'Launch on Warning' Nuclear Strategy and Its Insider Critics," (Washington: George Washington University, June 11, 2019), https://nsarchive.gwu.edu/briefing-book/nuclear-vault/2019-06-11/launch-warning-nuclear-strategy-its-insider-critics.

New START Treaty. Department of State. https://www.state.gov/new-start/.

Norris, Robert S., and Hans M. Kristensen. "The Cuban Missile Crisis: A Nuclear Order of Battle, October and November 1962." *Bulletin of the Atomic Scientists* 68, no. 6 (November 1, 2012): 85–91. doi.org/10.1177/0096340212464364.

Nuclear Threat Initiative. "India: Nuclear Weapons Program." November 2019. http://www.nti.org/learn/countries/india/nuclear.

Oppenheimer, J. Robert. "Physics in the Contemporary World." *Bulletin of the Atomic Scientists* 4, no. 3 (1948): 65–86.

Paul, T. V., Richard J. Harknett, and James J. Wirtz, eds. *The Absolute Weapon Revisited: Nuclear Arms and the Emerging International Order*. Ann Arbor: University of Michigan Press, 1998.

Pilkington, Ed, and Martin Pengally. "'Let It Be an Arms Race': Donald Trump Appears to Double Down on Nuclear Expansion." *The Guardian*, December 24, 2016. http://www.theguardian.com/us-news/2016/dec/23/donald-trump-nuclear-weapons-arms-race.

Pool, Robert. *Beyond Engineering: How Society Shapes Technology*. New York: Oxford University Press, 1997.

Reif, Kingston. "New START at a Glance." Arms Control Association, January 2020. https://www.armscontrol.org/factsheets/NewSTART.

Rhodes, Richard. *The Making of the Atomic Bomb*. New York: Simon & Schuster, 1986.

Robinson, Mary, Martti Ahtisaari, Ban Ki-moon, Ela Bhatt, Lakhdar Brahimi, Gro Harlem Brundtland, Fernando Henrique Cardoso, et al. "Nuclear Non-Proliferation and Disarmament." London: The Elders Foundation, November 2019. https://www.theelders.org/sites/default/files/The%20Elders%20Report%20Nuclear%20Position%20Paper%20A5%20Nov19.pdf.

Robock, Alan. "Nuclear Winter." *WIREs Climate Change* 1 (May/June 2010): 418–27. http://climate.envsci.rutgers.edu/pdf/WiresClimateChangeNW.pdf.Rotblat, Joseph. "The Early Days of Pugwash." *Physics Today* 54, no. 6 (June 2001): 50–55.

Rotblat, Joseph. *Scientists in the Quest for Peace: A History of the Pugwash Conferences*. Cambridge, MA: MIT Press, 1972.

Scharre, Paul. *Army of None: Autonomous Weapons and the Future of War*. New York: W.W. Norton, 2018.

Schell, Jonathan. "Nuclear Holocaust." In *The Nuclear Reader: Strategy, Weapons, War*. Edited by Charles W. Kegley, Jr., and Eugene R. Wittkopf. New York: St. Martin's Press, 1985, pp. 258–69. Shultz, George P., William J. Perry, Henry A. Kissinger, and Sam Nunn. "Toward a Nuclear-Free World." *Wall Street Journal*, January 16, 2008. https://www.wsj.com/articles/SB120036422673589947.

Shultz, George P., William J. Perry, Henry A. Kissinger, and Sam Nunn. "A World Free of Nuclear Weapons." *Wall Street Journal*, January 4, 2007. https://www.wsj.com/articles/SB116787515251566636.

Sime, Ruth Lewin. "Science and Politics: The Discovery of Nuclear Fission 75 Years Ago." *Annalen Der Physik* 526, no. 3–4 (2014): A27–31. doi.org/10.1002/andp.201400805.

Sokov, Nikolai N. "Why Russia Calls a Limited Nuclear Strike 'De-Escalation.'" *Bulletin of the Atomic Scientists* 70 (13 March 2014). https://thebulletin.org/2014/03/why-russia-calls-a-limited-nuclear-strike-de-escalation/.

Treaty on the Prohibition of Nuclear Weapons. New York: United Nations, 2017. https://www.icanw.org/full_text_of_the_treaty.US Air Force : *Nuclear Operations*. Doctrine Publication *3-72*. Curtis LeMay Center, December 18, 2020. https://www.doctrine.af.mil/Portals/61/documents/AFDP_3-72/3-72-AFDP-NUCLEAR-OPS.pdf.

Waltz, Kenneth N. The Spread of Nuclear Weapons: More May Be Better. Adelphi Papers, No. 171. London: International Institute for Strategic Studies, 1981.

Wellerstein, Alex. "Nukemap." The Nuclear Secrecy Blog. 2020. https://nuclearsecrecy.com/nukemap.

Wells, Samuel F., Jr. "The Origins of Massive Retaliation." *Political Science Quarterly* 96, no. 1 (1981): 31–52.

Annex A. A Primer on Nuclear Science and Technology

Ferguson, Charles. *Nuclear Energy: What Everyone Needs to Know*. Oxford: Oxford University Press, 2011.

World Nuclear Association. "Uranium Enrichment." 2016. http://www.world-nuclear.org/information-library/nuclear-fuel-cycle/conversion-enrichment-and-fabrication/uranium-enrichment.aspx.

Chapter 6. Global Health: Security and Inequality

Allam, Zaheer. "Chapter 1 - The First 50 Days of COVID-19: A Detailed Chronological Timeline and Extensive Review of Literature Documenting the Pandemic." In *Surveying the Covid-19 Pandemic and Its Implications: Urban Health, Data Technology and Political Economy (Elsevier Public Health Emergency Collection)*, 1–7. Elsevier, July 24, 2020. https://doi.org/10.1016/B978-0-12-824313-8.00001-2.

Birn, Ann-Emmanuelle, Yogan Pillay, and Timothy H. Holtz. *Textbook of International Health: Global Health in a Dynamic World*, 3rd ed. New York: Oxford University Press, 2009.

Buckley, Chris, David D. Kirkpatrick, Amy Qin, and Javier C. Hernández. "How Coronavirus Escaped from China's Grasp." *The New York Times*, December 31, 2020, A1.

Cha, Victor D. "Globalization and the Study of International Security." *Journal of Peace Research* 37, no. 3 (2000): 391–403.

Chuhan-Pole, Punam, and Angwaf Manka. *Yes, Africa Can: Success Stories from a Dynamic Continent*. Washington, DC: World Bank, 2011. https://openknowledge.worldbank.org/handle/10986/2335.

Cotlear, Daniel, Somil Nagpal, Owen Smith, Ajay Tandon, and Rafael Cortez. *Going Universal: How 24 Developing Countries Are Implementing Universal Health Coverage from the Bottom Up*. Washington, DC: World Bank, 2015.

Constitution of the World Health Organization. *United Nations Treaty Series* 14, (July 22, 1946): 186. https://treaties.un.org/Pages/showDetails.aspx?objid=080000028002d899&clang=_en.

Crane, Barbara B., Nils Daulaire, and Alex C. Ezeh. "Reproductive Health in Culture Wars Crossfire." *Science* 358, no. 6360 (October 13, 2017): 175–76. https://doi.org/10.1126/science.aao3940.

Cueto, Marcelo. "The Origins of Primary Health Care and Selective Primary Health Care." *American Journal of Public Health* 94, no. 11 (2004): 1864–74.

Farmer, Paul E., B. Nizeye, S. Stulac, S. Keshavjee. "Structural Violence and Clinical Medicine." *PLoS Medicine* 3, no. 10 (2006): e449.

Fidler, David P. *SARS, Governance and the Globalization of Disease*. New York: Palgrave Macmillan, 2004.

Fontaine, Olivier, Paul Garner, and M. K. Bahn. "Oral Rehydration Therapy: The Simple Solution for Saving Lives." *British Medical Journal* 334, (January, 2007): S14.

Fosdick, Raymond B. *The Story of the Rockefeller Foundation*. New York: Harper and Brothers, 1952.

Frenk, Julio. "Bridging the Divide: Global Lessons from Evidence-Based Health Policy in Mexico." *Lancet* 368, no. 9539 (2006): 954–61.

Frenk, Julio, Eduardo González-Pier, Octavio Gómez-Dantés, Miguel A. Lezana, and Felicia Marie Knaul. "Comprehensive Reform to Improve Health System Performance in Mexico." *Lancet* 368, no. 9546 (October 28, 2006): 1524–34. doi.org/10.1016/S0140-6736(06)69564-0.

Fried, Linda P., Margaret E. Bentley, Pierre Buekens, Donald S. Burke, Julio J. Frenk, Michael J. Klag, and Harrison C. Spencer. "Global Health Is Public Health." *Lancet* 375, no. 9714 (February 13, 2010): 535–37. doi.org/10.1016/S0140-6736(10)60203-6.

Garrett, Laurie. *The Coming Plague: Newly Emerging Diseases in a World Out of Balance*. New York: Penguin, 1995.

Garrett, Laurie. "Ebola's Lessons: How the WHO Mishandled the Crisis." *Foreign Affairs* 94, no. 5 (2017): 80–107.

Gaulkin, Thomas. "Pandemic Failure or Convenient Scapegoat: How Did WHO Get Here?" *Bulletin of the Atomic Scientists*, July 9, 2020. https://thebulletin.org/2020/07/pandemic-failure-or-convenient-scapegoat-how-did-who-get-here/.

Gebrekidan, Selam, Matt Apuzzo, Amy Qin, and Javier C. Hernández. "In Hunt for Virus Source, WHO Let China Take Charge." *The New York Times*, November 3, 2020, A1.

Gellman, Barton. "AIDS Is Declared Threat to Security." *Washington Post*, April 30, 2000. https://www.washingtonpost.com/archive/politics/2000/04/30/aids-is-declared-threat-to-security/c5e976e4-3fe8-411b-9734-ca44f3130b41/.

Gómez-Dantés, Octavio, Michael R. Reich, and Francisco Garrido-Latorre. "Political Economy of Pursuing the Expansion of Social Protection in Health in Mexico." *Health Systems & Reform* 1, no. 3 (2015): 207–16.

Gootnick, David. *Trends in U.S. Spending for Global HIV/AIDS and Other Health Assistance in Fiscal Years 2001–2008*. Washington, DC: U.S. Government Accountability Office, October 2010.

Gostin, Lawrence and Devi Srindha. "Global Health and the Law." *The New England Journal of Medicine* 370, no. 18 (2014): 1732.

Greene, Jeremy, Marguerite Thorp Basilico, Heidi Kim, and Paul Farmer. "Colonial Medicine and Its Legacies." In *Reimagining Global Health: An Introduction*. Edited by Paul Farmer, Arthur Kleinman, Jim Kim, and Matthew Basilico. Berkeley: University of California Press, 2013, pp. 33–73.

Haas, Ernst B. "Functionalism." *Beyond the Nation-State: Functionalism and International Organization*. Stanford, CA: Stanford University Press, 1964.

Hardon, Anita. "Consumers Versus Producers: Power Play Behind the Scenes." In *Drugs Policy in Developing Countries*. Edited by Najmi Kenjo, Anita Hardon, and Gill Walt. London: Zed Books, 1992, pp. 48–64.

Heaton, Matthew M. "Health and Medicine in Colonial Society." In *The Palgrave Handbook of African Colonial and Postcolonial History*. Edited by Martin S. Shanguhyia and Toyin Falola. New York: Palgrave Macmillan, 2018, pp. 303–18.

Horsely, Jamie P. "Let's End the COVID-19 Blame Game: Reconsidering China's Role in the Pandemic." Brookings. August 19, 2020. https://www.brookings.edu/blog/order-from-chaos/2020/08/19/lets-end-the-covid-19-blame-game-reconsidering-chinas-role-in-the-pandemic/.

Iwasaki, Akiko, and Patrick Wong. "The Immune Havoc of COVID-19." *Scientific American* 324, no. 1 (January 2021): 35–41.

Klausner, Jeffrey D., Mari-Rose Aplasca, Virginia P. Mesola, Gail Bolan, W. L. Whittington, and King K. Holmes. "Correlates of Gonococcal Infection and of Antimicrobial-Resistant Neisseria Gonorrhoeae Among Female Sex Workers, Republic of the Philippines, 1996–1997." *The Journal of Infectious Diseases* 179, no. 3 (March 1, 1999): 729–33. doi.org/10.1086/314625.

Knaul, Felicia Marie, Eduardo González-Pier, Octavio Gómez-Dantés, David García-Junco, Héctor Arreola-Ornelas, Mariana Barraza-Lloréns, Rosa Sandoval, et al. "The Quest for Universal Health Coverage: Achieving Social Protection for All in Mexico." *Lancet* 380, no. 9849 (October 6, 2012): 1259–79. doi.org/10.1016/S0140-6736(12)61068-X.

Lawn, Joy E. "The Child Survival Revolution: What Next?" *The Lancet* 384, no. 9947 (September 13, 2014): 931–33. doi.org/10.1016/S0140-6736(14)61057-6.

Lawrence, Susan V. "COVID-19 and China: A Chronology of Events (December 2019-January 2020)." R46354. Washington, DC: Congressional Research Service, May 13, 2020. https://fas.org/sgp/crs/row/R46354.pdf.

Lo, Nathan C., A. Lowe, and E. Bendavid. "Abstinence Funding Was Not Associated with Reductions in HIV Risk Behavior in Sub-Saharan Africa." *Health Affairs* 35, no. 5 (2016): 856–63.

Ma, Wenjun, Robert E Kahn, and Juergen A. Richt. "The Pig as a Mixing Vessel for Influenza Viruses: Human and Veterinary Implications." *Journal of Molecular and Genetic Medicine* 3, no. 1 (November 27, 2008): 158–66.

Manguvo, Angellar, and Benford Mafuvadze. "The Impact of Traditional and Religious Practices on the Spread of Ebola in West Africa: Time for a Strategic Shift." *The Pan African Medical Journal* 22, no. S1 (October 10, 2015). doi.org/10.11694/pamj.supp.2015.22.1.6190.

Moloo, Ashok. "River Blindness: Shifting from Prevention to Surveillance and Elimination." World Health Organization. November 12, 2017. http://www.who.int/neglected_diseases/news/shifting_from_prevention_to_surveillance_elimination/en/.

Nicholas Wade, "The Origin of COVID: Did People or Nature Open Pandora's Box at Wuhan?" *Bulletin of the Atomic Scientists* 5 (2021). https://thebulletin.org/2021/05/the-origin-of-covid-did-people-or-nature-open-pandoras-box-at-wuhan.

Nooruddin, Irfan, and Joel W. Simmons. "The Politics of Hard Choices: IMF Programs and Government Spending." *International Organization* 60, no. 4 (2006): 1001–33.

Norman, Leslie George. *Road Traffic Accidents: Epidemiology, Control, and Prevention*. Geneva: World Health Organization, 1962. https://apps.who.int/iris/handle/10665/39723.

Packard, Randall M. *A History of Global Health: Interventions into the Lives of Other Peoples*. Baltimore: Johns Hopkins Press, 2016.

Panisset, Ulysses B. *International Health Statecraft: Foreign Policy and Public Health in Peru's Cholera Epidemic*. Lanham, MD: University Press of America, 2000.

"A Plague of Finance." *The Economist*, September 29, 2001. https://www.economist.com/special-report/2001/09/29/a-plague-of-finance.

Quaye, Randolph. *Balancing Public and Private Health Care Systems: The Sub-Saharan African Experience*. Lanham, MD: University Press of America, 2010.

Qiu, Jane. "Chasing Plagues." *Scientific American, 322*, no.6, (June 1, 2020): 26–32.

Rauhala, Emily, Eva Dou, and Robyn Dixon. "Chinese and Russian Vaccines Remain Unproven — but Desperate Countries Plan to Use Them Anyway." *The Washington Post*, December 19, 2020. https://www.washingtonpost.com/world/vaccines-china-russia-safety/2020/12/18/8c87ba54-3637-11eb-9699-00d311f13d2d_story.html.

Reilley, Brigg, Michel Van Herp, Dan Sermand, and Nicoletta Dentico. "SARS and Carlo Urbani." *New England Journal of Medicine* 348 (2003): 1951–52.

Røttingen, John-Arne, Sadie Regmi, Mari Eide, Alison J. Young, Roderik F. Viergever, Christine Ardal, Javier Guzman, Danny Edwards, Stephen A. Matlin, and Robert F. Terry. "Mapping of Available Health Research and Development Data: What's There, What's Missing, and What Role Is There for a Global Observatory?" *Lancet* 382, no. 9900 (October 12, 2013): 1286–1307. doi.org/10.1016/S0140-6736(13)61046-6.

Saéz, Almudena Marí, Sabrina Weiss, Kathrin Nowak, Vincent Lapeyre, Fee Zimmermann, Ariane Düx, Hjalmar S Kühl, et al. "Investigating the Zoonotic Origin of the West African Ebola Epidemic." *EMBO Molecular Medicine* 7, no. 1 (January 1, 2015): 17–23. doi.org/10.15252/emmm.201404792.

Sarang, Anya, Lucy Platt, Inna Vyshemirskaya, and Tim Rhodes. "Prisons as a Source of Tuberculosis in Russia." *International Journal of Prisoner Health* 12 (March 14, 2016): 45–56. doi.org/10.1108/IJPH-07-2014-0022.

Sepúlveda, Jaime, Flavia Bustreo, Roberto Tapia, Juan Rivera, Rafael Lozano, Gustavo Oláiz, Virgilio Partida, Lourdes García-García, and José Luis Valdespino. "Improvement of Child Survival in Mexico: The Diagonal Approach." *Lancet* 368, no. 9551 (December 2, 2006): 2017–27. doi.org/10.1016/S0140-6736(06)69569-X.

Sharp, Paul M., and Beatrice H. Hahn. "Origins of HIV and the AIDS Pandemic." *Cold Spring Harbor Perspectives in Medicine* 1, no. 1 (September 1, 2011): a006841. doi.org/10.1101/cshperspect.a006841.

Shuman, Emily K. "Global Climate Change and Infectious Diseases." *New England Journal of Medicine* 362, no. 12 (March 25, 2010): 1061–63. doi.org/10.1056/NEJMp0912931.

Singh, K., and S. Mehta. "The Clinical Development Process for a Novel Preventive Vaccine: An Overview." *Journal of Postgraduate Medicine* 62, no. 1 (2016): 4–11. doi.org/10.4103/0022-3859.173187.

Thomas, Katie. "The Vaccines Will Probably Work. Making Them Fast Will Be the Hard Part." *The New York Times*, December 7, 2020, https://www.nytimes.com/2020/11/17/health/coronavirus-vaccine-operation-warp-speed.html.

Tilley, Helen. "Medicine, Empires, and Ethics in Colonial Africa." *AMA Journal of Ethics* 18, no. 7 (July 1, 2016): 743–53. doi.org/10.1001/journalofethics.2016.18.7.mhst1-1607.

United Nations. "Sustainable Development Goals".. https://www.un.org/sustainable-development/sustainable-development-goals/.

von Bertrab, Alejandra Xanic. "Part II: A Troubled Model for Reform in Mexico." Center for Public Integrity, November 15, 2010. https://publicintegrity.org/health/part-ii-a-troubled-model-for-reform-in-mexico/.

Walsh, J. A., and K. S. Warren. "Selective Primary Health Care: An Interim Strategy for Disease Control in Developing Countries." *New England Journal of Medicine* 301, no. 18 (November 1, 1979): 967–74. doi.org/10.1056/NEJM197911013011804.

Walt, Gil. "WHO Under Stress: Implications for Health Policy." *Health Policy* 24, no. 2 (1993). 125–44.

Weissman, R. "A Long, Strange TRIPS: The Pharmaceutical Industry Drive to Harmonize Global Intellectual Property Rules, and the Remaining WTO Legal Alternatives Available to Third World Countries." *University of Pennsylvania Journal of International Economic Law* 17, no. 4 (1996): 1069–1125.

Wenham, Clare. "The Oversecuritization of Global Health: Changing the Terms of Debate." *International Affairs* 95, no. 5 (September 1, 2019): 1093–1110. doi.org/10.1093/ia/iiz170.

World Bank. *World Development Report: Investing in Health*. New York: Oxford University Press, 1993.

World Health Organization. "10 Facts About Schistosomiasis." December 2017. www.who.int/features/factfiles/schistosomiasis/en/.

World Health Organization. "Health Security." http://www.who.int/health-security/en/.

World Health Organization. *International Health Regulations*, 3rd edition. Geneva: WHO, 2005. http://apps.who.int/iris/bitstream/10665/246107/1/9789241580496-eng.pdf?ua=1.

World Health Organization. *Malaria*. Global Health Observatory Data. 2018. www.who.int/gho/malaria/en/.

World Health Organization. *Report of the Ebola Interim Assessment Panel*. July 2015. http://www.who.int/csr/resources/publications/ebola/report-by-panel.pdf.

World Health Organization. "Voluntary Contributions." April 19, 2018. http://www.who.int/about/finances-accountability/funding/voluntary-contributions/en/.

World Health Organization. "WHO Global Health: Expenditure Atlas." Geneva: WHO, September 2014. https://www.who.int/health-accounts/atlas2014.pdf.

World Health Organization. *The World Health Report 2002 - Reducing Risks, Promoting Healthy Life*. Geneva: WHO, 2002. www.who.int/whr/2002/en/.

Zhong, Raymond, Paul Mozur, Jeff Kao, and Aaron Krolik. "No 'Negative' News: How China Censored the Coronavirus." *The New York Times*, January 13, 2021. https://www.nytimes.com/2020/12/19/technology/china-coronavirus-censorship.html.

Annex B. Science and Global Health

Fischetti, Mark, and Veronica Falconieri Hays. "Inside the Coronavirus: What Scientists Know About the Inner Workings of the Pathogen That Has Infected the World." *Scientific American* 323, no. 1 (July 2020): 32–37.

Herricks, Jennifer R., Peter J. Hotez, Valentine Wanga, Luc E. Coffeng, Juanita A. Haagsma, María-Gloria Basáñez, Geoffrey Buckle, et al. "The Global Burden of Disease Study 2013: What Does It Mean for the NTDs?" *PLOS Neglected Tropical Diseases* 11, no. 8 (August 3, 2017): e0005424. https://doi.org/10.1371/journal.pntd.0005424.

Jewell, Nicholas P., Joseph A. Lewnard, and Britta L. Jewell. "Caution Warranted: Using the Institute for Health Metrics and Evaluation Model for Predicting the Course of the COVID-19 Pandemic." *Annals of Internal Medicine* 173, no. 3 (April 14, 2020): 226–27. doi.org/10.7326/M20-1565.

Nájera, José A., Matiana González-Silva, and Pedro L. Alonso. "Some Lessons for the Future from the Global Malaria Eradication Programme (1955–1969)." *PLOS Medicine* 8, no. 1 (January 25, 2011): e1000412. https://doi.org/10.1371/journal.pmed.1000412.

Packard, Randall M. *A History of Global Health: Interventions into the Lives of Other Peoples*. Baltimore: Johns Hopkins Press, 2016.

World Health Organization. *The Top 10 Causes of Death*. Media Centre. May 24, 2018. https://www.who.int/news-room/fact-sheets/detail/the-top-10-causes-of-death.

Chapter 7. Globalization and the Burden of Disease

Akyeampong, Emmanuel Kwaku, Allan G. Hill, and Arthur Kleinman, eds. *The Culture of Mental Illness and Psychiatric Practice in Africa*. Bloomington, IN: Indiana University Press, 2015.

Alibek, Ken, and Stephen Handelman. *Biohazard: The Chilling True Story of the Largest Covert Biological Weapons Program in the World—Told from the inside by the Man Who Ran It*. New York: Random House, 1999.

Ben Ouagrham-Gormley, Sonia. "Potemkin or Real? North Korea's Biological Weapons Program." *Bulletin of the Atomic Scientists*, July 18, 2017. https://thebulletin.org/2017/07/potemkin-or-real-north-koreas-biological-weapons-program/.

Convention on Biological Diversity. United Nations (1992). https://www.cbd.int/doc/legal/cbd-en.pdf

Convention on the Prohibition of the Development, Production and Stockpiling of Bacteriological (Biological) and Toxin Weapons and on Their Destruction. United Nations Treaty Series 1015, no. 14860 (April 10, 1972): 163. https://www.state.gov/t/isn/bw/c48738.htm.

Curtis, Mark. "Gated Development: Is the Gates Foundation Always a Force for Good?" 2nd edition. (London: Global Justice Now, June 2016). https://www.globaljustice.org.uk/sites/default/files/files/resources/gjn_gates_report_june_2016_web_final_version_2.pdf.

Doughton, Sandi. "Not Many Speak Their Mind to Gates Foundation." *The Seattle Times*, August 2, 2008. https://www.seattletimes.com/seattle-news/not-many-speak-their-mind-to-gates-foundation/.

Fenn, Elizabeth A. "Biological Warfare in Eighteenth-Century North America: Beyond Jeffery Amherst." *The Journal of American History* 86, no. 4 (2000): 1552–80.

Feshbach, Murray. "Russia's Health and Demographic Crises: Policy Implications and Consequences." Washington, DC: Chemical and Biological Arms Control Institute, 2003.

Fisher, William W., and Cyrill P. Rigamonti. "The South Africa AIDS Controversy: A Case Study in Patent Law and Policy." Harvard Law School. February 10, 2005. boris.unibe.ch/69881/.

Gates, Bill. "Here's My Plan to Improve Our World—and How You Can Help." *Wired*, November 12, 2013. www.wired.com/2013/11/bill-gates-wired-essay/.

Gelling, Peter. "Indonesia Defiant on Refusal to Share Bird Flu Samples." *The New York Times*, March 25, 2007. www.nytimes.com/2007/03/26/world/asia/26cnd-flu.html.

Grenier, Marie-Lyne. "The Impact of Brain Drain in Underserved Countries: Implementing a Global Ethical Leadership Approach." *International Journal of Migration, Health and Social Care* 11, no. 3 (January 21, 2015): 218–22.

Institute of Complementary and Alternative Medicine. "Traditional Chinese Medicine: In Depth." National Center for Complementary and Integrative Health. 2012. https://nccih.nih.gov/health/whatiscam/chinesemed.htm.

Jackson, Joe. *The Thief at the End of the World: Rubber, Power, and the Seeds of Empire*. New York: Viking Press, 2008.

Jamison, Dean. "Global Health 2035: A World Converging Within a Generation." *The Lancet* 382, no. 9908 (December 13, 2013): 1898–1955.

Joy, Robert J. T. "Malaria in American Troops in the South and Southwest Pacific in World War II." *Medical History* 43, no. 2 (1999): 192–207.

Kim, Hyun-Kyung, Elizabeth Philipp, and Hattie Chung. "North Korea's Biological Weapons Program: The Known and Unknown." Cambridge, MA: Harvard University Kennedy School, 2017.

"Malaria Strategy Overview." Bill & Melinda Gates Foundation. 2018. www.gatesfoundation.org/What-We-Do/Global-Health/Malaria.

Mann, Charles C. *1491: New Revelations of the Americas Before Columbus*. New York: Knopf, 2005.

McNeill, John Robert. *Mosquito Empires: Ecology and War in the Greater Caribbean, 1620-1914*. New York: Cambridge University Press, 2010.

Mendenhall, Emily, Brandon A. Kohrt, Shane A. Norris, David Ndetei, and Dorairaj Prabhakaran. "Non-Communicable Disease Syndemics: Poverty, Depression, and

Diabetes Among Low-Income Populations." *Lancet* 389, no. 10072 (04 2017): 951–63. doi.org/10.1016/S0140-6736(17)30402-6.

Mukherjee, Ambarish, Mousumi Banerjee, Vivekananda Mandal, Amritesh C. Shukla, and Subhash C. Mandal. "Modernization of Ayurveda: A Brief Overview of Indian Initiatives." *Natural Product Communications* 9, no. 2 (February 1, 2014). doi.org/10–1177/1934578X1400900239.

Nagoya Protocol on Access to Genetic Resources and the Fair and Equitable Sharing of Benefits Arising from Their Utilization to The Convention on Biological Diversity, Text and Annex (Montreal, 2011). Secretariat of the Convention on Biological Diversity. https://www.cbd.int/abs/doc/protocol/nagoya-protocol-en.pdf

Nye, Joseph S. *Soft Power the Means to Success in World Politics* (New York: PublicAffairs, 2005).

People's Republic of China State Council Information Office. *Traditional Medicine in China*. December 6, 2016. http://english.gov.cn/archive/white_paper/2016/12/06/content_281475509333700.htm.

Prathapan, K. Divakaran, Rohan Pethiyagoda, Kamaljit S. Bawa, Peter H. Raven, and Priyadarsanan Dharma Rajan. "When the Cure Kills—CBD Limits Biodiversity Research." *Science* 360, no. 6396 (June 29, 2018): 1405–6. doi.org/10.1126/science.aat9844.

Protocol for the Prohibition of the Use in War of Asphyxiating, Poisonous or Other Gases, and of Bacteriological Methods of Warfare (Geneva Protocol). https://unoda-web.s3-accelerate.amazonaws.com/wp-content/uploads/assets/WMD/Bio/pdf/Status_Protocol.pdf.

Robbins, Anthony. "The CIA's Vaccination Ruse." *Journal of Public Health Policy* 33, no. 4 (2012): 387–89.

Sachs, Jeffrey D. *Macroeconomics and Health: Investing in Health for Economic Development*. Geneva: WHO, 2001.

Sarris, Peter. "The Justinianic Plague: Origins and Effects." *Continuity and Change* 17, no. 2 (2002): 169–82.

Stiffler, Lisa. "Bill Gates Sr. at 90: A Giant Impact on Technology, Philanthropy and the Seattle Region." *GeekWire*, November 26, 2015. www.geekwire.com/2015/life-and-times-of-bill-gates-sr-a-giant-impact-on-technology-philanthropy-and-the-seattle-region/.

"Thailand—Restrictions on Importation of and Internal Taxes on Cigarettes." *General Agreements on Tariffs and Trade (GATT) Basic Instruments and Selected Documents (BISD)* S37 (1991): 200.

The Lancet. "The Global Burden of Disease Study 2015: From Big Data to Meaningful Change." *The Lancet* 388, no. 10053 (October 8, 2016): 1447–1850.

"U.N. Biodiversity Summit Yields Welcome and Unexpected Progress." *Science* 330, no. 6605 (November 5, 2010): 742–43.

United Nations, Department of Economic and Social Affairs, Population Division. "The World Population Prospects: 2015 Revision." Working Paper ESA/P/WP.241. New York: United Nations, July 2015. www.un.org/en/development/desa/publications/world-population-prospects-2015-revision.html.

United Nations. *ESCAP Population Data Sheet*. 2016. https://www.unescap.org/sites/default/d8files/knowledge-products/SPPS%20PS%20data%20sheet%202016%20v15-2.pdf

United Nations, Population Division. "Population Ages 0–14 (% of Total)." World Bank. 2017. data.worldbank.org/indicator/SP.POP.0014.TO.ZS?end=2016&start=2016.

Weiss, Charles, and Thomas Eisner. "Partnerships for Bioprospecting." *Technology in Society* 20 (1998): 481–98.

Wenham, Clare. "The Oversecuritization of Global Health: Changing the Terms of Debate." *International Affairs* 95, no. 5 (September 1, 2019): 1093–1110. doi.org/10.1093/ia/iiz170.

Wheelis, Mark. "Biological Warfare at the 1346 Siege of Caffa." *Emerging Infectious Diseases* 8, no. 9 (2002): 971–75.

WHO Framework Convention on Tobacco Control. United Nations Treaty Series, 2302, no. 41032 (May 21, 2003), https://treaties.un.org/Pages/showDetails.aspx?objid=0800000280055410&clang=_en 2302, no. 41032 (May 21, 2003).

World Health Organization. "Social Determinants of Health." 2021. https://www.who.int/health-topics/social-determinants-of-health#tab=tab_1.

World Health Organization. "Ageing and Health." Media Centre. February 5, 2018. www.who.int/mediacentre/factsheets/fs404/en/.

World Health Organization. "NCD Mortality and Morbidity." August 24, 2018. www.who.int/gho/ncd/mortality_morbidity/en/.

World Health Organization and Calouste Gulbenkian Foundation. *Social Determinants of Mental Health*. Geneva: WHO, 2014.

Zinsser, Hans. *Rats, Lice and History*. New York: Little, Brown, 1935.

Chapter 8. The Internet and Social Media: Euphoria and Repression

Abatte, Janet. *Inventing the Internet*. Cambridge: MIT Press, 1991.

Barlow, John Perry. "A Declaration of the Independence of Cyberspace." Electronic Frontier Foundation. February 8, 1996. https://www.eff.org/cyberspace-independence.

Brown Weiss, Edith. *Establishing Norms in a Kaleidoscopic World*. General Course on Public International Law 396. Leiden: Brill/Nijhoff, 2018.

Castells, Manuel, Mireia Fernandez-Ardevol, Jack Linchuan Qiu, and Araba Sey. "Electronic Communication and Socio-Political Mobilisation: A New Form of Civil Society." In *Global Civil Society 2005/6*. Edited by Helmut K. Anheier, Mary Kaldor, and Marlies Glasius. London: SAGE Publications, 2006, pp. 112–31.

Cellan-Jones, Rory. "China to Track Cars in Terrorism Crackdown." *BBC News*, February 21, 2017. https://www.bbc.com/news/world-asia-china-39038364.

Curran, James, Natalie Fenton, and Des Freedman. *Misunderstanding the Internet*, 2nd edition. London: Routledge, 2016.

Economy, Elizabeth C. *The Third Revolution: Xi Jinping and the New Chinese State*. New York: Oxford University Press, 2018.

Fannin, Rebecca A. *Tech Titans of China*. Boston: Nicholas Brealey, 2019.

Fefer, Rachel F., and Kristin Archick. *E.U. Data Protection Rules and U.S. Implications*. Washington, DC: Congressional Research Service, 2019. https://crsreports.congress.gov/product/pdf/IF/IF10896.

Hafner, Katie, and Matthew Lyon. *Where the Wizards Stay Up Late: The Origins of the Internet*. New York: Touchstone, 1996.

Hill, Jonah Force. "The Growth of Data Localization Post-Snowden: Analysis and Recommendations for U.S. Policymakers and Industry Leaders." Lawfare Research Paper Series, 2, No. 3. July 21, 2014, 1–41. https://assets.documentcloud.org/documents/7276302/THE-GROWTH-OF-DATA-LOCALIZATION-POST-SNOWDEN.pdf.

Holderness, Mike. "It Could All Be So Much Better (Interview with Ted Nelson)." *New Scientist*, 191, 54–55 (September 16, 2006).

Inkster, Nigel. *China's Cyber Power*. Washington, DC: Routledge and the International Institute for Strategic Studies, 2018.

International Telecommunication Union. "Individuals Using the Internet (% of Population)." World Bank. October 2, 2019. https://data.worldbank.org/indicator/IT.NET.USER.ZS.

James, Randy. "A Brief History of: Chinese Internet Censorship." *Time*, March 18, 2009. http://content.time.com/time/world/article/0,8599,1885961,00.html.

Johnson, David R., and David Post. "Law and Borders: The Rise of Law in Cyberspace." *Stanford Law Review* 48, no. 5 (1996): 1367–1402. doi.org/10.2307/1229390.

Kaplan, Fred. *Dark Territory: The Secret History of Cyber War*. New York: Simon & Schuster, 2016.

Liang, Fan, Vishnupriya Das, Nadiya Kostyuk, and Muzammil M. Hussain. "Constructing a Data-Driven Society: China's Social Credit System as a State Surveillance Infrastructure." *Policy & Internet* 10, no. 4 (2018): 415–53.

Lim, Merlyna. "Clicks, Cabs, and Coffee Houses: Social Media and Oppositional Movements in Egypt, 2004–2011." *Journal of Communication* 62, no. 2 (2012): 231–48.

Morozov, Evgeny. *The Net Delusion: The Dark Side of Internet Freedom*. New York: PublicAffairs, 2011.

Negroponte, Nicholas. *Being Digital*. New York: Vintage Books, 1995.

Norris, Pippa. *Digital Divide: Civic Engagement, Information Poverty, and the Internet Worldwide*. New York: Cambridge University Press, 2001.

Qiang, Xiao. "President XI's Surveillance State." *Journal of Democracy* 30, no. 1 (2019): 53–67. doi.org/10.1353/jod.2019.0004.

Raustiala, K. "Governing the Internet." *American Journal of International Law* 110 (2016): 491–503.

Rubio, Marco, and Chris Smith. "2018 Annual Report: Xinjiang." Washington, DC: *Congressional-Executive Commission on China*, 2018. https://www.cecc.gov/publications/annual-reports/2018-annual-report.

Ryan, Johnny. *A History of the Internet and the Digital Future*. London: Reaktion Books, 2010.

Sanger, David E. *The Perfect Weapon: War, Sabotage and Fear in the Cyber Age*. New York: Crown Publishing, 2018.

Segal, Adam. "When China Rules the Internet." *Foreign Affairs* 85, no. 5 (2018): 10–18.

Snyder, Joel, Konstantinos Komaitis, and Andrei Robachevsky. "The History of IANA: An Extended Timeline with Citations and Commentary." Reston, VA: Internet Society, January 2017. https://www.internetsociety.org/wp-content/uploads/2016/05/IANA_Timeline_20170117.pdf.

Spar, Deborah. *Ruling the Waves: Cycles of Discovery, Chaos, and Wealth from the Compass to the Internet*. New York: Harcourt, 2001.

"State of Privacy Egypt." Privacy International. January 26, 2019. http://privacyinternational.org/state-privacy/1001/state-privacy-egypt.

Thompson, Stuart A., and Charlie Werzel. "Twelve Million Phones, One Dataset, Zero Privacy." *The New York Times*, December 19, 2019. https://www.nytimes.com/interactive/2019/12/19/opinion/location-tracking-cell-phone.html.

United Nations Population Division. "Population, Total." World Bank. October 16, 2019. https://data.worldbank.org/indicator/SP.POP.TOTL.

Weiss, Charles. "How Do Science and Technology Affect International Affairs?" *Minerva* 53 (2015): 411–30.

Weiss, Martin A., and Kristin Archick. *U.S.-EU Data Privacy: From Safe Harbor to Privacy Shield*. CRS 7–5700. Washington, DC: Congressional Research Service, 2016. https://epic.org/crs/R44257.pdf.

World Summit on the Information Society. *Tunis Agenda for the Information Society*. WSIS-05/TUNIS/DOC/6(Rev. 1)-E (2005). November 18, 2005. https://www.itu.int/net/wsis/docs2/tunis/off/6rev1.html.

Wray, Christopher. "Transcript: The Threat Posed by the Chinese Government and the Chinese Communist Party to the Economic and National Security of the United States." Hudson Institute. July 7, 2020. https://www.hudson.org/research/16201-transcript-the-threat-posed-by-the-chinese-government-and-the-chinese-communist-party-to-the-economic-and-national-security-of-the-united-states.

Zhong, Raymond. "China's Cryptocurrency Plan Has a Powerful Partner: Big Brother." *The New York Times*, October 18, 2019. https://www.nytimes.com/2019/10/18/technology/china-cryptocurrency-facebook-libra.html.

Zhong, Raymond, Paul Mozur, Jeff Kao, and Aaron Krolik. "No 'Negative' News: How China Censored the Coronavirus." *The New York Times*, December 19, 2020. https://www.nytimes.com/2020/12/19/technology/china-coronavirus-censorship.html.

Chapter 9. Cyberwarfare and Cybersecurity

Ahmed, Shazeda, Natasha E. Bajema, Samuel Bendett, Benjamin A. Chang, Rogier Creemers, Chris C. Demchak, Sarah W. Denton, et al. "AI, China, Russia, and the Global Order: Technological, Political, Global, and Creative Perspectives." Strategic Multilayer Assessment Periodic Publication. Boston, MA: Defense Technical Information Center, December 1, 2018. https://apps.dtic.mil/docs/citations/AD1066673.

Albright, David, Paul Brannan, and Christina Walrond. "Did Stuxnet Take Out 1,000 Centrifuges at the Natanz Enrichment Plant?" Washington, DC: Institute for Science and International Security, December 22, 2010. https://isis-online.org/isis-reports/detail/did-stuxnet-take-out-1000-centrifuges-at-the-natanz-enrichment-plant/.

Alford, Lionel D., Jr. "Cyber Warfare: Protecting Military Systems." *Acquisition Review Quarterly*, 99 (Spring 2000): 100–120.

Baezner, Marie, and Patrice Robin. "Hotspot Analysis: Stuxnet." Zurich: ETH CyberDefense Project, 2017. https://css.ethz.ch/content/dam/ethz/special-interest/gess/cis/center-for-securities-studies/pdfs/Cyber-Reports-2017-04.pdf.

Bartles, Charles K. "Getting Gerasimov Right." *Military Review* 96, no. 1 (January/February 2016): 30–38.

Brown Weiss, Edith. *Establishing Norms in a Kaleidoscopic World*. General Course on Public International Law 396. Leiden: Brill/Nijhoff, 2018.

Carpenter v. United States, 138 S. Ct. 2206, 2215–216 (2018).

Clapper, James R. *Facts and Fears: Hard Truths from a Life in Intelligence*. New York: Viking, 2018.

Department of Defense. *Cyber Strategy*. Arlington, VA: US Department of Defense, 2018. https://media.defense.gov/2018/Sep/18/2002041658/-1/-1/1/CYBER_STRATEGY_SUMMARY_FINAL.PDF.

Ephron, Dan, and Yuval Shany. "A rule book on the shelf? Tallinn Manual 2.0 on cyberoperations and subsequent state practice." *American Journal of International Law* 112, no. 4 (2019): 583–657.

Gerasimov, Valery. "The Value of Science is in the Foresight: New Challenges Demand Rethinking the Forms and Methods of Carrying Out Combat Operations." *Military Review,* 96, no. 1 (January/February 2016): 23–29. https://www.armyupress.army.mil/Portals/7/military-review/Archives/English/MilitaryReview_20160228_art008.pdf.

Geverd, Timothy. "Bulk Telephony Metadata Collection and the Fourth Amendment: The Case for Revisiting the Third-Party Disclosure Doctrine in the Digital Age." *John Marshall Journal of Information Technology & Privacy Law* 31, no. 2 (January 1, 2014): 191–235. https://repository.jmls.edu/jitpl/vol31/iss2/3.

Henriksen, Anders. "The End of the Road for the UNGGE Process: The Future Regulation of Cyberspace." *Journal of Cybersecurity* 5, no. 1 (January 1, 2019). doi.org/10.1093/cybsec/tyy009.

Kaplan, Fred. *Dark Territory: The Secret History of Cyber War.* New York: Simon & Schuster, 2016.

Kollars, Nina, and Jacqueline Schneider. "Defending Forward: The 2018 Cyber Strategy Is Here." War on the Rocks. September 20, 2018. https://warontherocks.com/2018/09/defending-forward-the-2018-cyber-strategy-is-here/.

Langner, Ralph. *To Kill a Centrifuge.* Arlington, VA: Langner Group, November 2013. https://www.langner.com/wp-content/uploads/2017/03/to-kill-a-centrifuge.pdf.

Lewis, Paul. "Snowden Documents Show NSA Gathering 5bn Cell Phone Records Daily." *The Guardian*, December 5, 2013. https://www.theguardian.com/world/2013/dec/04/nsa-storing-cell-phone-records-daily-snowden.

Martin, Diego A., and Jacob N. Shapiro. *Trends in Online Foreign Influence Efforts.* Princeton, NJ: ESOC Publications, 2019.

Mazanes, B. *The Evolution of Cyberwar: International Norms for Emerging Technology Weapons.* Lincoln: University of Nebraska Press, 2015.

Sarah McKune, "An Analysis of the International Code of Conduct for Information Security", Citizen Lab Report, Munk School of Global Affairs, University of Toronto, 2015. https://openeffect.ca/code-conduct/.

McMorrow, Ryan. "Huawei a Key Beneficiary of China Subsidies That US Wants Ended." *Phys.org,* May 30, 2019. https://phys.org/news/2019-05-huawei-key-beneficiary-china-subsidies.html.

Nakashima, Ellen. "U.S. Pushes Hard for a Ban on Huawei in Europe, but the Firm's 5G Prices Are Nearly Irresistible." *Washington Post*, May 29, 2019. https://www.washingtonpost.com/world/national-security/for-huawei-the-5g-play-is-in-europe--and-the-us-is-pushing-hard-for-a-ban-there/2019/05/28/582a8ff6-78d4-11e9-b7ae-390de4259661_story.html.

Office of the Press Secretary. "President Xi Jinping's State Visit to the United States." White House Archives. September 25, 2015. https://obamawhitehouse.archives.gov/the-press-office/2015/09/25/fact-sheet-president-xi-jinpings-state-visit-united-states.

Polyakova, Alina, and Chris Meserole. "Exporting Digital Authoritarianism." Democracy & Disorder policy brief. Brookings Institution. August 26, 2019. https://www.brookings.edu/wp-content/uploads/2019/08/FP_20190826_digital_authoritarianism_polyakova_meserole.pdf.

Sanger, David E. *The Perfect Weapon: War, Sabotage and Fear in the Cyber Age.* New York: Crown Publishing, 2018.

Sanger, David E., and Nicole Perlroth. "FireEye, a Top Cybersecurity Firm, Says It Was Hacked by a Nation-State." *The New York Times*, December 8, 2020, updated February 6, 2021. https://www.nytimes.com/2020/12/08/technology/fireeye-hacked-russians.html.

Sanger, Peter, Nicole Perlroth, and Julian E. Barnes. "As Understanding of Russian Hacking Grows, So Does Alarm." *The New York Times*, January 2, 2021, updated January 5, 2021. https://www.nytimes.com/2021/01/02/us/politics/russian-hacking-government.html.

Schmitt, M., ed. *Tallinn Manual 2.0 on the International Law Applicable to Cyber Operations*. NATO Cooperative Cyber Defense Center of Excellence. Cambridge: Cambridge University Press, 2017.

Schneier, Bruce. *Click Here to Kill Everybody: Security and Survival in a Hyperconnected World*. New York: W. W. Norton, 2018.

Segal, Adam. "When China Rules the Internet." *Foreign Affairs* 85, no. 5 (2018): 10–18.

Smith, Brad. "Transcript: The Need for a Digital Geneva Convention." Keynote Address, RSA Conference, San Francisco. February 14, 2017. https://blogs.microsoft.com/wp-content/uploads/2017/03/Transcript-of-Brad-Smiths-Keynote-Address-at-the-RSA-Conference-2017.pdf.

United Nations General Assembly. Chair's Summary, Open-ended Working Group on Developments in the Field of Information and Telecommunications in the Context of International Security, third substantive session 8 12 March 2021 (Document A/AC.290/2021/CRP.3*, Conference room paper 10 March 2021).

United Nations Secretary-General. *Group of Governmental Experts on Developments in the Field of Information and Telecommunications in the Context of International Security*. A/70/174. New York: UN, July 22, 2015.

Weiner, Tim. *Legacy of Ashes: The History of the CIA*. New York: Anchor Books, 2008.

Weiss, Charles. "The Coming Technology of Knowledge Discovery: A Final Blow to Privacy Protection?" *University of Illinois Journal of Law, Technology and Policy* 2004, no. 2 (2004): 253–81.

White, Sarah P. "Understanding Cyberwarfare: Lessons from the Russia-Georgia War." West Point, NY: Modern War Institute, 2018. https://mwi.usma.edu/wp-content/uploads/2018/03/Understanding-Cyberwarfare.pdf.

Wray, Christopher. "Transcript: The Threat Posed by the Chinese Government and the Chinese Communist Party to the Economic and National Security of the United States." Hudson Institute. July 7, 2020. https://www.hudson.org/research/16201-transcript-the-threat-posed-by-the-chinese-government-and-the-chinese-communist-party-to-the-economic-and-national-security-of-the-united-states.

Chapter 10. Frugal Innovations for the "Bottom of the Pyramid"

Bakalar, Nicholas. "For Neonatal Jaundice, a New Option in Resource-Poor Nations." *The New York Times*, September 26, 2011. https://www.nytimes.com/2011/09/27/health/27jaundice.html.

Baum, Warren. *Partners Against Hunger: The Consultative Group on International Agricultural Research*. Washington, DC: World Bank, 1986.

Bazzano, Alessandra N., Kaitlin S. Potts, Lydia A. Bazzano, and John B. Mason. "The Life Course Implications of Ready to Use Therapeutic Food for Children in Low-Income Countries." *International Journal of Environmental Research and Public Health* 14, no. 4 (April 2017): 403. doi.org/10.3390/ijerph14040403.

Bhattacharjee, Yudhijit. "How an Indian Innovator Reverse-Engineered the Making of Sanitary Pads." *The New York Times*, November 10, 2016. https://www.nytimes.com/interactive/2016/11/13/magazine/design-issue-sanitary-pads-india.html.

Bonvillian, William B., and Charles Weiss. *Technological Innovation in Legacy Sectors*. New York: Oxford University Press, 2015.

Christensen, Clayton. *The Innovator's Dilemma: When New Technologies Cause Great Firms to Fail*. Cambridge, MA: Harvard Business School Press, 1997.

CMAM. "Key Data." PlumpyField Network. 2018. http://www.plumpyfield.com/about/key-data.

Conway, Gordon. *One Billion Hungry*. Ithaca, NY: Cornell University Press, 2012.

DeGusta, Michael. "Are Smart Phones Spreading Faster than Any Technology in Human History?" *MIT Technology Review*, May 9, 2012. https://www.technologyreview.com/s/427787/are-smart-phones-spreading-faster-than-any-technology-in-human-history/.

Edwards, Timothy L., Christophe Cox, Bart Weetjens, Tesfazghi Tewelde, and Alan Poling. "Giant African Pouched Rats (Cricetomys Gambianus) That Work on Tilled Soil Accurately Detect Land Mines." *Journal of Applied Behavior Analysis* 48, no. 3 (2015): 696–700. doi.org/10.1002/jaba.214.

Ehrlich, Paul. *The Population Bomb*. New York: Sierra Club-Ballantine, 1969.

Greco, Chiara, Gaston Arnolda, Nem-Yun Boo, Iman F. Iskander, Angela A. Okolo, Rinawati Rohsiswatmo, Steven M. Shapiro, et al. "Neonatal Jaundice in Low- and Middle-Income Countries: Lessons and Future Directions from the 2015 Don Ostrow Trieste Yellow Retreat." *Neonatology* 110, no. 3 (2016): 172–80. doi.org/10.1159/000445708.

Gunnerson, Charles G. "Sanitation Systems for Developing Countries." In *Technology, Finance and Development*. Edited by Charles Weiss and Nicolas Jequier. Lexington, MA: D.C. Heath & Co., 1984, pp. 125–140. http://documents1.worldbank.org/curated/en/446461468173928768/pdf/UNN334000Techn0ance0and0development.pdf

ImpactMatters. "D-Rev Impact Audit: Newborn Health Program and Mobility Program (2013–2015)." New York: ImpactMatters, September 26, 2017. https://s3-us-west-2.amazonaws.com/d-rev/wordpress/wp-content/uploads/2019/06/D-Rev-Impact-Audit-Report-2017-1.pdf.

Kammen, Daniel M. "Cookstoves for the Developing World." *Scientific American* 273, no. 1 (July 1, 1995): 72–75. doi.org/10.1038/scientificamerican0795-72.

Kaplinsky, Raphael. *The Economies of Small: Appropriate Technology in a Changing World*. London: Intermediate Technology Press, 1990.

Khandelwal, Meena, Matthew E. Hill, Paul Greenough, Jerry Anthony, Misha Quill, Marc Linderman, and H. S. Udaykumar. "Why Have Improved Cook-Stove Initiatives in India Failed?" *World Development* 92 (April 1, 2017): 13–27. doi.org/10.1016/j.worlddev.2016.11.006.

Liu, Li, Hope L. Johnson, Simon Cousens, Jamie Perin, Susana Scott, Joy E. Lawn, Igor Rudan, et al. "Global, Regional, and National Causes of Child Mortality: An Updated Systematic Analysis for 2010 with Time Trends since 2000." *Lancet* 379, no. 9832 (June 9, 2012): 2151–61. doi.org/10.1016/S0140-6736(12)60560-1.

Maitland, Donald. *The Missing Link: Report of the Independent Commission for World-Wide Telecommunications Development*. Geneva: International Telecommunications Union, 1985. https://www.itu.int/en/history/Pages/MaitlandReport.aspx.

Malla, Sunil, and Govina R. Timilsina. "Household Cooking Fuel Choice and Adoption of Improved Cookstoves in Developing Countries: A Review." Policy Research Working Paper No. 6903. Washington, DC: World Bank, June 2014. http://documents1.worldbank.org/curated/en/542071468331032496/pdf/WPS6903.pdf.

McKinsey & Company. "Driving Down the Cost of High-Quality Care: Lessons from the Aravind Eye Care System." *Health International* 11 (2011): 19–27. https://www.mckinsey.com/~/media/mckinsey/dotcom/client_service/Healthcare%20Systems%20and%20Services/Health%20International/Issue%2011%20new%20PDFs/HI11_18%20AravindEyeCareSys_noprint.ashx.

McRobie, George. *Small Is Possible*. New York: Harper and Row, 1981.

Ozgediz, Selcuk. *The CGIAR at 40: Institutional Evolution of the World's Premier Agricultural Research Network*. Washington, DC: CGIAR Fund, 2011.

Özler, Berk. "One Laptop Per Child Is Not Improving Reading or Math. But, Are We Learning Enough from These Evaluations?" World Bank Blogs, *Development Impact* (blog). June 14, 2012. https://blogs.worldbank.org/impactevaluations/one-laptop-per-child-is-not-improving-reading-or-math-but-are-we-learning-enough-from-these-evaluati.

Paddock, William, and Paul Paddock. *Famine 1975! America's Decision: Who Will Survive?* Boston, MA: Little-Brown, 1967.

Pingali, Prabhu L. "Green Revolution: Impacts, Limits, and the Path Ahead." *Proceedings of the National Academy of Sciences* 109, no. 31 (July 31, 2012): 12302–8. doi.org/10.1073/pnas.0912953109.

Prahalad, C. K. *The Fortune at the Bottom of the Pyramid: Eradicating Poverty Through Profits*. Upper Saddle River, NJ: Wharton School Publishing, 2005.

Rivera, William M., and John W. Cary. "Privatizing Agricultural Extension." In *Improving Agricultural Extension: A Reference Manual*. Edited by Burton E. Swanson, Robert P. Bentz, and Andrew J. Sofranko. Rome: Food and Agricultural Organization, 1997, chapter 22. http://www.fao.org/3/W5830E/w5830e0o.htm#TopOfPage

Roy, Subir. "Tata Nano—Why Did the People Not Want the People's Car?" *The Wire*, July 17, 2018. https://thewire.in/business/tata-nano-why-did-the-people-not-want-the-peoples-car.

Saunders, Robert J., Jeremy J. Warford, and Bjorn Wellenius. *Telecommunications and Economic Development*. Baltimore: Johns Hopkins University Press, 1994. http://documents.worldbank.org/curated/en/244111468763783754/Telecommunications-and-economic-development.

Schumacher, E. F. *Small Is Beautiful: Economics as if People Mattered*. London: Blond & Briggs, 1973.

Sun, Irene Yuan. *The Next Factory of the World: How Chinese Investment Is Reshaping Africa*. Cambridge, MA: Harvard Business School Press, 2017.

Türker, Umut. "Alternative Sewerage Solution: Condominial Method and Its Application." *Physics and Chemistry of the Earth* 36, no. 5 (January 1, 2011): 179–86. doi.org/10.1016/j.pce.2010.03.038.

Wu, Felicia, and William P. Butz. "The Green Revolution." In *The Future of Genetically Modified Crops*, 11–38. Santa Monica, CA: RAND Corporation, 2004. https://www.jstor.org/stable/10.7249/mg161rc.11.

Chapter 11. Jobs, Competitiveness, and Inequality: Up the Down Escalators in Manufacturing

Amsden, Alice H. *Asia's Next Giant: South Korea and Late Industrialization*. New York: Oxford University Press, 1992.

Autor, David. "Why Are There Still So Many Jobs?" *Journal of Economic Perspectives* 29, no. 3 (2015): 3–30.

Autor, David H. "Work of the Past, Work of the Future." *AEA Papers and Proceedings* 109 (May 2019): 1–32. https://doi.org/10.1257/pandp.20191110.

Autor, David, David A. Mindell, and Elisabeth B. Reynolds. "The Work of the Future: Shaping Technology and Institutions - Fall 2019 Report." Cambridge, MA: MIT Work of the Future, 2019.

Baily, Martin Neil, and Barry P. Bosworth. "US Manufacturing: Understanding Its Past and Its Potential Future." *Journal of Economic Perspectives* 28, no. 1 (2014): 3–26. https://www.brookings.edu/wp-content/uploads/2016/06/us-manufacturing-past-and-potential-future-baily-bosworth.pdf.

Berger, Suzanne. *How We Compete*. Cambridge, MA: MIT Press, 2006.

Bivens, Josh. "Updated Employment Multipliers for the U.S. Economy." Economic Policy Institute. January 23, 2019. https://www.epi.org/publication/updated-employment-multipliers-for-the-u-s-economy/.

Bonvillian, William B., and Peter L. Singer. *Advanced Manufacturing: The New American Innovation Policies*. Cambridge, MA: MIT Press, 2017.

Bonvillian, William B., and Charles Weiss. *Technological Innovation in Legacy Sectors*. New York: Oxford University Press, 2015.

Brynjolfsson, Erik, and Andrew McAfee. *The Second Machine Age: Work, Progress and Prosperity in a Time of Brilliant Technologies*. New York: W.W. Norton, 2014.

Carnevale, Anthony P., Tamara Jayasundera, and Artem Gulish. "Good Jobs Are Back: College Graduates Are First in Line." Washington, DC; Georgetown Center for Education and the Workforce, 2015. https://1gyhoq479ufd3yna29x7ubjn-wpengine.netdna-ssl.com/wp-content/uploads/Good-Jobs_Full_Final.pdf.

China Power Team. "Is China a Global Leader in Research and Development?" Center for Strategic and International Studies. March 12, 2020. https://chinapower.csis.org/china-research-and-development-rnd/.

"China's Reboot of the Belt and Road Initiative." *The Economist Intelligence Unit*, April 29, 2019. https://country.eiu.com/article.aspx?articleid=597952843&Country=China&topic=Politics.

Choi, Hyung-Sup. "Technology for Development: The South Korean Experience." In *Mobilizing Technology for World Development*. Edited by Jairam Ramesh and Charles Weiss. New York: Praeger Publishers, 1979, pp. 130–39. https://www.wired.com/2000/04/joy-2/.

Dahlman, Carl. *World Under Pressure: How China and India Are Influencing the Global Economy and Environment*. Palo Alto, CA: Stanford University Press, 2012.

Dicken, Peter. *Global Shift: Mapping the Changing Contours of the World Economy*, 7th edition. New York: Guilford Press, 2015.

Economy, Elizabeth C. "The Great Leap Backward? The Costs of China's Environmental Crisis." *Foreign Affairs* 86, no. 5 (2007): 38–59.

Economy, Elizabeth C. *The Third Revolution: Xi Jinping and the New Chinese State*. New York: Oxford University Press, 2018.

Ernst, Dieter. "Advanced Manufacturing and China's Future for Jobs." In *Confronting Dystopia: The New Technological Revolution and the Future of Work*. Edited by Eva Paus. Ithaca, NY: Cornell University Press, 2018, pp. 181–206.

European Commission. *Turning Europe into a True Innovation Union*. MEMO/10/473, October 6, 2010. https://ec.europa.eu/commission/presscorner/detail/en/MEMO_10_473.

EY Greater China. "China Is Poised to Win the 5G Race. Are You Up to Speed?" EY. June 13, 2018. https://www.ey.com/en_cn/tmt/china-is-poised-to-win-the-5g-race-are-you-up-to-speed.

Fernandez, Eduard. "What the Belt and Road Initiative Can Learn from the AIIB." *Infrastructure Investor*, February 27, 2019. https://www.infrastructureinvestor.com/what-the-belt-and-road-can-learn-from-the-aiib/.

Frey, Karl-Benedikt, and Michael A. Osborne. "The Future of Employment: How Susceptible Are Jobs to Computerization?" *Technology Forecasting and Social Change* 114 (January 2014): 254–80. https://www.oxfordmartin.ox.ac.uk/downloads/academic/The_Future_of_Employment.pdf.

Fuller, Douglas B. *Paper Tigers, Hidden Dragons: Firms and the Political Economy of China's Technological Development*. New York: Oxford University Press, 2019.

Halegua, Aaron, and Jerome A. Cohen. "The Forgotten Victims of China's Belt and Road Initiative." *Washington Post*, April 23, 2019. https://www.washingtonpost.com/opinions/2019/04/23/forgotten-victims-chinas-belt-road-initiative/.

Hui, Li, and Ben Blanchard. "China Taps Tech Training to Tackle Labor Market Mismatch." *Reuters*, June 8, 2014. https://www.reuters.com/article/us-china-labour-education-idUSKBN0EJ01A20140608.

Inkster, Nigel. *China's Cyber Power*. Washington, DC: Routledge and the International Institute for Strategic Studies, 2018.

ITU Telecommunication Standardization Bureau, *Understanding Patents, Competition & Standardization in an Interconnected World*. Geneva: International Telecommunications Union, 2014. https://www.itu.int/en/ITU-T/Documents/Manual_Patents_Final_E.pdf.

Joy, Bill. "Why the Future Doesn't Need Us?" *Wired*, 1 April 2000. https://www.wired.com/2000/04/joy-2/

Kang, Cecilia, and Alan Rappeport. "Trump Blocks Broadcom's Bid for Qualcomm." *The New York Times*, March 12, 2018. https://www.nytimes.com/2018/03/12/technology/trump-broadcom-qualcomm-merger.html.

Kaplinsky, Raphael. *Globalization, Poverty and Inequality*. Cambridge, MA: Polity Press, 2005.

Katz, Cheryl. "Piling Up: How China's Ban on Importing Waste Has Stalled Global Recycling." *Yale Environment 360* (blog), March 7, 2019. https://e360.yale.edu/features/piling-up-how-chinas-ban-on-importing-waste-has-stalled-global-recycling.

Kenn, Garry Chua. *Education, Skills, and Decent Work in Low and Middle Income Countries: Trends and Results from an Adult Skills Survey*. UNESCO. 2016. https://unesdoc.unesco.org/ark:/48223/pf0000245622.

Keynes, John Maynard. "Economic Possibilities for Our Grandchildren." In *Essays in Persuasion*, 358–73. New York: W. W. Norton, [1930] 1963.

Kipnis, Andrew, and Shanfeng Li. "Is Chinese Education Underfunded?" *The China Quarterly*, 202 (18 June 2010): 327–43.

Kurzweil, Ray. *The Singularity Is Near*. New York: Penguin Books, 2006.

Lewis, Joanna I. *Green Innovation in China*. New York: Columbia University Press, 2010.

Lippert, Barbara, and Volker Perthes. "Strategic Rivalry Between United States and China: Causes, Trajectories, and Implications for Europe." Berlin: Stiftung Wissenschaft und Politik - German Institute for International and Security Affairs, April 2020. https://www.swp-berlin.org/10.18449/2020RP04/.

Lohr, Steve. "U.S. Moves to Ban Huawei from Government Contracts." *The New York Times*, August 7, 2019. https://www.nytimes.com/2019/08/07/business/huawei-us-ban.html.

Malkin, Anton. "Beyond 'Forced' Technology Transfers: Analysis of and Recommendations on Intangible Economy Governance in China." Waterloo, ON: Center for International Governance Innovation, March 2020.

Mazzucatto, Mariana. *The Entrepreneurial State*. New York: Public Affairs, 2014.

McGeachy, Hilary. *US-China Technology Competition: Impacting a Rules-Based Order*. United States Studies Centre at the University of Sydney. May 2, 2019. https://www.ussc.edu.au/analysis/us-china-technology-competition-impacting-a-rules-based-orderhttps://www.ussc.edu.au/analysis/us-china-technology-competition-impacting-a-rules-based-order.

McGregor, Richard. "How the State Runs Business in China." *The Guardian*, July 25, 2019. https://www.theguardian.com/world/2019/jul/25/china-business-xi-jinping-communist-party-state-private-enterprise-huawei.

Moretti, Enrico. *The New Geography of Jobs*. New York: Harcourt-Brace, 2012.

Muro, Mark, Sifan Liu, Jacob Whiton, and Siddharth Kulkarni. "Digitalization and the American Work Force." Washington, DC: Brookings Institution, 2017. https://www.brookings.edu/research/digitalization-and-the-american-workforce/.

Nahm, Jonas, and Edward S. Steinfeld. "Scale-Up Nation: China's Specialization in Innovative Manufacturing." *World Development* 54 (February 2014): 288–300.

Nellis, Stephen. "Apple's IPhone X Has Higher Margin than IPhone 8: Analysis." *Reuters*, November 6, 2017. https://www.reuters.com/article/us-apple-iphone-idUSKBN1D62RZ.

Putnam, Robert D. *Bowling Alone: The Collapse and Revival of American Community*. New York: Simon & Schuster, 2000.

Qin, Amy. "Fraud Scandals Sap China's Dream of Becoming a Science Superpower." *The New York Times*, October 13, 2017. https://www.nytimes.com/2017/10/13/world/asia/china-science-fraud-scandals.html.

Rodrik, Dani. "Premature Deindustrialization." *Journal of Economic Growth* 21, no. 1 (2016): 1–33.

Ross, Michael L. "The Political Economy of the Resource Curse." *World Politics* 51, no. 2 (January 1999): 297–322. doi.org/10.1017/S0043887100008200.

Schlossstein, Steven. *Asia's New Little Dragons: The Dynamic Emergence of Indonesia, Thailand, and Malaysia*. Chicago: Contemporary Books, 1991.

State Council of China. *Made in China 2025*. July 7, 2015. http://www.cittadellascienza.it/cina/wp-content/uploads/2017/02/IoT-ONE-Made-in-China-2025.pdf.

Sun, Irene Yuan. *The Next Factory of the World: How Chinese Investment Is Reshaping Africa*. Boston: Harvard Business School Press, 2017.

United Nations Department of Economic and Social Affairs, Population Division. *World Population Prospects 2019: Highlights*. New York: United Nations, 2019. https://population.un.org/wpp/Publications/Files/WPP2019_Highlights.pdf.

van Agtmael, Antoine, and F. Bakker. *The Smartest Places on Earth: Why Rustbelts Are the Emerging Hotspots of Global Innovation*. New York: PublicAffairs, 2018.

Vogel, Ezra F. *Japan as Number One: Lessons for America*. New York: Harper and Row, 1985.

Weiss, Martin A. and Kristin Archick, *U.S.-EU Data Privacy: From Safe Harbor to Privacy Shield*, CRS 7-5700 (Washington, DC: Congressional Research Service, 2016).

Wessner, Charles, ed. *21st Century Manufacturing: The Role of the Manufacturing Extension Partnership Program*. Washington, DC: National Academies Press, 2013.

Womack, Jack, Daniel T. Jones, and Daniel Roos. *The Machine that Changed the World*. New York: Rawson Associates, 1990.

Zilibotti, Fabrizio. "Growing and Slowing Down Like China." *Journal of the European Economic Association* 15, no. 5 (October 1, 2017): 943–88. doi.org/10.1093/jeea/jvx018.

Chapter 12. New Technologies, New Issues: Gene Drivers and Geoengineering

Allenby, Brad. "Geoengineering Redivivus." *Elementa: Science of the Anthropocene* 2:000023 (2014).

Asilomar International Conference on Climate Intervention Technologies: Conference Report. Pacific Grove, CA: Climate Institute, 2010. http://climate.org/archive/resources/climate-archives/conferences/asilomar/report.html.

Berg, Paul. "Asilomar 1975: DNA Modification Secured." *Nature* 455, no. 7211 (September 18, 2008): 290–91.

Cartagena Protocol on Biosafety to the Convention on Biological Diversity. Montreal: Secretariat of the Convention on Biological Diversity, 2000. https://bch.cbd.int/protocol/text/.

Doudna, Jennifer A., and Samuel H. Sternberg. *A Crack in Creation: Gene Editing and the Unthinkable Power to Control Evolution*. Boston: Houghton-Mifflin, 2017.

Dzau, Victor J., Marcia McNutt, and Chunli Bai. "Wake-Up Call from Hong Kong." *Science* 362, no. 6420 (December 14, 2018): 1215. doi.org/10.1126/science.aaw3127.

ETC Group and Heinrich Boll Foundation. "Forcing the Farm: How Gene Drive Organisms Could Entrench Industrial Agriculture and Threaten Food Sovereignty." October 2018. http://www.etcgroup.org/sites/www.etcgroup.org/files/files/etc_hbf_forcing_the_farm_web.pdf.

Fountain, Henry. "A Rogue Climate Experiment Outrages Scientists." *The New York Times*, October 18, 2012. https://www.nytimes.com/2012/10/19/science/earth/iron-dumping-experiment-in-pacific-alarms-marine-experts.html.

Greely, Henry T. "How Should Science Respond to CRISPR'd Babies?" *Issues in Science and Technology* 35, no. 3 (2019): 32–37.

Irvine, Peter J., and David W. Keith. "Halving Warming with Stratospheric Aerosol Geoengineering Moderates Policy-Relevant Climate Hazards." *Environmental Research Letters* 15, no. 4 (March 2020): 044011. https://doi.org/10.1088/1748-9326/ab76de.

Jasanoff, Sheila, and J. Benjamin Hurlbut. "A Global Observatory for Gene Editing." *Nature* 555, no. 7697 (March 2018): 435–37. doi.org/10.1038/d41586-018-03270-w.

Kenis, Mark, B. P. Hurley, F. Colombari, S. Lawson, J. Sun, C. Wilcken, R. Weeks, and S. Sathyapala. "Guide to the Classical Biological Control of Insect Pests in Planted and

Natural Forests." Rome: Food and Agricultural Organization, 2019. http://www.fao.org/documents/card/en/c/CA3677EN.

Krimsky, Sheldon. *Genetic Alchemy: The Social History of the Recombinant DNA Controversy*. Cambridge, MA: MIT Press, 1982.

Kuiken, Todd. "Governance: Learn from DIY Biologists." *Nature News* 531, no. 7593 (March 10, 2016): 167. doi.org/10.1038/531167a.

Morgan, J., and W. J. Whelan. *Recombinant DNA and Genetic Recombination*. New York: Pergamon Press, 1979.

National Academies of Sciences, Engineering and Medicine. *International Summit on Human Gene Editing: A Global Discussion*. Washington, DC: National Academies Press, 2015. doi.org/10.17226/21913.

National Academies of Sciences, Engineering, and Medicine. *Second International Summit on Human Genome Editing: Continuing the Global Discussion*. Washington, DC: National Academies Press, 2019. doi.org/10.17226/25343.

Paarlberg, Robert. *Starved for Science: How Biotechnology Is Being Kept Out of Africa*. Cambridge, MA: Harvard University Press, 2008.

Randall, John M., and Mandy Tu. "Biological Control." In *Weed Control Methods Handbook*. Edited by Mandy Tu, Callie Hurd, and John M. Randall. Arlington, VA: The Nature Conservancy, 2003, chapter 4, pp. 4.1–4.24. https://www.invasive.org/gist/products/handbook/06.BiologicalControl.pdf.

Rogers, Michael. "The Pandora's Box Congress." *Rolling Stone*, no. 189 (June 19, 1975): 37–40, 42, 74, 77–78, 82.

Shepherd, J. G. S., Ken Caldeira, Peter Cox, Joanna Haigh, David Keith, Brian Launder, Georgina Mace et al. *Geoengineering the Climate: Science, Governance and Uncertainty*. RS Policy Document 10/09. London: The Royal Society, 2009.

Travis, John. "Germline Editing Dominates DNA Summit." *Science* 350, no. 6266 (December 11, 2015): 1299–1300. doi.org/10.1126/science.350.6266.1299.

Yong, Ed. "The CRISPR Baby Scandal Gets Worse by the Day." *The Atlantic*, December 3, 2018. https://www.theatlantic.com/science/archive/2018/12/15-worrying-things-about-crispr-babies-scandal/577234/.

Chapter 13. Playing with Fire

Finnemore, M. and K. Skinnick. International Norm Dynamics and Political Change, *International Organization* 52, 887-917 (1998).

International Covenant on Economic, Social and Cultural Rights. Office of the High Commissioner, United Nations Human Rights. https://www.ohchr.org/EN/ProfessionalInterest/Pages/CESCR.aspx

Kristensen, Hans M., Pierce Corden, Tony Fainberg, David Hafemeister, and Allison Macfarlane. "Alert Status of Nuclear Weapons." *AIP Conference Proceedings* 1898, no. 1 (November 15, 2017): 020003. doi.org/10.1063/1.5009208.

Rawls, John. *A Theory of Justice*. Cambridge MA: Harvard University Press, 1971.

Weiss, Charles. "How Do Science and Technology Affect International Affairs?" *Minerva* 53, no. 4 (2015): 411–30.

Index

For the benefit of digital users, indexed terms that span two pages (e.g., 52–53) may, on occasion, appear on only one of those pages.

ABM (Anti-Ballistic Missile treaty), 72–73, 75, 78
abstinence-only campaigns, 117
Academy of Science (China), 243–44
accountability, 4–5
ACE-2 proteins, 132
adenovirus vaccines, 130
Advanced Projects Research Agency (ARPA), 158
Africa. *See also specific countries*
  appropriate technology movement and, 194, 196
  CGIAR and, 202–10
  climate disruption and, 25, 30–31
  clinical trials in, 149
  colonial health legacy in, 111–12
  disease burdens of, 107–8, 115, 119, 143, 146
  manufacturing advances and, 230, 235–36
  mobile finance in, 189–91
  orphan technologies and, 199–200
  WHO and, 112–13
African sleeping sickness (trypanosomiasis), 111, 117–18, 143, 288n.26
Agreed Framework (1994) for denuclearized Korean peninsula, 80–81
agriculture
  antibiotics and, 110–11, 140
  biofuels and, 55
  climate disruption and, 4, 25, 26–27, 30–31, 59–60
  Consultative Group on International Agricultural Research and, 202–10
  efficient energy use in, 59–60
  energy systems and, 59–60
  geoengineering and, 246
  global health initiatives compared to, 203, 206, 207
  influenza and, 105
  mechanization in, 217–18
  nuclear war and, 84
  vertical vs. horizontal systems of extension in, 108–9, 206, 207, 306–7n.35
AIIB (Asian Infrastructure Investment Bank), 311n.47
alcohol industry, 108–9, 146–47, 154
Alibaba, 162–63
Alipay, 156
Alma Ata Declaration (1978), 112–14
Amazon, 45, 161, 162–63, 166, 176–77
Amazon Valley, 1–2, 26
American Meteorological Society, 248
American Petroleum Institute, 38
Amherst, Jeffery, 150, 296n.32
Andlinger Center for Energy and Environment (Princeton University), 276n.7
Andrade, Maria, 208–9
Antarctica, 19, 21–22, 25, 28
Anti-Ballistic Missile treaty (ABM, 1972), 72–73, 75, 78
antibiotic resistance, 107, 110–11, 139–40
anti-missile systems, 72–73, 82, 86–87
antiretroviral therapy, 108, 117, 135, 147
Apple, 45, 161, 162–63, 215, 224
appropriate technology movement, 192–95, 201–2, 211, 307n.42
Arab Spring, 155, 162, 297n.1, 297n.2
Aravind Eye Care Center, 198–99
Arctic region, 25, 27, 30–31
Argentina, 77, 282n.27
Arkhipov, Vasili, 74

Arms Control and Disarmament Agency, 76
arms control and nonproliferation
autonomous weapons and, 91–92, 253
concepts and doctrines of, 71–74
for cyberwarfare, 183–85, 253–54
first- and second-strike capability, 70–72
global regimes for, 6–7, 12, 68, 74–77, 85–86, 92, 93–94, 251
hypersonic missiles and, 92
Iran and North Korea, 80–82
lessons from, 85–86
science and technology informing, 95–102
in second nuclear age, 82–85
ARPA (Advanced Projects Research Agency), 158
ARPANET, 158–59
Arrhenius, Svante, 35
artificial intelligence, 284n.55
agriculture and, 278n.29
China and, 162, 180, 231–32
facial recognition and, 156, 162
jobs and, 218, 220
manufacturing and, 213
weapons systems and, 2, 82, 87–88, 89–90, 93, 162, 253, 258
Asian Infrastructure Investment Bank (AIIB), 311n.47
Asilomar Conference (1975), 241–42, 248
asymmetric weapons
cyberweapons as, 183–84
nuclear weapons as, 82–83
ATMs (automatic teller machines), 220
atomic vs. chemical explosions, 95
Atoms for Peace initiative, 76
attribution problem in cyberwarfare, 184–85
authoritarianism
China's export of, 180
cyberwarfare and, 185
Internet and, 3, 156–57, 163–66, 168, 170, 171, 180, 251–52, 258
automatic teller machines (ATMs), 220
automation and robotization, 213–14, 219–20. *See also* jobs; manufacturing
automobiles and private vehicles, 37, 60–62, 163, 196–97, 226. *See also* electric vehicles; transportation
autonomous weapons systems
global regimes for, 13, 89–92, 253, 258
threat of, 2, 82, 88–89, 93

Baidu, 162–63
banking, 189–91, 220
batteries, 48, 53–54, 61–62, 226, 231, 276n.3
BECCS (bioenergy with carbon capture and sequestration). *See* carbon capture, use, and sequestration; geoengineering
Belarus, 77
Bellagio meeting on global health, 113–14, 152–53
Belt and Road Initiative (BRI), 180, 231, 311n.47
Berners-Lee, Tim, 157–58
Biden, Joseph, 126
bifurcated needles, 136
Bikini Atoll, 74–75
Bill and Melinda Gates Foundation, 116, 117–18, 126, 152–53, 199, 205–6
bin Laden, Osama, 149
bioenergy with carbon capture and sequestration (BECCS). *See* carbon capture, use, and sequestration (CCUS); geoengineering
biofuels, 54–55, 60, 61–62
biological control. *See* classical biological control
Biological Weapons Convention (1975), 150–51
bioprospecting and biopiracy, 148–49
Bioversity International, 169–70, 207
blockchain technologies, 211–12
Borlaug, Norman, 203
Bouis, Howarth, 208–9
brain drain, 147
Brazil
COVID-19 pandemic and, 103
hydroelectric energy in, 55
Montreal Protocol and, 20
nuclear weapons and, 77, 282n.27
sanitation technology and, 201

Breakthrough Energy Initiative, 44, 48–49
breastfeeding, 110–11, 113–14, 288n.24
Brexit referendum, 181
BRI (Belt and Road Initiative), 180, 231, 311n.47
Briend, Andre, 196
Brilliance phototherapy lamp, 198
buildings and building design, 2, 37, 48, 61
Bulletin of the Atomic Scientists, 70
Burmese military and Rohingya, 155–56. *See also* Myanmar
Bush, George W., 117

Canada, 27
cancer, 15, 99, 128, 141, 279n.6
cap and trade, 37–38, 43, 65–66
carbon capture, use, and sequestration (CCUS), 48, 51–52, 54–55, 56, 60, 245–46
carbon charges, taxes, and budgets, 32–33, 37–38, 59
carbon dioxide emissions. *See* greenhouse gas emissions
carbon dioxide removal (CDR), 63, 245–46
Cartagena Protocol to the CBD, 315n.18
Carter Center, 200
cassava and cassava pink mealybug, 202–3, 205
Castle Bravo thermonuclear test (1954), 74–75
Castro, Fidel, 74, 151
catalytic chain reactions, 18, 21–22
cataract surgeries, 198–99, 211
CBD (Convention on Biodiversity), 148–49, 315n.18
CCUS (carbon capture, use, and sequestration), 48, 51–52, 54–55, 56, 60, 245–46
CDR (carbon dioxide removal), 63, 245–46
cell phones, 11, 162, 163, 189–91, 210
censorship, 125, 157, 159, 164–66, 170–71, 182. *See also* cyber-authoritarianism
Centers for Disease Control and Prevention (CDC), 119, 137
centrifuge method of uranium enrichment, 100, 300n.3
CFCs. *See* chlorofluorocarbons
CGIAR (formerly Consultative Group on International Agricultural Research), 202–10, 306–7n.35, 307n.36
chae-bols, 221–22
chain reactions and chain terminators, 18, 21, 96–97
Chernobyl nuclear accident (1986), 58, 277n.24
chikungunya, 106
children. *See* infants and newborns
China
  autonomous weapons in, 89
  carbon emissions of, 26–27, 42, 45
  CFC and HCFC manufacture in, 22
  clean energy technologies and, 49, 53
  COVID-19 pandemic and, 104–5, 110, 118, 123–26
  cyber-authoritarianism and, 3, 156–57, 163–66, 168, 170, 171, 180
  cyberwarfare and, 176, 178–83, 302n.20
  efficiency initiatives in, 62
  frugal technology and, 196
  hypersonic missiles and, 92
  Internet development in, 162–66
  manufacturing in, 214, 226, 230–33
  Montreal Protocol and, 20
  nuclear arsenal of, 74, 93
  nuclear energy in, 57
  political control vs. innovation in, 233–34
  rise of, 230–33
  SARS and, 118
  in second nuclear age, 82
  and technical standards, 232, 311–12n.52
  technological ambitions and plans of, 232, 311n.49
  threat of, 256–57
  vaccine development and, 292n.69
  WHO and, 110
chlorofluorocarbons (CFCs)
  advent and impact of, 15, 18–19
  air conditioning and, 61
  climate change and, 22
  Montreal protocol and, 19–20, 32
  policy debates over, 11, 15–16, 23
Choi Hyung-Sup, 222
cholera, 103, 108, 140, 200, 287n.18

Chotukool, 196
Christensen, Clayton, 304n.7
chula wood stoves, 194
CIP (International Potato Center), 208
classical biological control, 202–3, 314n.14
Climate Action 100+, 44
climate disruption, 24, 47, 270n.2
adaptation to, 62–63
adaptive management and, 64–65
agriculture and, 4, 25, 26–27, 30–31, 59–60
budgets and scenarios for, 32–33
buildings and, 61
causes and effects of, 24–31
CCUS technology and, 52
damage from, 30–31
diseases from, 24, 25, 26, 106
disinformation and, 34, 38–40, 63–64, 252
effects of, 24–31
energy systems and, 48–50, 53–56
ethics and, 66
as existential danger, 1–2
fossil fuels and, 51
Framework Convention and Kyoto Protocol on, 35–36
geoengineering and, 63, 245–47, 248–49
glaciers and, 26
industry and, 62
institutionalized scientific advice and, 40–41
limitations of science and, 45–46
low-income countries and, 26–27, 36
market-based mechanisms and, 36–38
mitigation of, 48–62
budgets for, 32–33
ethics of, 33
targets for, 32, 36, 42–43
wedges and, 32–33
models for, 10, 28–30, 48
nuclear energy and, 56–59
origin of, 27–30
Paris Agreement and, 42–45
stratospheric ozone negotiations compared to, 34, 35
transportation and, 61
urbanization and, 60–61
Clinton, Hillary, 181–82, 303n.29
coal, 10–11, 28, 42, 48, 51–52, 53
coffee, 204, 215, 216
Cold War, 8–9, 69–72, 92–93, 255, 257. *See also* Soviet Union
command-and-control centers, 184–85, 280–81n.20
competition and competitiveness
Dutch disease and, 229–30
global value chains and, 215, 216, 217
Internet development and, 162–63
jobs and, 226–27, 310n.37
manufacturing and, 213–14, 222–23, 225–27, 228
renewable energy and, 53
research collaboration and, 242–44, 263
rise of China and, 230–31, 233, 234, 235, 236
technologies for low-income populations and, 191–92, 193
vaccine development and, 104
Comprehensive Nuclear Test Ban Treaty (CTBT, 1996), 78
Conference of Parties (COP) of Montreal Protocol, 19, 20
conferences and meetings
Alma Ata Declaration, 112–14
Asilomar Conference, 241–42
Bellagio meeting on global health, 113–14, 152–53
Conference of Parties (COP) of Montreal Protocol, 19, 20
Paris International Sanitary Conference, 109
Pugwash Conference, 70
Congo, 143
Consultative Group on International Agricultural Research (CGIAR), 202–10, 306–7n.35, 307n.36
Convention on Biodiversity (CBD), 148–49, 315n.18
Copenhagen Amendment to Montreal Protocol, 21
coral reefs, 24, 30–31
corruption, 58, 64, 147, 164–65, 230
counter-force and counter-city strategies, 73

COVID-19 pandemic, 7–8, 103–5
asymptomatic transmission of, 122–23, 130–31, 291n.63
clean energy investments and, 48–49
dangers of, 1, 122, 130–32, 291n.63
early Chinese and U.S. response to, 104, 123–26, 164, 165–66
equity issues of, 126–27
Gates Foundation and, 152
global health regimes and, 12–13, 121–23, 251
manufacturing and, 214–15, 217
models for, 9–10, 122–23, 133–34
preparation for, 135
researching cure for, 132–33
as security threat, 289n.43
Survival Nexus exemplified in, 7–8
vaccine development, 7, 104, 122–23, 126–27, 129, 133, 286n.7
vaccine diplomacy in, 292n.69
COVID-19 virus, 132–33
CRISPR-cas9, 239–40, 249–50, 313–14n.7
critical mass, 96–97, 98–99
Crutzen, Paul, 23
CTBT (Comprehensive Nuclear Test Ban Treaty), 78
Cuban Missile Crisis, 72–73, 74
cyber-authoritarianism, 3, 156–57, 163–66, 168, 170, 171, 180, 251–52, 258
cyberwarfare
autonomous weapons systems and, 88, 90
global regimes for, 13, 183–87, 253–54
Snowden revelations and, 176–78
Stuxnet and, 173–76
U.S. and democratic dominance challenged by, 178–83, 258
cytokines, 130, 131–32

DALYs (disability-adjusted life years), 141, 142, 198
data analytics, 162, 167–68, 227
data localization, 167–68
data sovereignty, 170, 179, 186
DDoS (distributed denial of service) attacks, 173
deep fake videos, 182
deep learning. *See* artificial intelligence; autonomous weapons systems; data analytics
de-escalatory first strike, 83
demographic transitions and burden of disease, 144–45, 293n.2
dengue fever, 24, 106, 117–18
Der Spiegel, 176–77
deterrence. *See* arms control and nonproliferation; mutually assured destruction
deuterium, 97–98
Digital Green, 210
Digital Silk Road, 180, 232
direct foreign investment, 221–22, 231
dirty bombs, 279n.3
disability-adjusted life years (DALYs), 141, 142, 198
Disarmament Committee of the United Nations, 185–86
disciplinary divide, 5–6
diseases. *See also specific diseases*
agricultural, 203–4, 207
antibiotic resistance and, 139–40
calculating global burden of, 141
from climate disruption, 24, 25, 26, 106
demographic transition and quadruple burden of, 143, 144–45
eradication of, 137
frugal technology and, 198–99
global health and, 106–7, 153–54
military security and, 149–51
orphan technologies and, 199–200
reporting obligations for, 118–19
research communities and, 117–18
trade and globalization and, 111, 145–49
wood stoves and, 54, 194
disinformation 159
climate disruption and, 34, 38–40, 63–64, 252
COVID-19 pandemic and, 121, 123, 126, 251–52
Facebook and, 160
Internet infrastructure and, 159, 251–52
Russian campaigns of, 181–83, 256
disruptive innovations, 304n.7

distributed denial of service (DDoS) attacks, 173
dot-com boom and bust, 161
dracontiasis (Guinea worm), 199, 200, 289n.38
D-REV (health technology company), 198, 306n.24
drones, 88, 89, 284n.51. *See also* weapons and weapons systems
drought, 27, 30–31, 55, 202, 206
Dupont, 15–16, 20–21, 22–23, 200, 289n.38
Dutch disease, 229–30, 310n.38

earthquakes, 281n.23
Ebola, 107–8, 114, 119, 151
Edison Electric Institute, 38
education and educational institutions, 220, 221, 223, 230, 232–33, 254–55
Egypt and Arab Spring, 155–56, 297n.1
Einstein, Albert, 69, 279n.9
Eisenhower, Dwight D., 82–83
elections and election interference, 110, 170–71, 181–82, 185–86
electric grid, 49–50
electric vehicles, 54, 61–62
elephantiasis, 117–18, 137, 192, 199
employment. *See* jobs
energy systems and energy industry
  agriculture and, 59–60
  alternative energy (*see* renewable energy; wind, solar, hydro and geothermal energy)
  buildings and, 61
  carbon dioxide emissions and, 48, 49
  demand fluctuations and, 50
  efficient use in, 59–62
  electric grid restructuring and, 49–50
  Intermediate Technology Development Group and, 193
  internal carbon emissions evaluations by, 43–44
  manufacturing and, 62, 224–25
  mitigation through, 50–59
  scenarios for, 47–48
  storage in, 53–56
  transformation in, 48–49
  transportation and, 61–62
  urbanization and, 60–61
Environmental Protection Agency (EPA), 38–39
EPI (Expanded Program of Immunization), 113–14
Epidemic Intelligence Open Source System, 119
epidemiological transition, 143, 144
epidemiology, 5, 8, 103–4, 122–23, 133–34, 141, 286n.2, 293n.5
epistemic communities 16, 22–23, 34
Estonia, 83, 185–86
Estrada, Joseph, 155
ETC Group, 244
ethics, 264–66, 313n.1
  autonomous weapons and, 91–92
  climate and, 66
  gene drivers and geoengineering and, 4, 237–38, 240, 242–44, 248–49, 313n.1
  Internet and, 2–4
  manufacturing and, 3–4
  orphan technologies and, 192, 201–2
  pandemic and, 104
  of scientific and technological discoveries, 2–4, 192, 264–66
  toilets/sanitation technology and, 200–2
  tropical medicine and, 111
  vaccine development and, 126–27
Ethiopia, 120
European Union (EU). *See also specific countries*
  carbon emissions of, 42, 43, 45
  General Data Protection Regulation (GDPR) of, 167
  Paris agreement and, 42–43
  Privacy Shield agreement and, 167
Expanded Program of Immunization (EPI), 113–14
extinction of plants and animals, 25, 270–71n.7
ExxonMobil, 274n.42

Facebook, 155, 160, 161, 182
facial recognition technology, 156, 163
FAO (Food and Agricultural Organization), 205
Farmer, Paul, 116, 117

fexinidazole, 288n.26
Fire-Eye (cybersecurity company), 182
first-strike capabilities, 71–73, 83–84, 86–87, 183. *See also* nuclear weapons; arms control
FISA (Foreign Intelligence Surveillance) court, 301n.14
5G technologies, 171, 179–80, 231–32
Food and Agricultural Organization (FAO), 205
Food and Drug Administration (FDA), 309n.26
Ford Foundation, 203
Foreign Intelligence Surveillance (FISA) court, 301n.14
fossil fuels. *See* coal; oil industry
Fourth Amendment to the U.S. Constitution, 177–78, 301n.15
Fox, Vicente, 120
fracking, 51
Framework Convention on Tobacco Control, 146
France, 56–57, 74
Franck, James, 70, 280n.13
Fraunhofer Gesellschaft, 227
freedom of expression
  communications technologies and, 2[illegible]3
  global regimes protecting, 13, 253
  Internet and, 156–57, 160, 166, 170
Frenk, Julio, 120–21, 290–91n.61
Fujimori, Alberto, 103
Fukushima nuclear accident (2011), 57–58
functionalism, 109, 287n.17
fusion, *See* nuclear fusion

GAFA (Google, Apple, Facebook, Amazon), 161. *See also individual companies*
gambling and online gambling, 166
gaseous diffusion method of uranium enrichment, 100
Gates, Bill, 44, 116, 151–52. *See also* Bill and Melinda Gates Foundation
GATT (General Agreement on Trade and Tariffs), 145
GAVI (Global Alliance for Vaccines and Immunization), 117, 119, 152, 290n.50
GBD (global burden of disease), 144–45, 154
GDPR (General Data Protection Regulation) of the European Union, 167
gene banks for major food crops, 207, 307n.36
gene drivers, 313–14n.7, 314n.8
  Asilomar Congress on, 241–42
  ethics and values issues of, 4, 237–38, 240, 242–44, 313n.1
  global regimes for, 13, 254
  research collaboration and, 247–48
  responsible innovation and, 249–50
  technology of, 238–40
General Agreement on Trade and Tariffs (GATT), 145
General Data Protection Regulation (GDPR) of the European Union, 167
Geneva Convention, 91
Geneva Convention on Cybersecurity (proposed), 186–87
geoengineering, 315n.22, 315–16n.25
  climate disruption and, 62–63
  ethics of, 248–49
  global regimes for, 4, 13, 63, 254
  governance of, 250
  research collaboration and, 247–48
  responsible innovation and, 249–50
  technologies of, 245–47
Georgia, Russian invasion of, 173, 185–86
geothermal energy, 55–56
Gerasimov Doctrine, 180–81
Germany
  Internet governance and, 166–67
  manufacturing and labor in, 214, 226–27
germ plasm collections, 169–70
giant Mozambique rats, 304n.2
glaciers and ice fields, 25, 26, 28
Global Alliance for Vaccines and Immunization (GAVI), 117, 119, 152, 290n.50
global burden of disease (GBD), 141, 144–45, 154
Global Fund to Fight AIDS, Tuberculosis and Malaria, 117

global health
agricultural initiatives compared to, 203, 206, 207
challenges of 1980s to, 115–16
colonial and postcolonial background for, 111–12
COVID-19 pandemic and, 103–4, 121–27
economics and, 116
gag rule and, 290n.49
global boundaries and, 104–5, 106–8, 253
global burden of disease statistics and, 141, 144–45
health security and, 116, 149–51, 153, 289n.43
HIV/AIDS and, 107, 116–18
as human right, 112, 119–21
influenza and, 105–6
pandemic responses and, 118–19
science of, 128–42
technological innovation and, 11–12, 199–200
trade and globalization and, 145–49
vertical and horizontal services of, 108–9, 111–12, 153–54
WHO and politics of, 109–11
globalization
country-specific strategies for, 47
global health and, 145–49
information and communication technologies and, 4
international and domestic issues intertwined with, 46, 146
Internet governance and, 3, 167–68
manufacturing and, 214–17
regimes and codes of conduct, 12–13, 258
global warming potential, 270–71n.7
glyphosate (Roundup), 278n.29
GOBI program for public health, 113–14, 152
Google, 45, 161, 162–63, 166, 176
Gore, Al, 160, 298n.15
graphite nuclear reactors, 80, 102
Great Firewall of China, 164–65
greenhouse gas emissions, 270–71n.7, *See also* climate disruption
agriculture and, 59–60
biofuels and, 54, 55
budgets and scenarios for, 32–33, 251, 273nn.30–32
buildings and, 61
geoengineering and, 245–47
greenhouse effect from, 27–30, 35
of low- vs. high-income countries, 26–27
market-based mechanisms to reduce, 36–38
measurement of, 28
negotiations on, 34
persistence of, 26, 30
shadow prices for, 43–44
transportation and, 61–62
Greenland ice sheet, 25, 26, 28
Green Revolution, 203, 204, 205, 208, 209, 210
Guardian, 176–77
Guinea worm (dracontiasis), 199, 200, 289n.38
Gulf War, 78
gun-type atomic bombs, 84–85, 98–99

Haas, Ernst, 109, 287n.17
Hallock, Ted, 15
halons, 20
HCFCs (hydrochlorofluorocarbons), 21, 22, 61
healthy life expectancy (HLE), 141
heating technologies, 54, 61
heavy-water reactors, 102
He Jiankui, 237, 313n.2
Herren, Hans Rudolf, 202–3
HFCs (hydrofluorocarbons), 21
High Performance Computing Act (199), 160
Hiroshima and Nagasaki, 67–68, 70, 76, 98–99, 279n.3
HIV/AIDS, 107–8, 116–18, 128, 135–36, 143, 147, 313n.2
HLE (healthy life expectancy), 141
Hong Kong, 164–65, 166, 309n.23
horizontal systems of service delivery. *See* vertical vs. horizontal systems of service delivery
Huawei, 176, 179–80, 232, 302n.20
Hubble Telescope, 8

human rights and human rights law, 112, 119–21, 186
hydrochlorofluorocarbons (HCFCs), 21, 22, 61
hydroelectric energy, 49–50, 55
hydrofluorocarbons (HFCs), 21
hydrogen bombs, 97–98
hydroxychloroquine, 286n.3
hypersonic missiles
  arms race in, 82, 92, 258
  existential threat of, 67, 82, 86–88, 93
  global regimes for, 2, 13, 92, 253–54
hypertext, 158, 298n.14
Hyren, John, 83–84

IAEA (International Atomic Energy Agency), 76, 78, 80–82, 92–93, 281–82n.24
IANA (Internet Assigned Numbers Authority), 297n.12
IBPGR (International Board on Plant Genetic Resources), 169–70
ICANN (International Corporation for Assigned Names and Numbers), 168–69
ICRISAT (International Crop Research Institute for the Semi-Arid Tropics), 205
ICT. *See* information and communication technologies
ideathons, 307n.43
IETF (Internet Engineering Task Force), 158–59
IFPRI (International Food Policy Research Institute), 208, 306–7n.35
IGF (Internet Governance Forum), 169
IITA (International Institute for Tropical Agriculture), 202, 203
immigration 235–36
ImpactMatters, 306n.24
India
  appropriate technology movement and, 194–95
  carbon emissions of, 26–27
  frugal technology and, 196–97, 198–99
  Montreal Protocol and, 20
  nuclear arsenal of, 2, 77, 78, 84, 86, 93
  Paris agreement and, 42–43
inequity
  in COVID-19 vaccinations, 126–27
  in benefits of technology, 212
infants and newborns
  formula for, 110–11, 288n.24
  frugal technology and, 198
  jaundice and, 198, 305n.21
  mortality rates of, 143
  oral rehydration therapy for, 140
influenza, 105–6, 125–26, 130, 149
informal governance
  of gene banks, 168, 169–70, 207
  of the Internet, 168, 169–70, 207
information and communication technologies (ICT). *See also* cell phones; *specific types and companies*
  blockchain and, 211–12
  Europe and, 225
  globalization and, 4
  information access and, 2[illegible]3
  Internet development and, 157–59, 161–62, 170
  technological determinism and, 156–57
  techno-optimism and, 163–64
  United States and, 226
innovation 10–12, 161–62. *See also* technologies
  in China, 232–34
  and climate, 276n.5
  distinguished from invention, 217
  and gene drivers and geoengineering, 249–50
  and global health, 11–12, 199–200
  and manufacturing, 225–28
  pipeline model of, 279n.8
  policy influence on, 10–11
  poverty-oriented, 11–12
insecticides. *See* pesticides
Institute for Viral Preparations (Moscow), 137
intellectual property rights, 110, 121, 162–63, 186–87, 205, 309n.26
interferons, 130–31
Intergovernmental Panel on Climate Change (IPCC), 29–30, 35, 40–41, 65, 255, 273n.31
Intermediate-Range Nuclear Forces Treaty (INF, 1988), 75–76

Intermediate Technology Development Group (ITDG), 193
international agreements. *See specific names of agreements*
International Atomic Energy Agency (IAEA), 76, 78, 80–82, 92–93, 281–82n.24
International Board on Plant Genetic Resources (IBPGR), 169–70
International Congress on Recombinant DNA Molecules (1975), 241–42
International Corporation for Assigned Names and Numbers (ICANN), 168–69
International Crop Research Institute for the Semi-Arid Tropics (ICRISAT), 205
International Food Policy Research Institute (IFPRI), 208, 306–7n.35
International Institute for Tropical Agriculture (IITA), 202, 203
International Monitoring System (seismograph station network), 281n.23
International Office of Public Hygiene, 109, 114
International Organization for Standardization (ISO), 62, 232, 311–12n.52
international organizations and institutions. *See specific organizations*
International Potato Center (CIP), 208
International Service for National Agricultural Research (ISNAR), 206, 306–7n.35
International Telecommunications Union (ITU), 179–80, 189–90
International Treaty on Plant Genetic Resources (2006), 207
Internet and social media. *See also* cyberwarfare
  China's mastery of, 163–66
  competition and, 162–63
  convenience and repression and, 155–57
  COVID-19 pandemic and, 125, 126
  development and infrastructure of, 157–59, 160, 298n.14
  ethics and values issues of, 2–4
  future of, 170–72, 251–52
  governance of, 168–70, 171
  privatization and regulation of, 160–62
  Western democracies and, 166–68
Internet Assigned Numbers Authority (IANA), 297n.12
Internet Engineering Task Force (IETF), 158–59
Internet Governance Forum (IGF), 169
Internet of Things, 171, 179, 231
Internet service providers (ISPs), 160
IPCC (Intergovernmental Panel on Climate Change), 29–30, 35, 40–41, 65, 255, 273n.31
Iran, 77, 80–82, 173–74
Iraq, 77, 78
ISIS, 79, 84
ISNAR (International Service for National Agricultural Research), 206, 306–7n.35
ISO (International Organization for Standardization), 62, 232, 311–12n.52
ISPs (Internet service providers), 160
Israel, 77, 81, 89, 93, 173–74
Italy, 103
ITDG (Intermediate Technology Development Group), 193
ITU (International Telecommunications Union), 179–80, 189–90
Ivermectin, 108, 192, 199–200, 289n.38

Jaipur Foot (prosthetic), 188, 198
Japan
  carbon-neutral commitment of, 45
  Fukushima nuclear accident in, 57–58
  manufacturing in, 226
  nuclear weapons and, 77
  South Korea and, 221–22
jaundice, 198, 305n.21
Jenner, Edward, 128–29
jiko stove, 194
job polarization, 218
jobs
  China and, 231
  cutting-edge manufacturing and, 224–25

future of, 234–36, 251
manufacturing technology and, 3–4, 227–28
productivity increases and, 217–19
skills in future of, 219–20
wind energy and, 53
worldwide scramble for, 223–24
Justinian (Byzantine emperor), 150

KAIST (Korea Advanced Institute of Science and Technology), 222
Kaplan, Fred, 280n.14
Kazakhstan, 77
Kellogg Foundation, 203
Kennedy, John F., 74
Kenya and mobile finance, 190–91
Khan, A. Q., 86
Khrushchev, Nikita, 74
Kigali Amendment to Montreal Protocol, 21
Korea Advanced Institute of Science and Technology (KAIST), 222
Korea Institute of Science and Technology (KIST), 222
Korea Technical Assistance Corporation (KTAC), 222
Krass, Allan, 278n.1
Kyoto Protocol, 20, 34, 36–37, 42, 249

labor and employment. *See* jobs
laser technologies, 8
launch-on-warning doctrine, 73, 93
laws and legal principles
accountability through, 264
autonomous weapons, 91
climate disruption, 38–39, 42
common but differentiated responsibility, 36
customary international law, 304n.42
cyber warfare, 185–86, 187
equity, 36
geoengineering, 63
global health, 112, 118–19, 147
human rights, 186
intellectual property rights, 110, 121, 162–63, 186–87, 205, 309n.26
Internet governance, 166–67, 168–69
national sovereignty, 107, 109, 118, 148–49, 167, 179, 258
privacy rights, 157, 161, 166–67, 301n.15
stratospheric ozone, 15
leapfrogging, 162
Lesotho, 189
Li Wenliang, 124–25
libertarianism, 157, 158
Libya, 77, 86, 155
Licklider, John, 158
Lifebuoy India, 195–96
life expectancy, 141, 142, 143
Limited Test Ban Treat (1963), 74–75
Lin Hai, 164
lobbying, 39–40
London Amendment to Montreal Protocol, 20
London Suppliers Group, 76
Low, Jan, 208–9
low-income populations and countries
biofuels and, 54
carbon tax and, 37–38
climate disruption and, 25, 26–27, 33, 34, 62–63
COVID-19 and, 104, 126–27
digital divide and, 162, 170–71
disease burdens of, 11, 106, 143, 144–45, 153–54, 253
economic power exerted over, 145–49
energy systems transformation and, 49
health initiatives for, 11–12, 114, 116–18, 119, 120–21, 143
intellectual property rules and, 110
Kyoto Protocol and, 36
manufacturing advances and, 214, 223–24, 228–30, 235
Montreal Protocol and, 20
technological innovation and, 11–12, 191–93, 195–97, 199–212, 257, 264
UNFCCC and, 36
value chain and production networks in, 216–17

MAD. *See* mutually assured destruction
Mahler, Halfden, 112–13

malaria
climate disruption and, 24, 30–31
disease burden to low-income countries of, 11–12, 143
eradication efforts against, 137–38, 152–53
Gates Foundation and, 152
gene drivers and, 239–40
vaccine development for, 106, 117–18, 128, 135–36, 138–39
World War II and, 150, 151, 295n.30
Mali, 143
Mandela, Nelson, 147
Manhattan Project, 69–70
Manning, Chelsea, 176–77
manufacturing, 213–14
China and, 230–33
cutting edge of, 224–25
ethics and values issues of, 3–4
global value chains and networks in, 214–17
innovation from, 225–28
jobs and productivity increases in, 217–19
in low- and middle-income countries, 228–30
skills in future of, 219–20, 234–36
South Korea and, 220–23
markets
environmental mechanisms by means of, 36–38, 64
global health and, 116
"magic" of, 10–12, 13–14
massive retaliation doctrine, 82–83
Max Planck Institutes, 227
McRobie, George, 192–93
meat consumption, 60, 107–8, 211–12
Megatons to Megawatts Program, 77
memory B cells, 129
mental health and mental illnesses, 108–9, 141, 143, 144–45, 149, 154
Merck, 199–200, 289n.38
MERS (Middle Eastern respiratory syndrome), 107–8, 122, 124
messenger RNA (mRNA) vaccines, 129–30
methane, 25
Mexico and health initiatives, 120–21
Microsoft, 152, 186–87, 311–12n.52
Middle Eastern respiratory syndrome (MERS), 107–8, 122, 124
middle-income trap, 228–29
Midgley, Thomas, 15
migration
climate disruption and, 25, 236
manufacturing advances and, 4, 230, 235–36, 258
urbanization and, 144, 235
MILNET, 159
MIRVs (multiple independently targeted vehicles), 75
missile defense systems, 72–73, 82, 86–87
Mission Innovation program, 44, 48–49
mitigation of global warming. *See* climate disruption
Mitticool, 196
mobile phones. *See* cell phones
models 8–9
for climate disruption, 10, 28–30, 48
COVID-19 and, 9–10, 122–23, 133–34
of stratospheric ozone, 16
uncertainty and, 9–10
Molina, Mario, 23
monsoons, 26
Montreal Protocol, 18–21, 22, 23, 32
Copenhagen and Kigali Amendments to 21
London Conference of the Parties (COP) of 20–21
Moonlight Maze, 175
moral hazard, 63, 246
mortality and morbidity data, 141
mosquito-borne illnesses, 24, 106, 108, 137–39. *See also* malaria; pesticides
motor vehicles. *See* automobiles and private vehicles
M-PESA (mobile finance), 190–91, 304n.6
mRNA vaccines. *See* messenger RNA vaccines
Mubarak, Hosni, 155
multiple independently targeted vehicles (MIRVs), 75
Muruganantham, Arunachalam, 195, 305n.14
mutual assured cyber-destruction, 183

mutually assured destruction (MAD), 69–71, 72–74, 78–80
Mwanga, Robert, 208–9
Myanmar
climate disruption and, 31
technology management and operation in, 224

Nagasaki. *See* Hiroshima and Nagasaki
Nano car, 196–97
NARSs (national agricultural research services), 205, 206, 209, 306–7n.35
National Academy of Sciences (U.S.), 70, 243–44
national agricultural research services (NARSs), 205, 206, 209, 306–7n.35
National Basketball Association, 166
National Coal Association, 38
National Institutes of Health (NIH), 117–18
National Science Foundation (NSF), 159, 160
National Security Agency (NSA), 167, 174–75, 176–78, 301n.14
national sovereignty, 107, 109, 118, 148–49, 167, 179, 258
natural gas, 48, 49–50, 51, 56–57
neem oil, 148
negative emissions. *See* geoengineering
net zero emissions, 33
Netherlands, 111
network economies, 161, 190
neural networks. *See* artificial intelligence; autonomous weapons systems
neutrophiles, 131
NIH (National Institutes of Health), 117–18
Nitro-Zeus (U.S. cyberwarfare plan), 174
nonproliferation. *See* arms control and nonproliferation
Nordhaus, William, 32
North Korea
biological warfare and, 150
cyberwarfare and, 178
nuclear arsenal of, 77, 78, 93
nuclear capabilities of, 58–59
nuclear weapons and, 80–82, 84–85
Norway, 55
no-till agricultural technology, 60, 278n.29
NPT. *See* Treaty on the Non-Proliferation of Nuclear Weapons
NSA (National Security Agency), 167, 174–75, 176–78, 301n.14
NSF (National Science Foundation), 159, 160
NSFNET, 159, 160
nuclear energy, 45, 49–50, 56–59, 68, 69, 76, 97, 277n.22, 277n.24
nuclear fusion, 97–98, 277n.22
nuclear weapons, 279n.5, 283n.41
arms control and, 71–74, 77–80
basic scientific research and, 69–70
command-and-control of, 280–81n.20
as existential danger, 2, 4, 67–69, 82–85
extended deterrence and, 282n.31
fully autonomous weapons and, 88–92
future of, 92–94
hypersonic missiles and, 86–88
international efforts against, 6–7
Iran and North Korea, 80–82
lessons from, 85–86
low-yield weapons and, 83
mutually assured destruction and, 69–71, 72–74, 78–80
no-first-use doctrine for, 83–84
nonproliferation and, 68, 74–77, 80–82
nuclear doctrine and, 283n.41
nuclear energy and, 58–59, 76
nuclear umbrella and, 68, 282n.31
science and technology of, 95–102
second nuclear age and, 82–85
terrorists and, 84–85
Nunn-Lugar Act, 77
nutrition and nutritional deficiencies, 141, 144, 145–47, 154, 287n.14

Obama, Barack, 185–86, 303n.36, 309n.32
obesity, 208
oceans and sea levels, 24–25, 26, 30–31, 271n.10
oil industry
agriculture and, 59–60
disinformation campaigns by, 38–39
electric grid restructuring and, 49–50
green energy commitments of, 45, 65–66

oil industry (*cont.*)
greenhouse gas emissions of, 2, 27–28, 273n.32
middle-income trap and, 229
phasing out of, 48, 54
subsidies for, 10–11
One Health framework, 108
One Laptop per Child (OLPC) project, 197
Operation Eligible Receiver, 174–75
Operation Solar Sunrise, 175
Operation Warp Speed, 104
Oppenheimer, J. Robert, 70, 280n.12
oral rehydration therapy (ORT), 108, 113–14, 140, 200, 211
Oregon, 15
orphan technologies, 11, 192, 199–200, 211, 253
Oxford Principles on geoengineering, 248–49
ozone hole and ozone molecules. *See* stratospheric ozone depletion

packet switching, 158
Pakistan
nuclear arsenal of, 2, 58–59, 77, 78, 84, 93, 280–81n.20
sanitation technology and, 201
palm oil, 54, 55, 60
pandemics, 1, 109–10, 134–35. *See also* COVID-19 pandemic; influenza
pangolins, 104–5, 107, 124
Paris Agreement, 30, 31, 34, 42–45, 52, 65
Paris International Sanitary Conference (1851), 109
passive house standards, 61
peer review for scientific research, 5–6
PEPFAR (President's Emergency Plan for AIDS Relief), 117, 119–20, 152–53
permafrost, 25
Peru, 103
pesticides, 60, 105, 108, 137–38, 199–200, 204, 210, 238–39, 295n.30
Petrov, Stanislav, 67
Phalanx system, 89
pharmaceutical industry
globalization and, 147–49, 294n.12
vaccine development and, 106, 286n.7
World Health Organization and, 110
Philippines, 155
photons, 17, 213
pipeline model of innovation, 279n.8
plankton, 17, 24
PLCs (programmable logic computers), 174
PlumpyNut, 196
plutonium. *See* uranium and plutonium
polar stratospheric clouds (PSCs), 21–22
policy learning, 47
polio, 108, 114, 137, 149
Postel, Jon, 158–59, 168, 297n.12
poverty. *See* low-income populations and countries
power-to-x technologies, 54
Prahalad, C. K., 195–96
precautionary principle, 16, 35–36, 244, 263, 269n.3, 274n.38
pre-competitive research, 247
premature deindustrialization, 229
President's Emergency Plan for AIDS Relief (PEPFAR), 117, 119–20, 152–53
PRISM, 177–78
privacy rights, 157, 161, 166–67, 301n.15
programmable logic computers (PLCs), 174
proliferation optimism, 78
PSCs (polar stratospheric clouds), 21–22
Pugwash Conference, 70
pumped storage, 53
Putin, Vladimir, 83

quantum computing, 297n.8

radiation from atomic bombs, 99, 279n.5, 279n.6
ransomware, 298n.14
Rawls criterion of equity, 263
recombinant protein vaccines, 128–29
regimes 12–13, 74, 251. *See also* specific topics
ReMotion (prosthetic knee), 198
renewable energy, 45, 48, 49–50, 53–56, 61, 253, 276n.7, 276n.13
reprocessing method for plutonium extraction, 101–2
resource curse. *See* Dutch Disease

rinderpest, 137
ring vaccination, 136, 137
river blindness (onchocerciasis), 108, 192, 199–200, 289n.38
Rockefeller Foundation, 113, 114, 203
Rohingya, 155–56
Roosevelt, Franklin D., 69
Roundup (glyphosate), 278n.29
Rowland, Sherwood, 23
Royal Society, 243–44
RT television network, 181–82
rubber plantations, 148
Russia. *See also* Soviet Union
  autonomous weapons in, 89
  climate disruption and, 27
  COVID-19 pandemic and, 103
  cyber-authoritarianism and, 3
  cyberwarfare and, 173, 178–83, 186–87
  global health and, 107
  hypersonic missiles and, 92
  Internet governance and, 170, 171
  nuclear weapons and, 2, 67–68, 69, 77–78, 93, 256
  in second nuclear age, 82–83
  threat of, 256–57
  vaccine development and, 292n.69
Rust Belt cities, 218–19

Safaricom, 304n.6
SALT I and SALT II treaties (1972, 1979), 75
sanitation technology, 200–2
SARS (severe acute respiratory syndrome), 118–19, 122, 125, 130–31, 164
Sasakawa Foundation, 200
Saudi Arabia, 77, 166
scale economies, 192–93
schistosomiasis, 106, 128, 199
Schumacher, E. F., 192–93
Science and Technology Studies (STS), 259–60
Science Diplomacy, 259–60
science, limitations of, 45–46
Science, Technology, and International Affairs (STIA), 260
Science, Technology and Society (STS), 259–60
scientific advice, 4–5, 6, 40–41, 123
scientific discoveries
  ethics/values issues of, 2–4, 264–66
  global health and, 108
  international affairs and, 4–7, 70, 74–75, 257–60
  norms for applying, 261–64
  peer review and, 5–6
  stimulation and guidance of, 252–55
  technology distinguished from, 8–9
scientific models. *See* models
scientific theories, 39
Sea Hunter autonomous weapons system, 89
sea levels. *See* oceans and sea levels
second-strike capabilities, 71–72
seismology, 74–75, 85, 257, 281n.23
semiconductor manufacturing and trade, 179, 213, 300n.40
Sethi, P. K., 188–89
severe acute respiratory syndrome (SARS), 118, 122, 125, 130–31, 164
Sharma, R. C., 188–89
Shi Zhengli, 291–92n.68
SIOP (Strategic Integrated Operations Plan), 73
slave trade, 111
small island states, 31
smallpox, 108, 128–29, 136–37, 150
smokeless chula wood stoves, 194
Snowden, Edward, 170, 174, 176–78
social credit systems, 163, 299n.26
social determinants of health, 120–21, 126–27, 144–45, 154, 293–94n.6
social distancing, 122–23, 125–26, 133–34
social media. *See* Internet and social media
solar energy. *See* wind and solar energy
solar radiation management (SRM), 63, 245
SolarWinds, 182
Sony, 226
South Africa, 77
South Korea
  industrial development in, 220–23
  nuclear weapons and, 77, 282n.27
Soviet Union. *See also* Russia
  biological warfare and, 149–51
  Cuban Missile Crisis and, 74
  nuclear arsenal of, 70, 71–72, 77

SRM (solar radiation management), 63, 245
STIA (Science, Technology, and International Affairs), 260
Strategic Arms Limitation Talk and SALT I treaty, 75
Strategic Arms Reduction Treaty (START, 1991), 75, 77
Strategic Integrated Operations Plan (SIOP), 73
Strategic vs. tactical strike, 68, 279n.5
stratospheric ozone depletion
  air conditioning and, 61
  international protection efforts and, 7, 11, 20–21
  lessons of success from, 15, 22–23
  Montreal Protocol and, 18–21, 22, 23, 32
  science behind, 17–18, 21–22
structural adjustment policies, 115, 307n.36
STS. *See* science and technology studies; science, technology and society
Stuxnet, 173–76, 183, 300nn.3–4
sub-Saharan Africa. *See* Africa
sugar cane, 54, 55
supply chain. *See* value chain
Survival Nexus, 266
  appropriate technology movement and, 193
  climate disruption and, 24, 46, 64–65
  COVID-19 pandemic and, 7–8, 103–4, 121–23
  global health and, 112–13, 145, 147
  globalism and, 13–14
  nuclear issues in, 69, 85
  regimes and codes of conduct for, 12–13, 261–64
  vertical and horizontal health systems in, 109, 206
sustainable development, 35
Sustainable Development Goals (UN), 120, 290n.56
syndemics, 145
Syria, 89, 155–56, 173
Szilard, Leo, 279n.9
Taiwan
  China's policy on, 309n.23
  manufacturing in, 226
  nuclear weapons and, 77, 282n.27
  semiconductor manufacturing and, 179, 216
  World Vegetable Center in, 208
Tallinn Manual, 185–87
Tata (Indian automobile manufacturer), 196–97
Taylor, Robert, 158
technologies, 269n.3, *See also* innovation; information and communication technologies
  appropriate technology moment and, 192–95
  Chinese initiatives for, 232–34
  COVID-19 and, 122
  ethics/values issues of, 2–4, 192, 264–66
  foreign investment and, 231, 232
  global health and, 108–9
  international affairs and, 4–7, 70, 76, 219, 257–60
  jobs and, 213, 217–19
  "lock-in" scenarios of, 276n.3
  markets and innovation in, 10–12
  NGOs and innovation in, 195–97
  norms for applying, 261–64
  "orphan technology" and, 11, 192, 199–200, 211, 253
  for the poor, 210–12
  responsible innovation and, 249–50
  science distinguished from, 8–9
  stimulation and guidance of, 252–55
  trickle-down impacts of, 191–92
telecommunications, 189–90, 301n.14. *See also* cell phones
TenCent, 162–63
terrorism
  cyberwarfare and, 185
  nuclear energy and, 58
  nuclear weapons and, 68, 79, 84–85, 93
tetraethyl lead, 15
Thailand
  Internet governance and, 166
  tobacco imports and, 145–46, 294n.9
  thermohaline circulation, 31, 271n.10

Three Gorges Dam, 55
Three Mile Island nuclear accident (1979), 58
Threshold Test Ban Treaty (1974), 74–75
tipping points in climate change, 26, 32, 65–66
tobacco industry, 110–11, 145–47, 154, 288n.24, 290–91n.61, 294n.9
trade agreements
  General Agreement on Trade and Tariffs, 145–46
  global health and, 145–49
  industrial development and, 222–23
  Montreal Protocol and, 19
  U.S./China rivalry and, 236
  value chains and production networks and, 217
Trade-Related Aspects of Intellectual Property (TRIPS), 110
traditional medicine, 104–5, 148–49, 294n.14, 295n.20
transportation. *See also* automobiles and private vehicles
  carbon dioxide emissions and, 48, 49
  efficient energy use in, 61–62
  electric vehicles, 54, 61–62
  manufacturing advances and, 229–30
  power-to-x technologies and, 54
  urbanization and, 60–61
Treaty Banning Nuclear Weapons Tests in the Atmosphere, in Outer Space and Under Water (1963), 74–75
Treaty on the Non-Proliferation of Nuclear Weapons (NPT, 1968), 75, 77, 78, 79, 80–81, 281–82n.24, 282n.27
Treaty on the Prohibition of Nuclear Weapons (2021), 79
trolls and trolling, 164, 181–82, 302n.28
Trump, Donald, 78, 125–26, 292n.4
trypanosomiasis. *See* African sleeping sickness
tuberculosis, 107, 116
Tunisia, 155
Twitter, 155, 181

Ukraine, 77, 173
ultraviolet radiation, 15, 17–18
uncertainty
  climate disruption and, 29–30
  IPCC and, 41
  precautionary principle and, 16, 35–36
  in scientific models, 9–10
Undersecretariat for Arms Control and International Security, 76
UN Framework Convention on Climate Change (UNFCCC), 34, 35–36, 65
UN Group of Governmental Experts (UNGGE), 186
unions and labor movements, 223, 226–27
United Kingdom
  carbon-neutral commitment of, 45
  COVID-19 pandemic and, 103
  manufacturing and, 214
  nuclear arsenal of, 74
  "tropical medicine" and, 111
United Nations Children's Fund (UNICEF), 113–14, 200
United Nations Development Program (UNDP), 205
United Nations Environmental Program (UNEP), 35
United Nations Sustainable Development Goals, 120
United States
  autonomous weapons and, 89
  carbon emissions of, 42
  CCUS technology and, 51–52
  climate change disinformation in, 2, 34, 38–40
  COVID-19 pandemic and, 103–4, 123–26
  cyberwarfare and, 172, 173–76, 181–83
  hypersonic missiles and, 92
  Internet governance and, 157, 166–68
  Iran denuclearization and, 81–82
  Kyoto Protocol and, 42
  manufacturing and, 214, 218–19, 224–26
  North Korea denuclearization and, 80–81
  nuclear arsenal of, 2, 67–68, 69, 70, 71–72, 93
  nuclear energy in, 57
  nuclear strategy of, 73, 75–76, 77, 78, 83, 282n.31

United States (*cont.*)
  Paris agreement and, 42–43
  Snowden and, 176–78
  WHO and, 110, 126, 288n.24
United States Army Medical Research Institute of Infectious Disease (USAMRIID), 150
Universal Postal Union, 109
uranium and plutonium, 95–97, 98–102, 278n.2, 279n.10
Urbani, Carlo, 118
urbanization, 60–61, 144, 235
URLs (universal resource locators), 158, 168
USAMRIID (United States Army Medical Research Institute of Infectious Disease), 150
Uyghurs, 156, 180

vaccines and vaccination
  CIA's false campaign of, 149
  COVID-19 and, 7, 104, 122–23, 126–27, 286n.7
  HIV/AIDS and, 135–36
  influenza and, 105
  in low-income countries, 143, 290n.50
  liability issues in manufacture of, 106, 286n.7
  malaria and, 106, 138–39, 199
  orphan technology challenge to, 199
  science of, 128–30
  smallpox and, 136–37
  vaccine hesitancy, 292n.70
value chains and production networks, 214–17
ventilated improved pit (VIP) latrines, 201, 306n.28
venture capital, 48, 56, 65–66, 161, 222, 259, 309n.26
vertical vs. horizontal systems of service delivery
  in agriculture, 206, 207, 306–7n.35
  in health, 108–9, 111–14, 119–20, 206
Vienna Convention on the Protection of the Stratospheric Ozone Layer, 16, 18–19
virtual private networks (VPN), 165–66
Vodaphone, 191, 304n.6

Washington Consensus, *See* structural adjustment
wasps and cassava mealybugs, 202–3
weapons and weapons systems. *See also* hypersonic missiles; nuclear weapons
  artificial intelligence and autonomy of, 2, 82, 87–90, 93, 162, 253, 258, 284n.51
  biological agents as, 150–51
  global regimes for, 13, 75–76
WeChat, 162–63
wedges, *See* under climate disruption
WikiLeaks, 176–77, 181–82
wildfires, 26, 27, 30–31
wind and solar energy, 48, 49–50, 53–56, 276n.13
work. *See* jobs; manufacturing
World Bank, 201, 205
World Health Organization (WHO), 12–13
  COVID-19 pandemic and, 104, 122–23, 124–25, 126
  Framework Convention on Tobacco Control and, 146
  Gates Foundation and, 152
  global health security and, 153, 289n.43, 296n.45
  on healthcare as human right, 112–14, 120
  international cooperation and, 109–11, 153
  malaria and, 137–38
  oral rehydration therapy and, 140, 200
  SARS epidemic and, 118–19
  smallpox eradication and, 111–12, 136
World Meteorological Organization (WMO), 35, 109, 248
World Summit on the Information Society (WSIS), 169, 170
World Trade Organization (WTO), 236
World Vegetable Center, 208
World Wide Web, 157–58. *See also* Internet and social media
Wuhan, China, 124–25, 291–92n.68

Xi Jingping, 165, 179, 233–34, 311n.47
Xiaomai, 162–63

years lost to disability (YLD), 141–42
years of life lost (YLL), 141
yellow fever, 106

zero-day vulnerabilities, 174, 175–76, 180–81, 300n.3
Zhong Nanshan, 125
Zika, 106
zoonoses, 104–5, 107–8, 287n.14, *See also* COVID-19 pandemic; influenza